KALMAN FILTERING AND NEURAL NETWORKS

KALMAN FILTERING AND NEURAL NETWORKS

Edited by

Simon Haykin

Communications Research Laboratory,
McMaster University, Hamilton, Ontario, Canada

A WILEY-INTERSCIENCE PUBLICATION

JOHN WILEY & SONS, INC.

New York / Chichester / Weinheim / Brisbane / Singapore / Toronto

This book is printed on acid-free paper. ∞

Published simultaneously in Canada.

For ordering and customer service, call 1-800-CALL-WILEY.

Library of Congress Cataloging-in-Publication Data:

CIP data is available.

ISBN 0-471-36998-5

Printed in the United States of America.

10 9 8 7 6 5 4 3 2 1

CONTENTS

PREFACE

This self-contained book, consisting of seven chapters, is devoted to Kalman filter theory applied to the training and use of neural networks, and some applications of learning algorithms derived in this way.

It is organized as follows:

- Chapter 1 presents an introductory treatment of Kalman filters, with emphasis on basic Kalman filter theory, the Rauch–Tung–Striebel smoother, and the extended Kalman filter.
- Chapter 2 presents the theoretical basis of a powerful learning algorithm for the training of feedforward and recurrent multilayered perceptrons, based on the decoupled extended Kalman filter (DEKF); the theory presented here also includes a novel technique called multistreaming.
- Chapters 3 and 4 present applications of the DEKF learning algorithm to the study of image sequences and the dynamic reconstruction of chaotic processes, respectively.
- Chapter 5 studies the dual estimation problem, which refers to the problem of simultaneously estimating the state of a nonlinear dynamical system and the model that gives rise to the underlying dynamics of the system.
- Chapter 6 studies how to learn stochastic nonlinear dynamics. This difficult learning task is solved in an elegant manner by combining two algorithms:
  1. The expectation-maximization (EM) algorithm, which provides an iterative procedure for maximum-likelihood estimation with missing hidden variables.
  2. The extended Kalman smoothing (EKS) algorithm for a refined estimation of the state.

- Chapter 7 studies yet another novel idea – the unscented Kalman filter – the performance of which is superior to that of the extended Kalman filter.

Except for Chapter 1, all the other chapters present illustrative applications of the learning algorithms described here, some of which involve the use of simulated as well as real-life data.

Much of the material presented here has not appeared in book form before. This volume should be of serious interest to researchers in neural networks and nonlinear dynamical systems.

SIMON HAYKIN

Communications Research Laboratory,
McMaster University, Hamilton, Ontario, Canada

Contributors

Sue Becker, Department of Psychology, McMaster University, 1280 Main Street West, Hamilton, ON, Canada L8S 4K1

Lee A. Feldkamp, Ford Research Laboratory, Ford Motor Company, 2101 Village Road, Dearborn, MI 48121-2053, U.S.A.

Simon Haykin, Communications Research Laboratory, McMaster University, 1280 Main Street West, Hamilton, ON, Canada L8S 4K1

Zoubin Ghahramani, Gatsby Computational Neuroscience Unit, University College London, Alexandra House, 17 Queen Square, London WC1N 3AR, U.K.

Alex T. Nelson, Department of Electrical and Computer Engineering, Oregon Graduate Institute of Science and Technology, 19600 N.W. von Neumann Drive, Beaverton, OR 97006-1999, U.S.A.

Gaurav S. Patel, 1553 Manton Blvd., Canton, MI 48187, U.S.A.

Gintaras V. Puskorius, Ford Research Laboratory, Ford Motor Company, 2101 Village Road, Dearborn, MI 48121-2053, U.S.A.

Ron Racine, Department of Psychology, McMaster University, 1280 Main Street West, Hamilton, ON, Canada L8S 4K1

Sam T. Roweis, Gatsby Computational Neuroscience Unit, University College London, Alexandra House, 17 Queen Square, London WC1N 3AR, U.K.

Rudolph van der Merwe, Department of Electrical and Computer Engineering, Oregon Graduate Institute of Science and Technology, 19600 N.W. von Neumann Drive, Beaverton, OR 97006-1999, U.S.A.

Eric A. Wan, Department of Electrical and Computer Engineering, Oregon Graduate Institute of Science and Technology, 19600 N.W. von Neumann Drive, Beaverton, OR 97006-1999, U.S.A.

KALMAN FILTERING AND NEURAL NETWORKS

1

KALMAN FILTERS

Simon Haykin
Communications Research Laboratory, McMaster University, Hamilton, Ontario, Canada
(haykin@mcmaster.ca)

1.1 INTRODUCTION

The celebrated *Kalman filter*, rooted in the state-space formulation of linear dynamical systems, provides a recursive solution to the linear optimal filtering problem. It applies to stationary as well as nonstationary environments. The solution is recursive in that each updated estimate of the state is computed from the previous estimate and the new input data, so only the previous estimate requires storage. In addition to eliminating the need for storing the entire past observed data, the Kalman filter is computationally more efficient than computing the estimate directly from the entire past observed data at each step of the filtering process.

In this chapter, we present an introductory treatment of Kalman filters to pave the way for their application in subsequent chapters of the book. We have chosen to follow the original paper by Kalman [1] for the

Kalman Filtering and Neural Networks, Edited by Simon Haykin
ISBN 0-471-36998-5

derivation; see also the books by Lewis [2] and Grewal and Andrews [3]. The derivation is not only elegant but also highly insightful.

Consider a *linear, discrete-time dynamical system* described by the block diagram shown in Figure 1.1. The concept of *state* is fundamental to this description. The *state vector* or simply *state*, denoted by $\mathbf{x}_k$, is defined as the minimal set of data that is sufficient to uniquely describe the unforced dynamical behavior of the system; the subscript k denotes discrete time. In other words, the state is the least amount of data on the past behavior of the system that is needed to predict its future behavior. Typically, the state $\mathbf{x}_k$ is unknown. To estimate it, we use a set of observed data, denoted by the vector $\mathbf{y}_k$.

In mathematical terms, the block diagram of Figure 1.1 embodies the following pair of equations:

1. *Process equation*

$$\mathbf{x}_{k+1} = \mathbf{F}_{k+1,k}\mathbf{x}_k + \mathbf{w}_k, \tag{1.1}$$

where $\mathbf{F}_{k+1,k}$ is the *transition matrix* taking the state $\mathbf{x}_k$ from time k to time $k+1$. The process noise $\mathbf{w}_k$ is assumed to be additive, white, and Gaussian, with zero mean and with covariance matrix defined by

$$E[\mathbf{w}_n\mathbf{w}_k^T] = \begin{cases} \mathbf{Q}_k & \text{for } n = k, \\ \mathbf{0} & \text{for } n \neq k, \end{cases} \tag{1.2}$$

where the superscript T denotes matrix transposition. The dimension of the state space is denoted by M.

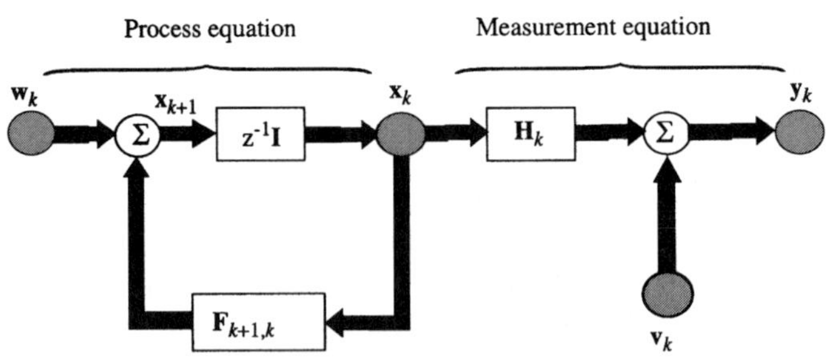

Figure 1.1 Signal-flow graph representation of a linear, discrete-time dynamical system.

2. *Measurement equation*

$$\mathbf{y}_k = \mathbf{H}_k \mathbf{x}_k + \mathbf{v}_k, \tag{1.3}$$

where $\mathbf{y}_k$ is the *observable* at time k and $\mathbf{H}_k$ is the *measurement matrix*. The measurement noise $\mathbf{v}_k$ is assumed to be additive, white, and Gaussian, with zero mean and with covariance matrix defined by

$$E[\mathbf{v}_n \mathbf{v}_k^T] = \begin{cases} \mathbf{R}_k & \text{for } n = k, \\ \mathbf{0} & \text{for } n \neq k. \end{cases} \tag{1.4}$$

Moreover, the measurement noise $\mathbf{v}_k$ is uncorrelated with the process noise $\mathbf{w}_k$. The dimension of the measurement space is denoted by N.

The Kalman filtering problem, namely, the problem of jointly solving the process and measurement equations for the unknown state in an optimum manner may now be formally stated as follows:

- Use the entire observed data, consisting of the vectors $\mathbf{y}_1, \mathbf{y}_2, \ldots, \mathbf{y}_k$, to find for each $k \geq 1$ the minimum mean-square error estimate of the state $\mathbf{x}_i$.

The problem is called *filtering* if $i = k$, *prediction* if $i > k$, and *smoothing* if $1 \leq i < k$.

1.2 OPTIMUM ESTIMATES

Before proceeding to derive the Kalman filter, we find it useful to review some concepts basic to optimum estimation. To simplify matters, this review is presented in the context of scalar random variables; generalization of the theory to vector random variables is a straightforward matter. Suppose we are given the observable

$$y_k = x_k + v_k,$$

where x_k is an unknown signal and v_k is an additive noise component. Let $\hat{x}_k$ denote the a posteriori estimate of the signal x_k, given the observations $y_1, y_2, \ldots, y_k$. In general, the estimate $\hat{x}_k$ is different from the unknown signal x_k. To derive this estimate in an optimum manner, we need a *cost (loss) function* for incorrect estimates. The cost function should satisfy two requirements:

- The cost function is nonnegative.
- The cost function is a nondecreasing function of the *estimation error* $\tilde{x}_k$ defined by

$$\tilde{x}_k = x_k - \hat{x}_k.$$

These two requirements are satisfied by the *mean-square error* defined by

$$\begin{aligned} J_k &= E[(x_k - \hat{x}_k)^2] \\ &= E[\tilde{x}_k^2], \end{aligned}$$

where E is the expectation operator. The dependence of the cost function J_k on time k emphasizes the nonstationary nature of the recursive estimation process.

To derive an optimal value for the estimate $\hat{x}_k$, we may invoke two theorems taken from stochastic process theory [1, 4]:

Theorem 1.1 Conditional mean estimator *If the stochastic processes $\{x_k\}$ and $\{y_k\}$ are jointly Gaussian, then the optimum estimate $\hat{x}_k$ that minimizes the mean-square error J_k is the conditional mean estimator:*

$$\hat{x}_k = E[x_k | y_1, y_2, \ldots, y_k].$$

Theorem 1.2 Principle of orthogonality *Let the stochastic processes $\{x_k\}$ and $\{y_k\}$ be of zero means; that is,*

$$E[x_k] = E[y_k] = 0 \quad \text{for all } k.$$

Then:

(i) *the stochastic process $\{x_k\}$ and $\{y_k\}$ are jointly Gaussian; or*

(ii) *if the optimal estimate $\hat{x}_k$ is restricted to be a linear function of the observables and the cost function is the mean-square error,*

(iii) *then the optimum estimate $\hat{x}_k$, given the observables y_1, $y_2, \ldots, y_k$, is the orthogonal projection of x_k on the space spanned by these observables.*

With these two theorems at hand, the derivation of the Kalman filter follows.

1.3 KALMAN FILTER

Suppose that a measurement on a linear dynamical system, described by Eqs. (1.1) and (1.3), has been made at time k. The requirement is to use the information contained in the new measurement $\mathbf{y}_k$ to update the estimate of the unknown state $\mathbf{x}_k$. Let $\hat{\mathbf{x}}_k^-$ denote a priori estimate of the state, which is already available at time k. With a linear estimator as the objective, we may express the a posteriori estimate $\hat{\mathbf{x}}_k$ as a linear combination of the a priori estimate and the new measurement, as shown by

$$\hat{\mathbf{x}}_k = \mathbf{G}_k^{(1)}\hat{\mathbf{x}}_k^- + \mathbf{G}_k\mathbf{y}_k, \tag{1.5}$$

where the multiplying matrix factors $\mathbf{G}_k^{(1)}$ and $\mathbf{G}_k$ are to be determined. To find these two matrices, we invoke the principle of orthogonality stated under Theorem 1.2. The *state-error vector* is defined by

$$\tilde{\mathbf{x}}_k = \mathbf{x}_k - \hat{\mathbf{x}}_k. \tag{1.6}$$

Applying the principle of orthogonality to the situation at hand, we may thus write

$$E[\tilde{\mathbf{x}}_k\mathbf{y}_i^T] = \mathbf{0} \quad \text{for } i = 1, 2, \ldots, k-1. \tag{1.7}$$

Using Eqs. (1.3), (1.5), and (1.6) in (1.7), we get

$$E[(\mathbf{x}_k - \mathbf{G}_k^{(1)}\hat{x}_k^- - \mathbf{G}_k\mathbf{H}_k\mathbf{x}_k - \mathbf{G}_k\mathbf{w}_k)\mathbf{y}_i^T] = \mathbf{0} \quad \text{for } i = 1, 2, \ldots, k-1. \tag{1.8}$$

Since the process noise $\mathbf{w}_k$ and measurement noise $\mathbf{v}_k$ are uncorrelated, it follows that

$$E[\mathbf{w}_k\mathbf{y}_i^T] = \mathbf{0}.$$

Using this relation and rearranging terms, we may rewrite Eq. (8) as

$$E[(\mathbf{I} - \mathbf{G}_k\mathbf{H}_k - \mathbf{G}_k^{(1)})\mathbf{x}_k\mathbf{y}_i^T + \mathbf{G}_k^{(1)}(\mathbf{x}_k - \hat{\mathbf{x}}_k^-)\mathbf{y}_i^T] = \mathbf{0}, \tag{1.9}$$

where $\mathbf{I}$ is the identity matrix. From the principle of orthogonality, we now note that

$$E[(\mathbf{x}_k - \hat{\mathbf{x}}_k^-)\mathbf{y}_i^T] = \mathbf{0}.$$

Accordingly, Eq. (1.9) simplifies to

$$(\mathbf{I} - \mathbf{G}_k\mathbf{H}_k - \mathbf{G}_k^{(1)})E[\mathbf{x}_k\mathbf{y}_i^T] = \mathbf{0} \quad \text{for } i = 1, 2, \ldots, k-1. \tag{1.10}$$

For arbitrary values of the state $\mathbf{x}_k$ and observable $\mathbf{y}_i$, Eq. (1.10) can only be satisfied if the scaling factors $\mathbf{G}_k^{(1)}$ and $\mathbf{G}_k$ are related as follows:

$$\mathbf{I} - \mathbf{G}_k\mathbf{H}_k - \mathbf{G}_k^{(1)} = \mathbf{0},$$

or, equivalently, $\mathbf{G}_k^{(1)}$ is defined in terms of $\mathbf{G}_k$ as

$$\mathbf{G}_k^{(1)} = \mathbf{I} - \mathbf{G}_k\mathbf{H}_k. \tag{1.11}$$

Substituting Eq. (1.11) into (1.5), we may express the a posteriori estimate of the state at time k as

$$\hat{\mathbf{x}}_k = \hat{\mathbf{x}}_k^- + \mathbf{G}_k(\mathbf{y}_k - \mathbf{H}_k\hat{\mathbf{x}}_k^-), \tag{1.12}$$

in light of which, the matrix $\mathbf{G}_k$ is called the *Kalman gain*.

There now remains the problem of deriving an explicit formula for $\mathbf{G}_k$. Since, from the principle of orthogonality, we have

$$E[(\mathbf{x}_k - \hat{\mathbf{x}}_k)\mathbf{y}_k^T] = \mathbf{0}, \tag{1.13}$$

it follows that

$$E[(\mathbf{x}_k - \hat{\mathbf{x}}_k)\hat{\mathbf{y}}_k^T] = \mathbf{0}, \tag{1.14}$$

where $\hat{\mathbf{y}}_k^T$ is an estimate of $\mathbf{y}_k$ given the previous measurement $\mathbf{y}_1, \mathbf{y}_2, \ldots, \mathbf{y}_{k-1}$. Define the *innovations* process

$$\tilde{\mathbf{y}}_k = \mathbf{y}_k - \hat{\mathbf{y}}_k. \tag{1.15}$$

The innovation process represents a measure of the "new" information contained in $\mathbf{y}_k$; it may also be expressed as

$$\begin{aligned} \tilde{\mathbf{y}}_k &= \mathbf{y}_k - \mathbf{H}_k\hat{\mathbf{x}}_k^- \\ &= \mathbf{H}_k\mathbf{x}_k + \mathbf{v}_k - \mathbf{H}_k\hat{\mathbf{x}}_k^- \\ &= \mathbf{H}_k\tilde{\mathbf{x}}_k^- + \mathbf{v}_k. \end{aligned} \tag{1.16}$$

Hence, subtracting Eq. (1.14) from (1.13) and then using the definition of Eq. (1.15), we may write

$$E[(\mathbf{x}_k - \hat{\mathbf{x}}_k)\tilde{\mathbf{y}}_k^T] = \mathbf{0}. \tag{1.17}$$

Using Eqs. (1.3) and (1.12), we may express the state-error vector $\mathbf{x}_k - \hat{\mathbf{x}}_k$ as

$$\begin{aligned} \mathbf{x}_k - \hat{\mathbf{x}}_k &= \tilde{\mathbf{x}}_k^- - \mathbf{G}_k(\mathbf{H}_k\tilde{\mathbf{x}}_k^- + \mathbf{v}_k) \\ &= (\mathbf{I} - \mathbf{G}_k\mathbf{H}_k)\tilde{\mathbf{x}}_k^- - \mathbf{G}_k\mathbf{v}_k. \end{aligned} \tag{1.18}$$

Hence, substituting Eqs. (1.16) and (1.18) into (1.17), we get

$$E[\{(\mathbf{I} - \mathbf{G}_k\mathbf{H}_k)\tilde{\mathbf{x}}_k^- - \mathbf{G}_k\mathbf{v}_k\}(\mathbf{H}_k\tilde{\mathbf{x}}_k^- + \mathbf{v}_k)] = \mathbf{0}. \tag{1.19}$$

Since the measurement noise $\mathbf{v}_k$ is independent of the state $\mathbf{x}_k$ and therefore the error $\tilde{\mathbf{x}}_k^-$, the expectation of Eq. (1.19) reduces to

$$(\mathbf{I} - \mathbf{G}_k\mathbf{H}_k)E[\tilde{\mathbf{x}}_k\tilde{\mathbf{x}}_k^{T-}]\mathbf{H}_k^T - \mathbf{G}_kE[\mathbf{v}_k\mathbf{v}_k^T] = \mathbf{0}. \tag{1.20}$$

Define the *a priori covariance matrix*

$$\begin{aligned} \mathbf{P}_k^- &= E[(\mathbf{x}_k - \hat{\mathbf{x}}_k^-)(\mathbf{x}_k - \hat{\mathbf{x}}_k^-)^T] \\ &= E[\tilde{\mathbf{x}}_k^- \cdot \tilde{\mathbf{x}}_k^{T-}]. \end{aligned} \tag{1.21}$$

Then, invoking the covariance definitions of Eqs. (1.4) and (1.21), we may rewrite Eq. (1.20) as

$$(\mathbf{I} - \mathbf{G}_k \mathbf{H}_k)\mathbf{P}_k^- \mathbf{H}_k^T - \mathbf{G}_k \mathbf{R}_k = \mathbf{0}.$$

Solving this equation for $\mathbf{G}_k$, we get the desired formula

$$\mathbf{G}_k = \mathbf{P}_k^- \mathbf{H}_k^T [\mathbf{H}_k \mathbf{P}_k^- \mathbf{H}_k^T + \mathbf{R}_k]^{-1}, \tag{1.22}$$

where the symbol $[\cdot]^{-1}$ denotes the inverse of the matrix inside the square brackets. Equation (1.22) is the desired formula for computing the Kalman gain $\mathbf{G}_k$, which is defined in terms of the a priori covariance matrix $\mathbf{P}_k^-$.

To complete the recursive estimation procedure, we consider the *error covariance propagation*, which describes the effects of time on the covariance matrices of estimation errors. This propagation involves two stages of computation:

1. The a priori covariance matrix $\mathbf{P}_k^-$ at time k is defined by Eq. (1.21). Given $\mathbf{P}_k^-$, compute the a posteriori covariance matrix $\mathbf{P}_k$, which, at time k, is defined by

$$\begin{aligned} \mathbf{P}_k &= E[\tilde{\mathbf{x}}_k \tilde{\mathbf{x}}_k^T] \\ &= E[(\mathbf{x}_k - \hat{\mathbf{x}}_k)(\mathbf{x}_k - \hat{\mathbf{x}}_k)^T]. \end{aligned} \tag{1.23}$$

2. Given the "old" a posteriori covariance matrix, $\mathbf{P}_{k-1}$, compute the "updated" a priori covariance matrix $\mathbf{P}_k^-$.

To proceed with stage 1, we substitute Eq. (1.18) into (1.23) and note that the noise process $\mathbf{v}_k$ is independent of the a priori estimation error $\tilde{x}_k^-$. We thus obtain[1]

$$\begin{aligned} \mathbf{P}_k &= (\mathbf{I} - \mathbf{G}_k \mathbf{H}_k) E[\tilde{\mathbf{x}}_k^- \tilde{\mathbf{x}}_k^{T-}](\mathbf{I} - \mathbf{G}_k \mathbf{H}_k)^T + \mathbf{G}_k E[\mathbf{v}_k \mathbf{v}_k^T] \mathbf{G}_k^T \\ &= (\mathbf{I} - \mathbf{G}_k \mathbf{H}_k) \mathbf{P}_k^- (\mathbf{I} - \mathbf{G}_k \mathbf{H}_k)^T + \mathbf{G}_k \mathbf{R}_k \mathbf{G}_k^T. \end{aligned} \tag{1.24}$$

[1]Equation (1.24) is referred to as the "Joseph" version of the covariance update equation [5].

Expanding terms in Eq. (1.24) and then using Eq. (1.22), we may reformulate the dependence of the a posteriori covariance matrix $\mathbf{P}_k$ on the a priori covariance matrix $\mathbf{P}_k^-$ in the simplified form

$$\begin{aligned}\mathbf{P}_k &= (\mathbf{I} - \mathbf{G}_k\mathbf{H}_k)\mathbf{P}_k^- - (\mathbf{I} - \mathbf{G}_k\mathbf{H}_k)\mathbf{P}_k^-\mathbf{H}_k^T\mathbf{G}_k^T + \mathbf{G}_k\mathbf{R}_k\mathbf{G}_k^T \\ &= (\mathbf{I} - \mathbf{G}_k\mathbf{H}_k)\mathbf{P}_k^- - \mathbf{G}_k\mathbf{R}_k\mathbf{G}_k^T + \mathbf{G}_k\mathbf{R}_k\mathbf{G}_k^T \\ &= (\mathbf{I} - \mathbf{G}_k\mathbf{H}_k)\mathbf{P}_k^-. \end{aligned} \tag{1.25}$$

For the second stage of error covariance propagation, we first recognize that the a priori estimate of the state is defined in terms of the "old" a posteriori estimate as follows:

$$\hat{\mathbf{x}}_k^- = \mathbf{F}_{k,k-1}\hat{\mathbf{x}}_{k-1}. \tag{1.26}$$

We may therefore use Eqs. (1.1) and (1.26) to express the a priori estimation error in yet another form:

$$\begin{aligned}\tilde{\mathbf{x}}_k^- &= \mathbf{x}_k - \hat{\mathbf{x}}_k^- \\ &= (\mathbf{F}_{k,k-1}\mathbf{x}_{k-1} + \mathbf{w}_{k-1}) - (\mathbf{F}_{k,k-1}\hat{\mathbf{x}}_{k-1}) \\ &= \mathbf{F}_{k,k-1}(\mathbf{x}_{k-1} - \hat{\mathbf{x}}_{k-1}) + \mathbf{w}_{k-1} \\ &= \mathbf{F}_{k,k-1}\tilde{\mathbf{x}}_{k-1} + \mathbf{w}_{k-1}. \end{aligned} \tag{1.27}$$

Accordingly, using Eq. (1.27) in (1.21) and noting that the process noise $\mathbf{w}_k$ is independent of $\tilde{\mathbf{x}}_{k-1}$, we get

$$\begin{aligned}\mathbf{P}_k^- &= \mathbf{F}_{k,k-1}E[\tilde{\mathbf{x}}_{k-1}\tilde{\mathbf{x}}_{k-1}^T]\mathbf{F}_{k,k-1}^T + E[\mathbf{w}_{k-1}\mathbf{w}_{k-1}^T] \\ &= \mathbf{F}_{k,k-1}\mathbf{P}_{k-1}\mathbf{F}_{k,k-1}^T + \mathbf{Q}_{k-1}, \end{aligned} \tag{1.28}$$

which defines the dependence of the a priori covariance matrix $\mathbf{P}_k^-$ on the "old" a posteriori covariance matrix $\mathbf{P}_{k-1}$.

With Eqs. (1.26), (1.28), (1.22), (1.12), and (1.25) at hand, we may now summarize the recursive estimation of state as shown in Table 1.1. This table also includes the initialization. In the absence of any observed data at time $k = 0$, we may choose the initial estimate of the state as

$$\hat{\mathbf{x}}_0 = E[\mathbf{x}_0], \tag{1.29}$$

Table 1.1 Summary of the Kalman filter

State-space model

$$\mathbf{x}_{k+1} = \mathbf{F}_{k+1,k}\mathbf{x}_k + \mathbf{w}_k,$$
$$\mathbf{y}_k = \mathbf{H}_k\mathbf{x}_k + \mathbf{v}_k,$$

where $\mathbf{w}_k$ and $\mathbf{v}_k$ are independent, zero-mean, Gaussian noise processes of covariance matrices $\mathbf{Q}_k$ and $\mathbf{R}_k$, respectively.
Initialization: For $k = 0$, set

$$\hat{\mathbf{x}}_0 = E[\mathbf{x}_0],$$
$$\mathbf{P}_0 = E\left[(\mathbf{x}_0 - E[\mathbf{x}_0])(\mathbf{x}_0 - E[\mathbf{x}_0])^T\right].$$

Computation: For $k = 1, 2, \ldots$, compute:
State estimate propagation

$$\hat{\mathbf{x}}_k^- = \mathbf{F}_{k,k-1}\hat{\mathbf{x}}_{k-1}^-;$$

Error covariance propagation

$$\mathbf{P}_k^- = \mathbf{F}_{k,k-1}\mathbf{P}_{k-1}\mathbf{F}_{k,k-1}^T + \mathbf{Q}_{k-1};$$

Kalman gain matrix

$$\mathbf{G}_k = \mathbf{P}_k^-\mathbf{H}_k^T\left[\mathbf{H}_k\mathbf{P}_k^-\mathbf{H}_k^T + \mathbf{R}_k\right]^{-1};$$

State estimate update

$$\hat{\mathbf{x}}_k = \hat{\mathbf{x}}_k^- + \mathbf{G}_k\left(\mathbf{y}_k - \mathbf{H}_k\hat{\mathbf{x}}_k^-\right);$$

Error covariance update

$$\mathbf{P}_k = (\mathbf{I} - \mathbf{G}_k\mathbf{H}_k)\mathbf{P}_k^-.$$

---

and the initial value of the a posteriori covariance matrix as

$$\mathbf{P}_0 = E[(\mathbf{x}_0 - E[\mathbf{x}_0])(\mathbf{x}_0 - E[\mathbf{x}_0])^T]. \tag{1.30}$$

This choice for the initial conditions not only is intuitively satisfying but also has the advantage of yielding an *unbiased* estimate of the state $\mathbf{x}_k$.

1.4 DIVERGENCE PHENOMENON: SQUARE-ROOT FILTERING

The Kalman filter is prone to serious numerical difficulties that are well documented in the literature [6]. For example, the a posteriori covariance matrix $\mathbf{P}_k$ is defined as the difference between two matrices $\mathbf{P}_k^-$ and

$\mathbf{G}_k\mathbf{H}_k\mathbf{P}_k^-$; see Eq. (1.25). Hence, unless the numerical accuracy of the algorithm is high enough, the matrix $\mathbf{P}_k$ resulting from this computation may *not* be nonnegative-definite. Such a situation is clearly unacceptable, because $\mathbf{P}_k$ represents a covariance matrix. The unstable behavior of the Kalman filter, which results from numerical inaccuracies due to the use of finite-wordlength arithmetic, is called the *divergence phenomenon.*

A refined method of overcoming the divergence phenomenon is to use numerically stable unitary transformations at every iteration of the Kalman filtering algorithm [6]. In particular, the matrix $\mathbf{P}_k$ is propagated in a square-root form by using the *Cholesky factorization*:

$$\mathbf{P}_k = \mathbf{P}_k^{1/2}\mathbf{P}_k^{T/2}, \tag{1.31}$$

where $\mathbf{P}_k^{1/2}$ is reserved for a lower-triangular matrix, and $\mathbf{P}_k^{T/2}$ is its transpose. In linear algebra, the Cholesky factor $\mathbf{P}_k^{1/2}$ is commonly referred to as the square root of the matrix $\mathbf{P}_k$. Accordingly, any variant of the Kalman filtering algorithm based on the Cholesky factorization is referred to as square-root filtering. The important point to note here is that the matrix product $\mathbf{P}_k^{1/2}\mathbf{P}_k^{T/2}$ is much less likely to become indefinite, because the product of any square matrix and its transpose is always positive-definite. Indeed, even in the presence of roundoff errors, the numerical conditioning of the Cholesky factor $\mathbf{P}_k^{1/2}$ is generally much better than that of $\mathbf{P}_k$ itself.

1.5 RAUCH–TUNG–STRIEBEL SMOOTHER

In Section 1.3, we addressed the optimum linear filtering problem. The solution to the linear prediction problem follows in a straightforward manner from the basic theory of Section 1.3. In this section, we consider the *optimum smoothing problem.*

To proceed, suppose that we are given a set of data over the time interval $0 < k \leq N$. Smoothing is a non-real-time operation in that it involves estimation of the state $\mathbf{x}_k$ for $0 < k \leq N$, using all the available data, past as well as future. In what follows, we assume that the final time N is *fixed.*

To determine the optimum state estimates $\hat{\mathbf{x}}_k$ for $0 < k \leq N$, we need to account for past data $\mathbf{y}_j$ defined by $0 < j \leq k$, and future data $\mathbf{y}_j$ defined by $k < j \leq N$. The estimation pertaining to the past data, which we refer to as *forward filtering* theory, was presented in Section 1.3. To deal with the issue of state estimation pertaining to the future data, we use *backward filtering*, which starts at the final time N and runs backwards. Let $\hat{\mathbf{x}}_k^f$ and $\hat{\mathbf{x}}_k^b$ denote the state estimates obtained from the forward and backward recursions, respectively. Given these two estimates, the next issue to be considered is how to combine them into an overall smoothed estimate $\hat{\mathbf{x}}_k$, which accounts for data over the entire time interval. Note that the symbol $\hat{\mathbf{x}}_k$ used for the smoothed estimate in this section is not to be confused with the filtered (i.e., a posteriori) estimate used in Section 1.3.

We begin by rewriting the process equation (1.1) as a recursion for decreasing k, as shown by

$$\mathbf{x}_k = \mathbf{F}_{k+1,k}^{-1}\mathbf{x}_{k+1} - \mathbf{F}_{k+1,k}^{-1}\mathbf{w}_k, \tag{1.32}$$

where $\mathbf{F}_{k+1,k}^{-1}$ is the inverse of the transition matrix $\mathbf{F}_{k+1,k}$. The rationale for backward filtering is depicted in Figure 1.2*a*, where the recursion begins at the final time N. This rationale is to be contrasted with that of forward filtering depicted in Figure 1.2*b*. Note that the a priori estimate $\hat{\mathbf{x}}_k^{b-}$ and the a posteriori estimate $\hat{\mathbf{x}}_k^b$ for backward filtering occur to the right and left of time k in Figure 1.2*a*, respectively. This situation is the exact opposite to that occurring in the case of forward filtering depicted in Figure 1.2*b*.

To simplify the presentation, we introduce the two definitions:

$$\mathbf{S}_k = [\mathbf{P}_k^b]^{-1}, \tag{1.33}$$

$$\mathbf{S}_k^- = [\mathbf{P}_k^{b-}]^{-1}, \tag{1.34}$$

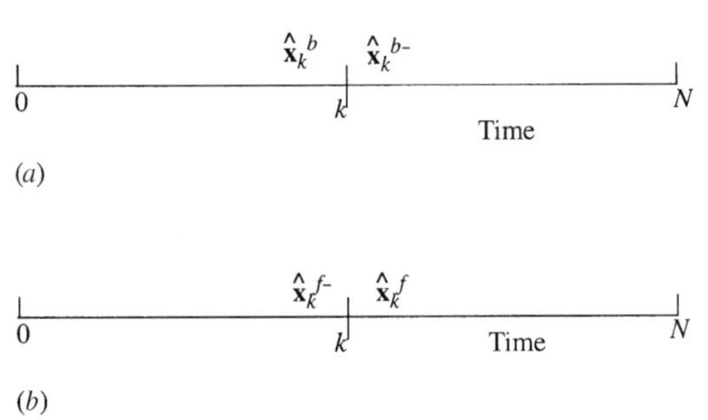

Figure 1.2 Illustrating the smoother time-updates for (*a*) backward filtering and (*b*) forward filtering.

and the two intermediate variables

$$\hat{\mathbf{z}}_k = [\mathbf{P}_k^b]^{-1}\hat{\mathbf{x}}_k^b = \mathbf{S}_k\hat{\mathbf{x}}_k^b, \tag{1.35}$$

$$\hat{\mathbf{z}}_k^- = [\mathbf{P}_k^{b-}]^{-1}\hat{\mathbf{x}}_k^{b-} = \mathbf{S}_k^-\hat{\mathbf{x}}_k^{b-}. \tag{1.36}$$

Then, building on the rationale of Figure 1.2*a*, we may derive the following updates for the backward filter [2]:

1. *Measurement updates*

$$\mathbf{S}_k = \mathbf{S}_k^- + \mathbf{H}_k\mathbf{R}_k^{-1}\mathbf{H}_k, \tag{1.37}$$

$$\hat{\mathbf{z}}_k = \hat{\mathbf{z}}_k^- + \mathbf{H}_k^T\mathbf{R}_k^{-1}\mathbf{y}_k, \tag{1.38}$$

   where $\mathbf{y}_k$ is the observable defined by the measurement equation (1.3), $\mathbf{H}_k$ is the measurement matrix, and $\mathbf{R}_k^{-1}$ is the inverse of the covariance matrix of the measurement noise $\mathbf{v}_k$.

2. *Time updates*

$$\mathbf{G}_k^b = \mathbf{S}_{k+1}[\mathbf{S}_{k+1} + \mathbf{Q}_k^{-1}]^{-1}, \tag{1.39}$$

$$\mathbf{S}_k^- = \mathbf{F}_{k+1,k}^T(\mathbf{I} - \mathbf{G}_k^b)\mathbf{S}_{k+1}\mathbf{F}_{k+1,k}, \tag{1.40}$$

$$\hat{\mathbf{z}}_k^- = \mathbf{F}_{k+1,k}^T(\mathbf{I} - \mathbf{G}_k^b)\hat{\mathbf{z}}_{k+1}, \tag{1.41}$$

   where $\mathbf{G}_k^b$ is the Kalman gain for backward filtering and $\mathbf{Q}_k^{-1}$ is the inverse of the covariance matrix of the process noise $\mathbf{w}_k$. The backward filter defined by the measurement and time updates of Eqs. (1.37)–(1.41) is the *information formulation* of the Kalman filter. The information filter is distinguished from the basic Kalman filter in that it propagates the inverse of the error covariance matrix rather than the error covariance matrix itself.

Given observable data over the interval $0 < k \leq N$ for fixed N, suppose we have obtained the following two estimates:

- The forward a posteriori estimate $\hat{\mathbf{x}}_k^f$ by operating the Kalman filter on data $\mathbf{y}_j$ for $0 < j \leq k$.
- The backward a priori estimate $\hat{\mathbf{x}}_k^{b-}$ by operating the information filter on data $\mathbf{y}_j$ for $k < j \leq N$.

With these two estimates and their respective error covariance matrices at hand, the next issue of interest is how to determine the smoothed estimate $\hat{\mathbf{x}}_k$ and its error covariance matrix, which incorporate the overall data over the entire time interval $0 < k \leq N$.

Recognizing that the process noise $\mathbf{w}_k$ and measurement noise $\mathbf{v}_k$ are independent, we may formulate the error covariance matrix of the a posteriori smoothed estimate $\hat{\mathbf{x}}_k$ as follows:

$$\begin{aligned} \mathbf{P}_k &= [[\mathbf{P}_k^f]^{-1} + [\mathbf{P}_k^{b-}]^{-1}]^{-1} \\ &= [[\mathbf{P}_k^f]^{-1} + \mathbf{S}_k^-]^{-1}. \end{aligned} \tag{1.42}$$

To proceed further, we invoke the *matrix inversion lemma*, which may be stated as follows [7]. Let $\mathbf{A}$ and $\mathbf{B}$ be two positive-definite matrices related by

$$\mathbf{A} = \mathbf{B}^{-1} + \mathbf{C}\mathbf{D}^{-1}\mathbf{C}^T,$$

where $\mathbf{D}$ is another positive-definite matrix and $\mathbf{C}$ is a matrix with compatible dimensions. The matrix inversion lemma states that we may express the inverse of the matrix $\mathbf{A}$ as follows:

$$\mathbf{A}^{-1} = \mathbf{B} - \mathbf{B}\mathbf{C}[\mathbf{D} + \mathbf{C}^T\mathbf{B}\mathbf{C}]^{-1}\mathbf{C}^T\mathbf{B}.$$

For the problem at hand, we set

$$\begin{aligned} \mathbf{A} &= \mathbf{P}_k^{-1}, \\ \mathbf{B} &= \mathbf{P}_k^f, \\ \mathbf{C} &= \mathbf{I}, \\ \mathbf{D} &= [\mathbf{S}_k^-]^{-1}, \end{aligned}$$

where $\mathbf{I}$ is the identity matrix. Then, applying the matrix inversion lemma to Eq. (1.42), we obtain

$$\begin{aligned} \mathbf{P}_k &= \mathbf{P}_k^f - \mathbf{P}_k^f[\mathbf{P}_k^{b-} + \mathbf{P}_k^f]^{-1}\mathbf{P}_k^f \\ &= \mathbf{P}_k^f - \mathbf{P}_k^f\mathbf{S}_k^-[\mathbf{I} + \mathbf{P}_k^f\mathbf{S}_k^-]^{-1}\mathbf{P}_k^f. \end{aligned} \tag{1.43}$$

From Eq. (1.43), we find that the a posteriori smoothed error covariance matrix $\mathbf{P}_k$ is smaller than or equal to the a posteriori error covariance

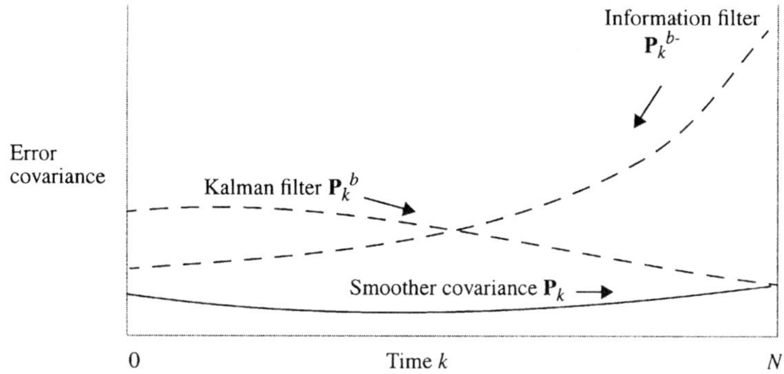

Figure 1.3 Illustrating the error covariance for forward filtering, backward filtering, and smoothing.

matrix $\mathbf{P}_k^f$ produced by the Kalman filter, which is naturally due to the fact that smoothing uses additional information contained in the future data. This point is borne out by Figure 1.3, which depicts the variations of $\mathbf{P}_k$, $\mathbf{P}_k^f$, and $\mathbf{P}_k^{b-}$ with k for a one-dimensional situation.

The a posteriori smoothed estimate of the state is defined by

$$\hat{\mathbf{x}}_k = \mathbf{P}_k([\mathbf{P}_k^f]^{-1}\hat{\mathbf{x}}_k^f + [\mathbf{P}_k^{b-}]^{-1}\hat{\mathbf{x}}_k^{b-}). \tag{1.44}$$

Using Eqs. (1.36) and (1.43) in (1.44) yields, after simplification,

$$\hat{\mathbf{x}}_k = \hat{\mathbf{x}}_k^f + (\mathbf{P}_k\mathbf{z}_k^- - \mathbf{G}_k\hat{\mathbf{x}}_k^f), \tag{1.45}$$

where the *smoother gain* is defined by

$$\mathbf{G}_k = \mathbf{P}_k^f\mathbf{S}_k^-[\mathbf{I} + \mathbf{P}_k^f\mathbf{S}_k^-]^{-1}, \tag{1.46}$$

which is not to be confused with the Kalman gain of Eq. (1.22).

The optimum smoother just derived consists of three components:

- A forward filter in the form of a Kalman filter.
- A backward filter in the form of an information filter.
- A separate smoother, which combines results embodied in the forward and backward filters.

The *Rauch–Tung–Striebel* smoother, however, is more efficient than the three-part smoother in that it incorporates the backward filter and separate smoother into a single entity [8, 9]. Specifically, the measurement update of the Rauch–Tung–Striebel smoother is defined by

$$\mathbf{P}_k = \mathbf{P}_k^f - \mathbf{A}_k(\mathbf{P}_{k+1}^{f-} - \mathbf{P}_{k+1})\mathbf{A}_k^T, \tag{1.47}$$

where $\mathbf{A}_k$ is the new gain matrix:

$$\mathbf{A}_k = \mathbf{P}_k^f \mathbf{F}_{k+1,k}^T [\mathbf{P}_{k+1}^{f-}]^{-1}. \tag{1.48}$$

The corresponding time update is defined by

$$\hat{\mathbf{x}}_k = \hat{\mathbf{x}}_k^f + \mathbf{A}_k(\hat{\mathbf{x}}_{k+1} - \hat{\mathbf{x}}_{k+1}^{f-}) \tag{1.49}$$

The Rauch–Tung–Striebel smoother thus proceeds as follows:

1. The Kalman filter is applied to the observable data in a forward manner, that is, $k = 0, 1, 2, \ldots$, in accordance with the basic theory summarized in Table 1.1.
2. The recursive smoother is applied to the observable data in a backward manner, that is, $k = N - 1, N - 2, \ldots$, in accordance with Eqs. (1.47)–(1.49).
3. The initial conditions are defined by

$$\mathbf{P}_N = \mathbf{P}_N^f, \tag{1.50}$$

$$\hat{\mathbf{x}}_k = \hat{\mathbf{x}}_k^f. \tag{1.51}$$

Table 1.2 summarizes the computations involved in the Rauch–Tung–Striebel smoother.

1.6 EXTENDED KALMAN FILTER

The Kalman filtering problem considered up to this point in the discussion has addressed the estimation of a state vector in a linear model of a dynamical system. If, however, the model is *nonlinear*, we may extend the use of Kalman filtering through a linearization procedure. The resulting filter is referred to as the *extended Kalman filter* (*EKF*) [10–12]. Such an

Table 1.2 Summary of the Rauch–Tung–Striebel smoother

State-space model

$$\mathbf{x}_{k+1} = \mathbf{F}_{k+1,k}\mathbf{x}_k + \mathbf{w}_k$$
$$\mathbf{y}_k = \mathbf{H}_k\mathbf{x}_k + \mathbf{v}_k$$

where $\mathbf{w}_k$ and $\mathbf{v}_k$ are independent, zero-mean, Gaussian noise processes of covariance matrices $\mathbf{Q}_k$ and $\mathbf{R}_k$, respectively.

Forward filter

Initialization: For $k = 0$, set

$$\hat{\mathbf{x}}_0 = E[\mathbf{x}_0],$$
$$\mathbf{P}_0 = E[(\mathbf{x}_0 - E[\mathbf{x}_0](\mathbf{x}_0 - E[\mathbf{x}_0])^T].$$

Computation: For $k = 1, 2, \ldots,$ compute

$$\hat{\mathbf{x}}_k^{f-} = \mathbf{F}_{k,k-1}\hat{\mathbf{x}}_{k-1}^{f-},$$
$$\mathbf{P}_k^{f-} = \mathbf{F}_{k,k-1}\mathbf{P}_{k-1}^{f}\mathbf{F}_{k,k-1}^T + \mathbf{Q}_{k-1},$$
$$\mathbf{G}_k^f = \mathbf{P}_k^{f-}\mathbf{H}_k^T[\mathbf{H}_k\mathbf{P}_k^{f-}\mathbf{H}_k^T + \mathbf{R}_k]^{-1},$$
$$\hat{\mathbf{x}}_k^f = \hat{\mathbf{x}}_k^{f-} + \mathbf{G}_k^f(\mathbf{y}_k - \mathbf{H}_k\hat{\mathbf{x}}_k^{f-}).$$

Recursive smoother

Initialization: For $k = N$, set

$$\mathbf{P}_N = \mathbf{P}_N^f,$$
$$\hat{\mathbf{x}}_k = \hat{\mathbf{x}}_k^f.$$

Computation: For $k = N - 1, N - 2,$ compute

$$\mathbf{A}_k = \mathbf{P}_k^f\mathbf{F}_{k+1,k}^T[\mathbf{P}_{k+1}^{f-}]^{-1},$$
$$\mathbf{P}_k = \mathbf{P}_k^f - \mathbf{A}_k(\mathbf{P}_{k+1}^{f-} - \mathbf{P}_{k+1})\mathbf{A}_k^T,$$
$$\hat{\mathbf{x}}_k = \hat{\mathbf{x}}_k^f + \mathbf{A}_k\left(\hat{\mathbf{x}}_{k+1} - \hat{\mathbf{x}}_{k+1}^{f-}\right).$$

---

extension is feasible by virtue of the fact that the Kalman filter is described in terms of difference equations in the case of discrete-time systems.

To set the stage for a development of the extended Kalman filter, consider a nonlinear dynamical system described by the state-space model

$$\mathbf{x}_{k+1} = \mathbf{f}(k, \mathbf{x}_k) + \mathbf{w}_k, \tag{1.52}$$

$$\mathbf{y}_k = \mathbf{h}(k, \mathbf{x}_k) + \mathbf{v}_k, \tag{1.53}$$

where, as before, $\mathbf{w}_k$ and $\mathbf{v}_k$ are independent zero-mean white Gaussian noise processes with covariance matrices $\mathbf{R}_k$ and $\mathbf{Q}_k$, respectively. Here, however, the functional $\mathbf{f}(k, \mathbf{x}_k)$ denotes a nonlinear transition matrix function that is possibly time-variant. Likewise, the functional $\mathbf{h}(k, \mathbf{x}_k)$ denotes a *nonlinear measurement matrix* that may be time-variant, too.

The basic idea of the extended Kalman filter is to *linearize* the state-space model of Eqs. (1.52) and (1.53) at each time instant around the most recent state estimate, which is taken to be either $\hat{\mathbf{x}}_k$ or $\hat{\mathbf{x}}_k^-$, depending on which particular functional is being considered. Once a linear model is obtained, the standard Kalman filter equations are applied.

More explicitly, the approximation proceeds in two stages.

Stage 1 The following two matrices are constructed:

$$\mathbf{F}_{k+1,k} = \left. \frac{\partial \mathbf{f}(k, \mathbf{x})}{\partial \mathbf{x}} \right|_{\mathbf{x}=\hat{\mathbf{x}}_k}, \tag{1.54}$$

$$\mathbf{H}_k = \left. \frac{\partial \mathbf{h}(k, \mathbf{x}_k)}{\partial \mathbf{x}} \right|_{\mathbf{x}=\hat{\mathbf{x}}_k^-}. \tag{1.55}$$

That is, the ijth entry of $\mathbf{F}_{k+1,k}$ is equal to the partial derivative of the ith component of $\mathbf{F}(k, \mathbf{x})$ with respect to the jth component of $\mathbf{x}$. Likewise, the ijth entry of $\mathbf{H}_k$ is equal to the partial derivative of the ith component of $\mathbf{H}(k, \mathbf{x})$ with respect to the jth component of $\mathbf{x}$. In the former case, the derivatives are evaluated at $\hat{\mathbf{x}}_k$, while in the latter case, the derivatives are evaluated at $\hat{\mathbf{x}}_k^-$. The entries of the matrices $\mathbf{F}_{k+1,k}$ and $\mathbf{H}_k$ are all known (i.e., computable), by having $\hat{\mathbf{x}}_k$ and $\hat{\mathbf{x}}_k^-$ available at time k.

Stage 2 Once the matrices $\mathbf{F}_{k+1,k}$ and $\mathbf{H}_k$ are evaluated, they are then employed in a *first-order Taylor approximation* of the nonlinear functions $\mathbf{F}(k, \mathbf{x}_k)$ and $\mathbf{H}(k, \mathbf{x}_k)$ around $\hat{\mathbf{x}}_k$ and $\hat{\mathbf{x}}_k^-$, respectively. Specifically, $\mathbf{F}(k, \mathbf{x}_k)$ and $\mathbf{H}(k, \mathbf{x}_k)$ are approximated as follows

$$\mathbf{F}(k, \mathbf{x}_k) \approx \mathbf{F}(\mathbf{x}, \hat{\mathbf{x}}_k) + \mathbf{F}_{k+1,k}(\mathbf{x}, \hat{\mathbf{x}}_k), \tag{1.56}$$

$$\mathbf{H}(k, \mathbf{x}_k) \approx \mathbf{H}(\mathbf{x}, \hat{\mathbf{x}}_k^-) + \mathbf{H}_{k+1,k}(\mathbf{x}, \hat{\mathbf{x}}_k^-). \tag{1.57}$$

With the above approximate expressions at hand, we may now proceed to approximate the nonlinear state equations (1.52) and (1.53) as shown by, respectively,

$$\mathbf{x}_{k+1} \approx \mathbf{F}_{k+1,k}\mathbf{x}_k + \mathbf{w}_k + \mathbf{d}_k,$$

$$\bar{\mathbf{y}}_k \approx \mathbf{H}_k\mathbf{x}_k + \mathbf{v}_k,$$

where we have introduced two new quantities:

$$\bar{\mathbf{y}}_k = \mathbf{y}_k - \{\mathbf{h}(\mathbf{x}, \hat{\mathbf{x}}_k^-) - \mathbf{H}_k \hat{\mathbf{x}}_k^-\}, \tag{1.58}$$

$$\mathbf{d}_k = \mathbf{f}(\mathbf{x}, \hat{\mathbf{x}}_k) - \mathbf{F}_{k+1,k} \hat{\mathbf{x}}_k. \tag{1.59}$$

The entries in the term $\bar{\mathbf{y}}_k$ are all known at time k, and, therefore, $\bar{\mathbf{y}}_k$ can be regarded as an observation vector at time n. Likewise, the entries in the term $\mathbf{d}_k$ are all known at time k.

Table 1.3 Extended Kalman filter

State-space model

$$\mathbf{x}_{k+1} = \mathbf{f}(k, \mathbf{x}_k) + \mathbf{w}_k,$$
$$\mathbf{y}_k = \mathbf{h}(k, \mathbf{x}_k) + \mathbf{v}_k,$$

where $\mathbf{w}_k$ and $\mathbf{v}_k$ are independent, zero mean, Gaussian noise processes of covariance matrices $\mathbf{Q}_k$ and $\mathbf{R}_k$, respectively.

Definitions

$$\mathbf{F}_{k+1,k} = \frac{\partial \mathbf{f}(k, \mathbf{x})}{\partial \mathbf{x}} |_{\mathbf{x}=\mathbf{x}_k},$$
$$\mathbf{H}_k = \frac{\partial \mathbf{h}(k, \mathbf{x})}{\partial \mathbf{x}} |_{\mathbf{x}=\mathbf{x}_k^-}.$$

Initialization: For $k = 0$, set

$$\hat{\mathbf{x}}_0 = E[\mathbf{x}_0],$$
$$\mathbf{P}_0 = E[(\mathbf{x}_0 - E[\mathbf{x}_0])(\mathbf{x}_0 - E[\mathbf{x}_0])^T].$$

Computation: For $k = 1, 2, \ldots,$ compute:

State estimate propagation

$$\hat{\mathbf{x}}_k^- = \mathbf{f}(k, \hat{\mathbf{x}}_{k-1});$$

Error covariance propagation

$$\mathbf{P}_k^- = \mathbf{F}_{k,k-1} \mathbf{P}_{k-1} \mathbf{F}_{k,k-1}^T + \mathbf{Q}_{k-1};$$

Kalman gain matrix

$$\mathbf{G}_k = \mathbf{P}_k^- \mathbf{H}_k^T \left[\mathbf{H}_k \mathbf{P}_k^- \mathbf{H}_k^T + \mathbf{R}_k\right]^{-1};$$

State estimate update

$$\hat{\mathbf{x}}_k = \hat{\mathbf{x}}_k^- + \mathbf{G}_k \mathbf{y}_k - \mathbf{h}(k, \hat{\mathbf{x}}_k^-);$$

Error covariance update

$$\mathbf{P}_k = (\mathbf{I} - \mathbf{G}_k \mathbf{H}_k) \mathbf{P}_k^-.$$

Given the linearized state-space model of Eqs. (1.58) and (1.59), we may then proceed and apply the Kalman filter theory of Section 1.3 to derive the extended Kalman filter. Table 1.2 summarizes the recursions involved in computing the extended Kalman filter.

1.7 SUMMARY

The basic Kalman filter is a linear, discrete-time, finite-dimensional system, which is endowed with a recursive structure that makes a digital computer well suited for its implementation. A key property of the Kalman filter is that it is the *minimum mean-square* (*variance*) *estimator of the state* of a linear dynamical system.

The Kalman filter, summarized in Table 1.1, applies to a linear dynamical system, the state space model of which consists of two equations:

- The process equation that defines the evolution of the state with time.
- The measurement equation that defines the observable in terms of the state.

The model is stochastic owing to the additive presence of process noise and measurement noise, which are assumed to be Gaussian with zero mean and known covariance matrices.

The Rauch–Tung–Striebel smoother, summarized in Table 1.2, builds on the Kalman filter to solve the optimum smoothing problem in an efficient manner. This smoother consists of two components: a forward filter based on the basic Kalman filter, and a combined backward filter and smoother.

Applications of Kalman filter theory may be extended to nonlinear dynamical systems, as summarized in Table 1.3. The derivation of the extended Kalman filter hinges on linearization of the nonlinear state-space model on the assumption that deviation from linearity is of first order.

REFERENCES

[1] R.E. Kalman, "A new approach to linear filtering and prediction problems," *Transactions of the ASME, Ser. D, Journal of Basic Engineering*, **82**, 34–45 (1960).

[2] F.H. Lewis, *Optical Estimation with an Introduction to Stochastic Control Theory.* New York: Wiley, 1986.

[3] M.S. Grewal and A.P. Andrews, *Kalman Filtering: Theory and Practice.* Englewood Cliffs, NJ: Prentice-Hall, 1993.

[4] H.L. Van Trees, *Detection, Estimation, and Modulation Theory*, Part I. New York: Wiley, 1968.

[5] R.S. Bucy and P.D. Joseph, *Filtering for Stochastic Processes, with Applications to Guidance.* New York: Wiley, 1968.

[6] P.G. Kaminski, A.E. Bryson, Jr., and S.F. Schmidt, "Discrete square root filtering: a survey of current techniques," *IEEE Transactions on Automatic Control*, **16**, 727–736 (1971).

[7] S. Haykin, *Adaptive Filter Theory*, 3rd ed. Upper Saddle River, NJ: Prentice-Hall, 1996.

[8] H.E. Rauch, "Solutions to the linear smoothing problem," *IEEE Transactions on Automatic Control*, **11**, 371–372 (1963).

[9] H.E. Rauch, F. Tung, and C.T. Striebel, "Maximum likelihood estimates of linear dynamic systems," *AIAA Journal*, **3**, 1445–1450 (1965).

[10] A.H. Jazwinski, *Stochastic Processes and Filtering Theory.* New York: Academic Press, 1970.

[11] P.S. Maybeck, *Stochastic Models, Estimation and Control*, Vol. 1. New York: Academic Press, 1979.

[12] P.S. Maybeck, *Stochastic Models, Estimation, and Control*, Vol. 2. New York: Academic Press, 1982.

2

PARAMETER-BASED KALMAN FILTER TRAINING: THEORY AND IMPLEMENTATION

Gintaras V. Puskorius and Lee A. Feldkamp

Ford Research Laboratory, Ford Motor Company, Dearborn, Michigan, U.S.A.
(gpuskori@ford.com, lfeldkam@ford.com)

2.1 INTRODUCTION

Although the rediscovery in the mid 1980s of the backpropagation algorithm by Rumelhart, Hinton, and Williams [1] has long been viewed as a landmark event in the history of neural network computing and has led to a sustained resurgence of activity, the relative ineffectiveness of this simple gradient method has motivated many researchers to develop enhanced training procedures. In fact, the neural network literature has been inundated with papers proposing alternative training

Kalman Filtering and Neural Networks, Edited by Simon Haykin
ISBN 0-471-36998-5

methods that are claimed to exhibit superior capabilities in terms of training speed, mapping accuracy, generalization, and overall performance relative to standard backpropagation and related methods.

Amongst the most promising and enduring of enhanced training methods are those whose weight update procedures are based upon second-order derivative information (whereas standard backpropagation exclusively utilizes first-derivative information). A variety of second-order methods began to be developed and appeared in the published neural network literature shortly after the seminal article on backpropagation was published. The vast majority of these methods can be characterized as batch update methods, where a single weight update is based on a matrix of second derivatives that is approximated on the basis of many training patterns. Popular second-order methods have included weight updates based on quasi-Newton, Levenburg–Marquardt, and conjugate gradient techniques. Although these methods have shown promise, they are often plagued by convergence to poor local optima, which can be partially attributed to the lack of a stochastic component in the weight update procedures. Note that, unlike these second-order methods, weight updates using standard backpropagation can either be performed in batch or instance-by-instance mode.

The extended Kalman filter (EKF) forms the basis of a second-order neural network training method that is a practical and effective alternative to the batch-oriented, second-order methods mentioned above. The essence of the recursive EKF procedure is that, during training, in addition to evolving the weights of a network architecture in a sequential (as opposed to batch) fashion, an approximate error covariance matrix that encodes second-order information about the training problem is also maintained and evolved. The global EKF (GEKF) training algorithm was introduced by Singhal and Wu [2] in the late 1980s, and has served as the basis for the development and enhancement of a family of computationally effective neural network training methods that has enabled the application of feedforward and recurrent neural networks to problems in control, signal processing, and pattern recognition.

In their work, Singhal and Wu developed a second-order, sequential training algorithm for static multilayered perceptron networks that was shown to be substantially more effective (orders of magnitude) in terms of number of training epochs than standard backpropagation for a series of pattern classification problems. However, the computational complexity of GEKF scales as the square of the number of weights, due to the development and use of second-order information that correlates every pair of network weights, and was thus found to be impractical for all but the simplest network architectures, given the state of standard computing hardware in the early 1990s.

In response to the then-intractable computational complexity of GEKF, we developed a family of training procedures, which we named the *decoupled* EKF algorithm [3]. Whereas the GEKF procedure develops and maintains correlations between each pair of network weights, the DEKF family provides an approximation to GEKF by developing and maintaining second-order information only between weights that belong to mutually exclusive groups. We have concentrated on what appear to be some relatively natural groupings; for example, the *node-decoupled* (NDEKF) procedure models only the interactions between weights that provide inputs to the same node. In one limit of a separate group for each network weight, we obtain the *fully decoupled* EKF procedure, which tends to be only slightly more effective than standard backpropagation. In the other extreme of a single group for all weights, DEKF reduces exactly to the GEKF procedure of Singhal and Wu.

In our work, we have successfully applied NDEKF to a wide range of network architectures and classes of training problems. We have demonstrated that NDEKF is extremely effective at training feedforward as well as recurrent network architectures, for problems ranging from pattern classification to the on-line training of neural network controllers for engine idle speed control [4, 5]. We have demonstrated the effective use of dynamic derivatives computed by both forward methods, for example those based on real-time-recurrent learning (RTRL) [6, 7], as well as by truncated backpropagation through time (BPTT(h)) [8] with the parameter-based DEKF methods, and have extended this family of methods to optimize cost functions other than sum of squared errors [9], which we describe below in Sections 2.7.2 and 2.7.3.

Of the various extensions and enhancements of EKF training that we have developed, perhaps the most enabling is one that allows for EKF procedures to perform a single update of a network's weights on the basis of more than a single training instance [10–12]. As mentioned above, EKF algorithms are intrinsically sequential procedures, where, at any given time during training, a network's weight values are updated on the basis of one and only one training instance. When EKF methods or any other sequential procedures are used to train networks with distributed representations, as in the case of multilayered perceptrons and time-lagged recurrent neural networks, there is a tendency for the training procedure to concentrate on the most recently observed training patterns, to the detriment of training patterns that had been observed and processed a long time in the past. This situation, which has been called the *recency* phenomenon, is particularly troublesome for training of recurrent neural networks and/or neural network controllers, where the temporal order of presentation of data during training must be respected. It is likely that sequential training procedures will perform greedily for these systems, for example by merely changing a network's output bias during training to accommodate a new region of operation. On the other hand, the off-line training of static networks can circumvent difficulties associated with the recency effect by employing a scrambling of the sequence of data presentation during training.

The recency phenomenon can be at least partially mitigated in these circumstances by providing a mechanism that allows for multiple training instances, preferably from different operating regions, to be simultaneously considered for each weight vector update. *Multistream* EKF training is an extension of EKF training methods that allows for multiple training instances to be batched, while remaining consistent with the Kalman methods.

We begin with a brief discussion of the types of feedforward and recurrent network architectures that we are going to consider for training by EKF methods. We then discuss the *global* EKF training method, followed by recommendations for setting of parameters for EKF methods, including the relationship of the choice of learning rate to the initialization of the error covariance matrix. We then provide treatments of the decoupled extended Kalman filter (DEKF) method as well as the multistream procedure that can be applied with any level of decoupling. We discuss at length a variety of issues related to computer implementation, including derivative calculations, computationally efficient formulations, methods for avoiding matrix inversions, and square-root filtering for computational stability. This is followed by a number of special topics, including training with constrained weights and alternative cost functions. We then provide an overview of applications of EKF methods to a series of problems in control, diagnosis, and modeling of automotive powertrain systems. We conclude the chapter with a discussion of the virtues and limitations of EKF training methods, and provide a series of guidelines for implementation and use.

2.2 NETWORK ARCHITECTURES

We consider in this chapter two types of network architecture: the well-known feedforward layered network and its dynamic extension, the recurrent multilayered perceptron (RMLP). A block-diagram representa-

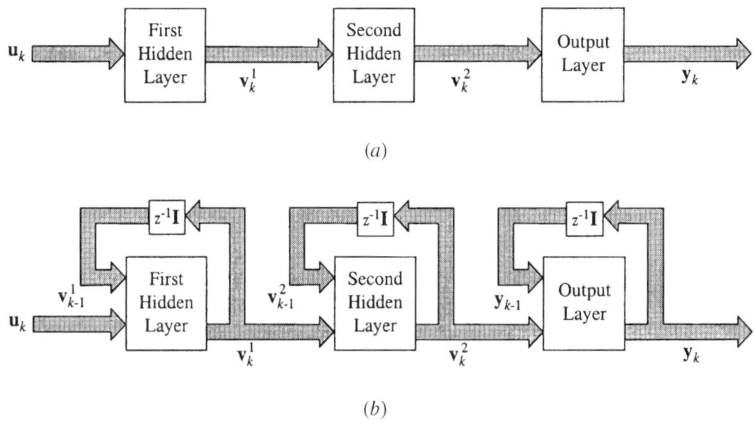

Figure 2.1 Block-diagram representation of two hidden layer networks. (*a*) depicts a feedforward layered neural network that provides a static mapping between the input vector $\mathbf{u}_k$ and the output vector $\mathbf{y}_k$. (*b*) depicts a recurrent multilayered perceptron (RMLP) with two hidden layers. In this case, we assume that there are time-delayed recurrent connections between the outputs and inputs of all nodes within a layer. The signals $\mathbf{v}_k^i$ denote the node activations for the *i*th layer. Both of these block representations assume that bias connections are included in the feedforward connections.

tion of these types of networks is given in Figure 2.1. Figure 2.2 shows an example network, denoted as a 3-3-3-2 network, with three inputs, two hidden layers of three nodes each, and an output layer of two nodes. Figure 2.3 shows a similar network, but modified to include interlayer, time-delayed recurrent connections. We denote this as a 3-3R-3R-2R RMLP, where the letter "R" denotes a recurrent layer. In this case, both hidden layers as well as the output layer are recurrent. The essential difference between the two types of networks is the recurrent network's ability to encode temporal information. Once trained, the feedforward

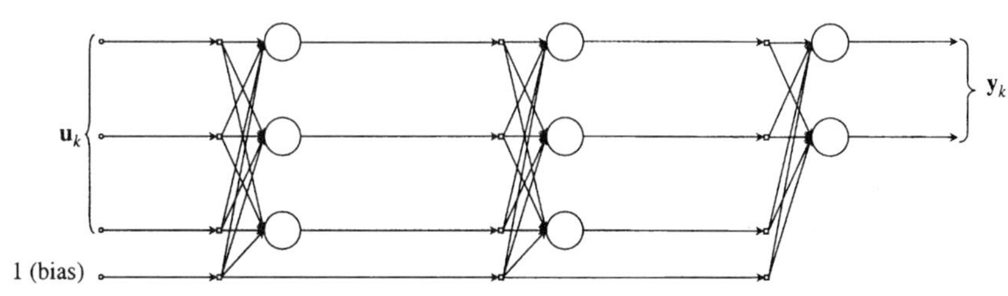

Figure 2.2 A schematic diagram of a 3-3-3-2 feedforward network architecture corresponding to the block diagram of Figure 2.1*a*.

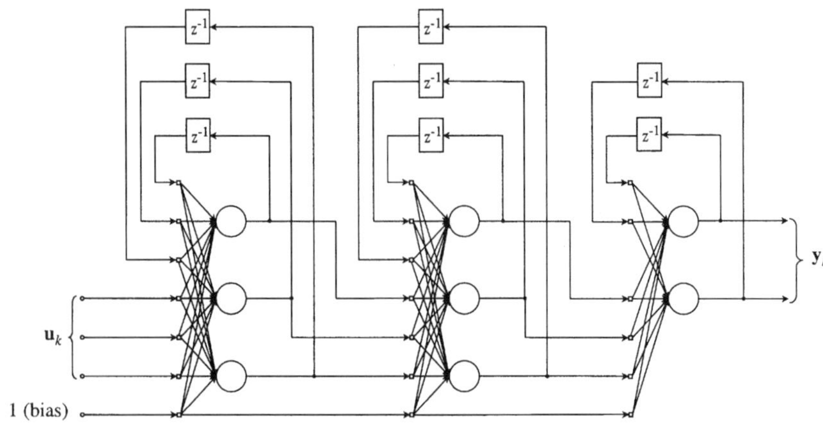

Figure 2.3. A schematic diagram of a 3-3R-3R-2R recurrent network architecture corresponding to the block diagram of Figure 2.1*b*. Note the presence of time delay operators and recurrent connections between the nodes of a layer.

network merely carries out a static mapping from input signals $\mathbf{u}_k$ to outputs $\mathbf{y}_k$, such that the output is independent of the history in which input signals are presented. On the other hand, a trained RMLP provides a dynamic mapping, such that the output $\mathbf{y}_k$ is not only a function of the current input pattern $\mathbf{u}_k$, but also implicitly a function of the entire history of inputs through the time-delayed recurrent node activations, given by the vectors $\mathbf{v}_{k-1}^i$, where i indexes layer number.

2.3 THE EKF PROCEDURE

We begin with the equations that serve as the basis for the derivation of the EKF family of neural network training algorithms. A neural network's behavior can be described by the following nonlinear discrete-time system:

$$\mathbf{w}_{k+1} = \mathbf{w}_k + \boldsymbol{\omega}_k \tag{2.1}$$

$$\mathbf{y}_k = \mathbf{h}_k(\mathbf{w}_k, \mathbf{u}_k, \mathbf{v}_{k-1}) + \boldsymbol{\nu}_k. \tag{2.2}$$

The first of these, known as the process equation, merely specifies that the state of the ideal neural network is characterized as a stationary process corrupted by process noise $\boldsymbol{\omega}_k$, where the state of the system is given by the network's weight parameter values $\mathbf{w}_k$. The second equation, known as the observation or measurement equation, represents the network's desired response vector $\mathbf{y}_k$ as a nonlinear function of the input vector $\mathbf{u}_k$, the weight parameter vector $\mathbf{w}_k$, and, for recurrent networks, the recurrent node activations $\mathbf{v}_k$; this equation is augmented by random measurement noise $\boldsymbol{\nu}_k$. The measurement noise $\boldsymbol{\nu}_k$ is typically characterized as zero-mean, white noise with covariance given by $E[\boldsymbol{\nu}_k \boldsymbol{\nu}_l^T] = \delta_{k,l}\mathbf{R}_k$. Similarly, the process noise $\boldsymbol{\omega}_k$ is also characterized as zero-mean, white noise with covariance given by $E[\boldsymbol{\omega}_k \boldsymbol{\omega}_l^T] = \delta_{k,l}\mathbf{Q}_k$.

2.3.1 Global EKF Training

The training problem using Kalman filter theory can now be described as finding the minimum mean-squared error estimate of the state $\mathbf{w}$ using all observed data so far. We assume a network architecture with M weights and N_o output nodes and cost function components. The EKF solution to the training problem is given by the following recursion (see Chapter 1):

$$\mathbf{A}_k = [\mathbf{R}_k + \mathbf{H}_k^T \mathbf{P}_k \mathbf{H}_k]^{-1}, \tag{2.3}$$

$$\mathbf{K}_k = \mathbf{P}_k \mathbf{H}_k \mathbf{A}_k, \tag{2.4}$$

$$\hat{\mathbf{w}}_{k+1} = \hat{\mathbf{w}}_k + \mathbf{K}_k \boldsymbol{\xi}_k, \tag{2.5}$$

$$\mathbf{P}_{k+1} = \mathbf{P}_k - \mathbf{K}_k \mathbf{H}_k^T \mathbf{P}_k + \mathbf{Q}_k. \tag{2.6}$$

The vector $\hat{\mathbf{w}}_k$ represents the estimate of the state (i.e., weights) of the system at update step k. This estimate is a function of the Kalman gain matrix $\mathbf{K}_k$ and the error vector $\boldsymbol{\xi}_k = \mathbf{y}_k - \hat{\mathbf{y}}_k$, where $\mathbf{y}_k$ is the target vector and $\hat{\mathbf{y}}_k$ is the network's output vector for the kth presentation of a training pattern. The Kalman gain matrix is a function of the approximate error covariance matrix $\mathbf{P}_k$, a matrix of derivatives of the network's outputs with respect to all trainable weight parameters $\mathbf{H}_k$, and a global scaling matrix $\mathbf{A}_k$. The matrix $\mathbf{H}_k$ may be computed via static backpropagation or backpropagation through time for feedforward and recurrent networks, respectively (described below in Section 2.6.1). The scaling matrix $\mathbf{A}_k$ is a function of the measurement noise covariance matrix $\mathbf{R}_k$, as well as of the matrices $\mathbf{H}_k$ and $\mathbf{P}_k$. Finally, the approximate error covariance matrix $\mathbf{P}_k$ evolves recursively with the weight vector estimate; this matrix encodes second derivative information about the training problem, and is augmented by the covariance matrix of the process noise $\mathbf{Q}_k$. This algorithm attempts to find weight values that minimize the sum of squared error $\sum_k \boldsymbol{\xi}_k^T \boldsymbol{\xi}_k$. Note that the algorithm requires that the measurement and process noise covariance matrices, $\mathbf{R}_k$ and $\mathbf{Q}_k$, be specified for all training instances. Similarly, the approximate error covariance matrix $\mathbf{P}_k$ must be initialized at the beginning of training. We consider these issues below in Section 2.3.3.

GEKF training is carried out in a sequential fashion as shown in the signal flow diagram of Figure 2.4. One step of training involves the following steps:

1. An input training pattern $\mathbf{u}_k$ is propagated through the network to produce an output vector $\hat{\mathbf{y}}_k$. Note that the forward propagation is a function of the recurrent node activations $\mathbf{v}_{k-1}$ from the previous time step for RMLPs. The error vector $\boldsymbol{\xi}_k$ is computed in this step as well.
2. The derivative matrix $\mathbf{H}_k$ is obtained by backpropagation. In this case, there is a separate backpropagation for each component of the output vector $\hat{\mathbf{y}}_k$, and the backpropagation phase will involve a time history of recurrent node activations for RMLPs.
3. The Kalman gain matrix is computed as a function of the derivative matrix $\mathbf{H}_k$, the approximate error covariance matrix $\mathbf{P}_k$, and the measurement covariance noise matrix $\mathbf{R}_k$. Note that this step includes the computation of the global scaling matrix $\mathbf{A}_k$.
4. The network weight vector is updated using the Kalman gain matrix $\mathbf{K}_k$, the error vector $\boldsymbol{\xi}_k$, and the current values of the weight vector $\hat{\mathbf{w}}_k$.

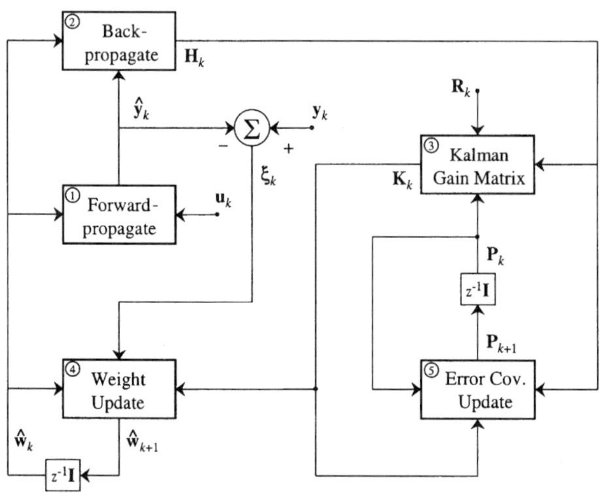

Figure 2.4 Signal flow diagram for EKF neural network training. The first two steps, comprising the forward- and backpropagation operations, will depend on whether or not the network being trained has recurrent connections. On the other hand, the EKF calculations encoded by steps (3)–(5) are independent of network type.

5. The approximate error covariance matrix is updated using the Kalman gain matrix $\mathbf{K}_k$, the derivative matrix $\mathbf{H}_k$, and the current values of the approximate error covariance matrix $\mathbf{P}_k$. Although not shown, this step also includes augmentation of the error covariance matrix by the covariance matrix of the process noise $\mathbf{Q}_k$.

2.3.2 Learning Rate and Scaled Cost Function

We noted above that $\mathbf{R}_k$ is the covariance matrix of the measurement noise and that this matrix must be specified for each training pattern. Generally speaking, training problems that are characterized by noisy measurement data usually require that the elements of $\mathbf{R}_k$ be scaled larger than for those problems with relatively noise-free training data. In [5, 7, 12], we interpret this measurement error covariance matrix to represent an inverse learning rate: $\mathbf{R}_k = \eta_k^{-1}\mathbf{S}_k^{-1}$, where the training cost function at time step k is now given by $\mathscr{E}_k = \frac{1}{2}\boldsymbol{\xi}_k^T\mathbf{S}_k\boldsymbol{\xi}_k$, and $\mathbf{S}_k$ allows the various network output components to be scaled nonuniformly. Thus, the global scaling matrix $\mathbf{A}_k$ of equation (2.3) can be written as

$$\mathbf{A}_k = \left[\frac{1}{\eta_k}\mathbf{S}_k^{-1} + \mathbf{H}_k^T\mathbf{P}_k\mathbf{H}_k\right]^{-1}. \tag{2.7}$$

The use of the weighting matrix $\mathbf{S}_k$ in Eq. (2.7) poses numerical difficulties when the matrix is singular.[1] We reformulate the GEKF algorithm to eliminate this difficulty by distributing the square root of the weighting matrix into both the derivative matrices as $\mathbf{H}_k^* = \mathbf{H}_k\mathbf{S}_k^{1/2}$ and the error vector as $\boldsymbol{\xi}_k^* = \mathbf{S}_k^{1/2}\boldsymbol{\xi}_k$. The matrices $\mathbf{H}_k^*$ thus contain the scaled derivatives of network outputs with respect to the weights of the network. The rescaled extended Kalman recursion is then given by

$$\mathbf{A}_k^* = \left[\frac{1}{\eta_k}\mathbf{I} + (\mathbf{H}_k^*)^T\mathbf{P}_k\mathbf{H}_k^*\right]^{-1}, \tag{2.8}$$

$$\mathbf{K}_k^* = \mathbf{P}_k\mathbf{H}_k^*\mathbf{A}_k^*, \tag{2.9}$$

$$\hat{\mathbf{w}}_{k+1} = \hat{\mathbf{w}}_k + \mathbf{K}_k^*\boldsymbol{\xi}_k^*, \tag{2.10}$$

$$\mathbf{P}_{k+1} = \mathbf{P}_k - \mathbf{K}_k^*(\mathbf{H}_k^*)^T\mathbf{P}_k + \mathbf{Q}_k. \tag{2.11}$$

Note that this rescaling does not change the evolution of either the weight vector or the approximate error covariance matrix, and eliminates the need

[1]This may occur when we utilize penalty functions to impose explicit constraints on network outputs. For example, when a constraint is not violated, we set the corresponding diagonal element of $\mathbf{S}_k$ to zero, thereby rendering the matrix singular.

to compute the inverse of the weighting matrix $\mathbf{S}_k$ for each training pattern. For the sake of clarity in the remainder of this chapter, we shall assume a uniform scaling of output signals, $\mathbf{S}_k = \mathbf{I}$, which implies $\mathbf{R}_k = \eta_k^{-1}\mathbf{I}$, and drop the asterisk notation.

2.3.3 Parameter Settings

EKF training algorithms require the setting of a number of parameters. In practice, we have employed the following rough guidelines. First, we typically assume that the input–output data have been scaled and transformed to reasonable ranges (e.g., zero mean, unit variance for all continuous input and output variables). We also assume that weight values are initialized to small random values drawn from a zero-mean uniform or normal distribution. The approximate error covariance matrix is initialized to reflect the fact that no a priori knowledge was used to initialize the weights; this is accomplished by setting $\mathbf{P}_0 = \epsilon^{-1}\mathbf{I}$, where ϵ is a small number (of the order of 0.001–0.01). As noted above, we assume uniform scaling of outputs: $\mathbf{S}_k = \mathbf{I}$. Then, training data that are characterized by noisy measurements usually require small values for the learning rate η_k to achieve good training performance; we typically bound the learning rate to values between 0.001 and 1. Finally, the covariance matrix $\mathbf{Q}_k$ of the process noise is represented by a scaled identity matrix $q_k\mathbf{I}$, with the scale factor q_k ranging from as small as zero (to represent no process noise) to values of the order of 0.1. This factor is generally annealed from a large value to a limiting value of the order of 10^{-6}. This annealing process helps to accelerate convergence and, by keeping a nonzero value for the process noise term, helps to avoid divergence of the error covariance update in Eqs. (2.6) and (2.11).

We show here that the setting of the learning rate, the process noise covariance matrix, and the initialization of the approximate error covariance matrix are interdependent, and that an arbitrary scaling can be applied to $\mathbf{R}_k$, $\mathbf{P}_k$, and $\mathbf{Q}_k$ without altering the evolution of the weight vector $\hat{\mathbf{w}}$ in Eqs. (2.5) and (2.10). First consider the Kalman gain of Eqs. (2.4) and (2.9). An arbitrary positive scaling factor μ can be applied to $\mathbf{R}_k$ and $\mathbf{P}_k$ without altering the contents of $\mathbf{K}_k$:

$$\begin{aligned}\mathbf{K}_k &= \mathbf{P}_k\mathbf{H}_k[\mathbf{R}_k + \mathbf{H}_k^T\mathbf{P}_k\mathbf{H}_k]^{-1}\\ &= \mu\mathbf{P}_k\mathbf{H}_k[\mu\mathbf{R}_k + \mathbf{H}_k^T\mu\mathbf{P}_k\mathbf{H}_k]^{-1}\\ &= \mathbf{P}_k^\dagger\mathbf{H}_k[\mathbf{R}_k^\dagger + \mathbf{H}_k^T\mathbf{P}_k^\dagger\mathbf{H}_k]^{-1}\\ &= \mathbf{P}_k^\dagger\mathbf{H}_k\mathbf{A}_k^\dagger,\end{aligned}$$

where we have defined $\mathbf{R}_k^\dagger = \mu \mathbf{R}_k$, $\mathbf{P}_k^\dagger = \mu \mathbf{P}_k$, and $\mathbf{A}_k^\dagger = \mu^{-1}\mathbf{A}_k$. Similarly, the approximate error covariance update becomes

$$\begin{aligned}\mathbf{P}_{k+1}^\dagger &= \mu \mathbf{P}_{k+1} \\ &= \mu \mathbf{P}_k - \mathbf{K}_k \mathbf{H}_k^T \mu \mathbf{P}_k + \mu \mathbf{Q}_k \\ &= \mathbf{P}_k^\dagger - \mathbf{K}_k \mathbf{H}_k^T \mathbf{P}_k^\dagger + \mathbf{Q}_k^\dagger.\end{aligned}$$

This implies that a training trial characterized by the parameter settings $\mathbf{R}_k = \eta^{-1}\mathbf{I}$, $\mathbf{P}_0 = \epsilon^{-1}\mathbf{I}$, and $\mathbf{Q}_k = q\mathbf{I}$, would behave identically to a training trial with scaled versions of these parameter settings: $\mathbf{R}_k = \mu\eta^{-1}\mathbf{I}$, $\mathbf{P}_0 = \mu\epsilon^{-1}\mathbf{I}$, and $\mathbf{Q}_k = \mu q\mathbf{I}$. Thus, for any given EKF training problem, there is no one best set of parameter settings, but a continuum of related settings that must take into account the properties of the training data for good performance. This also implies that only two effective parameters need to be set. Regardless of the training problem considered, we have typically chosen the initial error covariance matrix to be $\mathbf{P}_0 = \epsilon^{-1}\mathbf{I}$, with $\epsilon = 0.01$ and 0.001 for sigmoidal and linear activation functions, respectively. This leaves us to specify values for η_k and $\mathbf{Q}_k$, which are likely to be problem-dependent.

2.4 DECOUPLED EKF (DEKF)

The computational requirements of GEKF are dominated by the need to store and update the approximate error covariance matrix $\mathbf{P}_k$ at each time step. For a network architecture with N_o outputs and M weights, GEKF's computational complexity is $\mathcal{O}(N_o M^2)$ and its storage requirements are $\mathcal{O}(M^2)$. The parameter-based DEKF algorithm is derived from GEKF by assuming that the interactions between certain weight estimates can be ignored. This simplification introduces many zeroes into the matrix $\mathbf{P}_k$. If the weights are decoupled so that the weight groups are mutually exclusive of one another, then $\mathbf{P}_k$ can be arranged into block-diagonal form. Let g refer to the number of such weight groups. Then, for group i, the vector $\hat{\mathbf{w}}_k^i$ refers to the estimated weight parameters, $\mathbf{H}_k^i$ is the submatrix of derivatives of network outputs with respect to the ith group's weights, $\mathbf{P}_k^i$ is the weight group's approximate error covariance matrix, and $\mathbf{K}_k^i$ is its Kalman gain matrix. The concatenation of the vectors $\hat{\mathbf{w}}_k^i$ forms the vector $\hat{\mathbf{w}}_k$. Similarly, the global derivative matrix $\mathbf{H}_k$ is composed via concatenation of the individual submatrices $\mathbf{H}_k^i$. The DEKF algorithm for the ith weight group is given by

$$\mathbf{A}_k = \left[\mathbf{R}_k + \sum_{j=1}^{g} (\mathbf{H}_k^j)^T \mathbf{P}_k^j \mathbf{H}_k^j \right]^{-1}, \tag{2.12}$$

$$\mathbf{K}_k^i = \mathbf{P}_k^i \mathbf{H}_k^i \mathbf{A}_k, \tag{2.13}$$

$$\hat{\mathbf{w}}_{k+1}^i = \hat{\mathbf{w}}_k^i + \mathbf{K}_k^i \boldsymbol{\xi}_k, \tag{2.14}$$

$$\mathbf{P}_{k+1}^i = \mathbf{P}_k^i - \mathbf{K}_k^i (\mathbf{H}_k^i)^T \mathbf{P}_k^i + \mathbf{Q}_k^i. \tag{2.15}$$

A single global sealing matrix $\mathbf{A}_k$, computed with contributions from all of the approximate error covariance matrices and derivative matrices, is used to compute the Kalman gain matrices, $\mathbf{K}_k^i$. These gain matrices are used to update the error covariance matrices for all weight groups, and are combined with the global error vector $\boldsymbol{\xi}_k$ for updating the weight vectors. In the limit of a single weight group ($g = 1$), the DEKF algorithm reduces exactly to the GEKF algorithm.

The computational complexity and storage requirements for DEKF can be significantly less than those of GEKF. For g disjoint weight groups, the computational complexity of DEKF becomes $\mathcal{O}(N_o^2 M + N_o \sum_{i=1}^{g} M_i^2)$, where M_i is the number of weights in group i, while the storage requirements become $\mathcal{O}(\sum_{i=1}^{g} M_i^2)$. Note that this complexity analysis does not include the computational requirements for the matrix of derivatives, which is independent of the level of decoupling. It should be noted that in the case of training recurrent networks or networks as feedback controllers, the computational complexity of the derivative calculations can be significant.

We have found that decoupling of the weights of the network by node (i.e., each weight group is composed of a single node's weight) is rather natural and leads to compact and efficient computer implementations. Furthermore, this level of decoupling typically exhibits substantial computational savings relative to GEKF, often with little sacrifice in network performance after completion of training. We refer to this level of decoupling as *node-decoupled* EKF or NDEKF. Other forms of decoupling considered have been *fully decoupled* EKF, in which each individual weight constitutes a unique group (thereby resulting in an error covariance matrix that has diagonal structure), and *layer-decoupled* EKF, in which weights are grouped by the layer to which they belong [13]. We show an example of the effect of all four levels of decoupling on the structure of

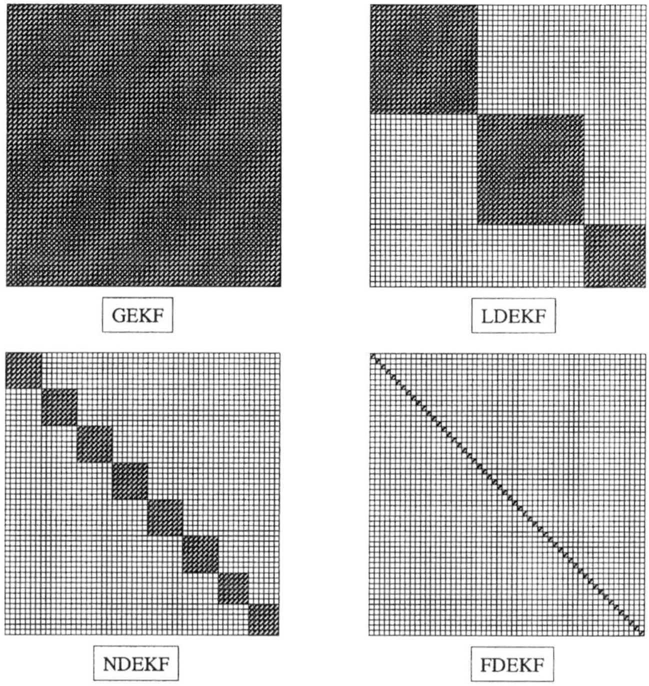

Figure 2.5 Block-diagonal representation of the approximate error covariance matrix $\mathbf{P}_k$ for the RMLP network shown in Figure 2.3 for four different levels of decoupling. This network has two recurrent layers with three nodes each and each node with seven incoming connections. The output layer is also recurrent, but its two nodes only have six connections each. Only the shaded portions of these matrices are updated and maintained for the various forms of decoupling shown. Note that we achieve a reduction by nearly a factor of 8 in computational complexity for the case of node decoupling relative to GEKF in this example.

the approximate error covariance matrix in Figure 2.5. For the remainder of this chapter, we explicitly consider only two different levels of decoupling for EKF training: global and node-decoupled EKF.

2.5 MULTISTREAM TRAINING

Up to this point, we have considered forms of EKF training in which a single weight-vector update is performed on the basis of the presentation of a single input–output training pattern. However, there may be situations for which a *coordinated* weight update, on the basis of multiple training patterns, would be advantageous. We consider in this section an abstract example of such a situation, and describe the means by which the EKF method can be naturally extended to simultaneously handle multiple training instances for a single weight update.[2]

Consider the standard recurrent network training problem: training on a sequence of input–output pairs. If the sequence is in some sense homogeneous, then one or more linear passes through the data may well produce good results. However, in many training problems, especially those in which external inputs are present, the data sequence is heterogeneous. For example, regions of rapid variation of inputs and outputs may be followed by regions of slow change. Alternatively, a sequence of outputs that centers about one level may be followed by one that centers about a different level. In any case, the tendency always exists in a straightforward training process for the network weights to be adapted unduly in favor of the currently presented training data. This *recency effect* is analogous to the difficulty that may arise in training feedforward networks if the data are repeatedly presented in the same order.

In this latter case, an effective solution is to scramble the order of presentation; another is to use a batch update algorithm. For recurrent networks, the direct analog of scrambling the presentation order is to present randomly selected subsequences, making an update only for the last input–output pair of the subsequence (when the network would be expected to be independent of its initialization at the beginning of the sequence). A full batch update would involve running the network through the entire data set, computing the required derivatives that correspond to each input–output pair, and making an update based on the entire set of errors.

The multistream procedure largely circumvents the recency effect by combining features of both scrambling and batch updates. Like full batch methods, multistream training [10–12] is based on the principle that each weight update should attempt to satisfy simultaneously the demands from multiple input–output pairs. However, it retains the useful stochastic aspects of sequential updating, and requires much less computation time between updates. We now describe the mechanics of multistream training.

[2]In the case of purely linear systems, there is no advantage in batching up a collection of training instances for a single weight update via Kalman filter methods, since all weight updates are completely consistent with previously observed data. On the other hand, derivative calculations and the extended Kalman recursion for nonlinear networks utilize first-order approximations, so that weight updates are no longer guaranteed to be consistent with all previously processed data.

In a typical training problem, we deal with one or more files, each of which contains a sequence of data. Breaking the overall data into multiple files is typical in practical problems, where the data may be acquired in different sessions, for distinct modes of system operation, or under different operating conditions.

In each cycle of training, we choose a specified number N_s of randomly selected starting points in a chosen set of files. Each such starting point is the beginning of a *stream*. In the multistream procedure we progress sequentially through each stream, carrying out weight updates according to the set of current points. Copies of recurrent node outputs must be maintained separately for each stream. Derivatives are also computed separately for each stream, generally by truncated backpropagation through time (BPTT(h)) as discussed in Section 2.6.1 below. Because we generally have no prior information with which to initialize the recurrent network, we typically set all state nodes to values of zero at the start of each stream. Accordingly, the network is executed but updates are suspended for a specified number N_p of time steps, called the *priming length*, at the beginning of each stream. Updates are performed until a specified number N_t of time steps, called the *trajectory length*, have been processed. Hence, $N_t - N_p$ updates are performed in each training cycle.

If we take $N_s = 1$ and $N_t - N_p = 1$, we recover the order-scrambling procedure described above; N_t may be identified with the subsequence length. On the other hand, we recover the batch procedure if we take N_s equal to the number of time steps for which updates are to be performed, assemble streams systematically to end at the chosen N_s steps, and again take $N_t - N_p = 1$.

Generally speaking, apart from the computational overhead involved, we find that performance tends to improve as the number of streams is increased. Various strategies are possible for file selection. If the number of files is small, it is convenient to choose N_s equal to a multiple of the number of files and to select each file the same number of times. If the number of files is too large to make this practical, then we tend to select files randomly. In this case, each set of $N_t - N_p$ updates is based on only a subset of the files, so it seems reasonable not to make the trajectory length N_t too large.

An important consideration is how to carry out the EKF update procedure. If gradient updates were being used, we would simply average the updates that would have been performed had the streams been treated separately. In the case of EKF training, however, averaging separate updates is incorrect. Instead, we treat this problem as that of training a single, shared-weight network with N_oN_s outputs. From the standpoint of the EKF method, we are simply training a multiple-output network in which the number of original outputs is multiplied by the number of streams. The nature of the Kalman recursion, because of the global scaling matrix $\mathbf{A}_k$, is then to produce weight updates that are not a simple average of the weight updates that would be computed separately for each output, as is the case for a simple gradient descent weight update. Note that we are still minimizing the same sum of squared error cost function.

In single-stream EKF training, we place derivatives of network outputs with respect to network weights in the matrix $\mathbf{H}_k$ constructed from N_o column vectors, each of dimension equal to the number of trainable weights, N_w. In multistream training, the number of columns is correspondingly increased to $N_o N_s$. Similarly, the vector of errors $\boldsymbol{\xi}_k$ has $N_o N_s$ elements. Apart from these augmentations of $\mathbf{H}_k$ and $\boldsymbol{\xi}_k$, the form of the Kalman recursion is unchanged.

Given these considerations, we define the decoupled multistream EKF recursion as follows. We shall alter the temporal indexing by specifying a range of training patterns that indicate how the multi-stream recursion should be interpreted. We define $l = k + N_s - 1$ and allow the range $k : l$ to specify the batch of training patterns for which a single weight vector update will be performed. Then, the matrix $\mathbf{H}^i_{k:l}$ is the concatenation of the derivative matrices for the ith group of weights and for training patterns that have been assigned to the range $k : l$. Similarly, the augmented error vector is denoted by $\boldsymbol{\xi}_{k:l}$. We construct the derivative matrices and error vector, respectively, by

$$\mathbf{H}_{k:l} = (\mathbf{H}_k \mathbf{H}_{k+1} \mathbf{H}_{k+2} \cdots \mathbf{H}_{l-1} \mathbf{H}_l),$$

$$\boldsymbol{\xi}_{k:l} = (\boldsymbol{\xi}_k^T \boldsymbol{\xi}_{k+1}^T \boldsymbol{\xi}_{k+2}^T \cdots \boldsymbol{\xi}_{l-1}^T \boldsymbol{\xi}_l^T)^T.$$

We use a similar notation for the measurement error covariance matrix $\mathbf{R}_{k:l}$ and the global scaling matrix $\mathbf{A}_{k:l}$, both square matrices of dimension $N_o N_s$, and for the Kalman gain matrices $\mathbf{K}^i_{k:l}$, with size $M_i \times N_o N_s$. The multistream DEKF recursion is then given by

$$\mathbf{A}_{k:l} = \left[\mathbf{R}_{k:l} + \sum_{j=1}^{g} (\mathbf{H}^j_{k:l})^T \mathbf{P}^j_k \mathbf{H}^j_{k:l} \right]^{-1}, \tag{2.16}$$

$$\mathbf{K}^i_{k:l} = \mathbf{P}^i_k \mathbf{H}^i_{k:l} \mathbf{A}_{k:l}, \tag{2.17}$$

$$\hat{\mathbf{w}}^i_{k+N_s} = \hat{\mathbf{w}}^i_k + \mathbf{K}^i_{k:l} \boldsymbol{\xi}_{k:l}, \tag{2.18}$$

$$\mathbf{P}^i_{k+N_s} = \mathbf{P}^i_k - \mathbf{K}^i_{k:l} (\mathbf{H}^i_{k:l})^T \mathbf{P}^i_k + \mathbf{Q}^i_k. \tag{2.19}$$

Note that this formulation reduces correctly to the original DEKF recursion in the limit of a single stream, and that multistream GEKF is given in the case of a single weight group. We provide a block diagram representation of the multistream GEKF procedure in Figure 2.6. Note that the steps of training are very similar to the single-stream case, with the exception of multiple forward-propagation and backpropagation steps, and the concatenation operations for the derivative matrices and error vectors.

Let us consider the computational implications of the multistream method. The sizes of the approximate error covariance matrices $\mathbf{P}_k^i$ and the weight vectors $\mathbf{w}_k^i$ are independent of the chosen number of streams. On the other hand, we noted above the increase in size for the derivative matrices $\mathbf{H}_{k:l}^i$, as well as of the Kalman gain matrices $\mathbf{K}_{k:l}^i$. However, the computation required to obtain $\mathbf{H}_{k:l}^i$ and to compute updates to $\mathbf{P}_k^i$ is the same as for N_s separate updates. The major additional computational burden is the inversion required to obtain the matrix $\mathbf{A}_{k:l}$ whose dimension is N_s times larger than in the single-stream case. Even this cost tends to be small compared with that associated with the $\mathbf{P}_k^i$ matrices, as long as

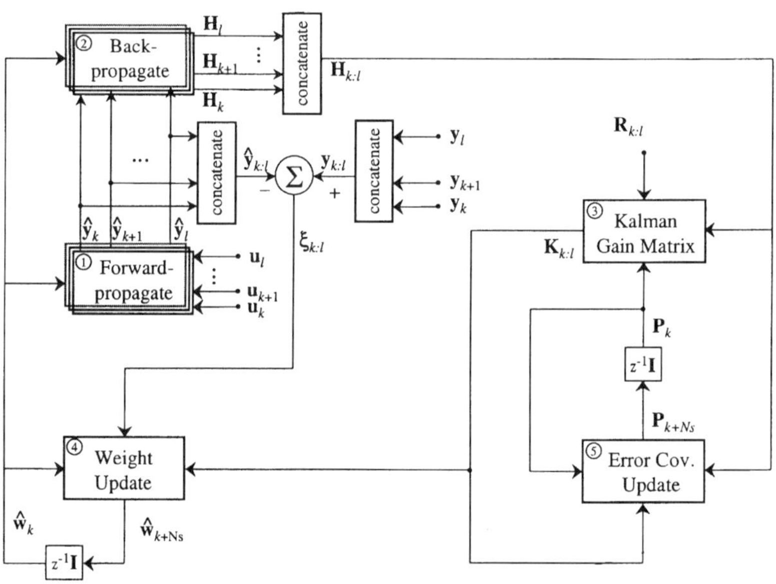

Figure 2.6 Signal flow diagram for multistream EKF neural network training. The first two steps are comprised of multiple forward- and backpropagation operations, determined by the number of streams N_s selected; these steps also depend on whether or not the network being trained has recurrent connections. On the other hand, once the derivative matrix $\mathbf{H}_{k:l}$ and error vector $\xi_{k:l}$ are formed, the EKF steps encoded by steps (3)–(5) are independent of number of streams and network type.

$N_o N_s$ is smaller than the number of network weights (GEKF) or the maximum number of weights in a group (DEKF).

If the number of streams chosen is so large as to make the inversion of $\mathbf{A}_{k:l}$ impractical, the inversion may be avoided by using one of the alternative EKF formulations described below in Section 2.6.3.

2.5.1 Some Insight into the Multistream Technique

A simple means of motivating how multiple training instances can be used simultaneously for a single weight update via the EKF procedure is to consider the training of a single linear node. In this case, the application of EKF training is equivalent to that of the recursive least-squares (RLS) algorithm. Assume that a training data set is represented by m unique training patterns. The kth training pattern is represented by a d-dimensional input vector $\mathbf{u}_k$, where we assume that all input vectors include a constant bias component of value equal to 1, and a 1-dimensional output target y_k. The simple linear model for this system is given by

$$\hat{y}_k = \mathbf{u}_k^T \mathbf{w}_f, \tag{2.20}$$

where $\mathbf{w}_f$ is the single node's d-dimensional weight vector. The weight vector $\mathbf{w}_f$ can be found by applying m iterations of the RLS procedure as follows:

$$a_k = [1 + \mathbf{u}_k^T \mathbf{P}_k \mathbf{u}_k]^{-1}, \tag{2.21}$$

$$\mathbf{k}_k = \mathbf{P}_k \mathbf{u}_k a_k, \tag{2.22}$$

$$\mathbf{w}_{k+1} = \mathbf{w}_k + \mathbf{k}_k (y_k - \hat{y}_k), \tag{2.23}$$

$$\mathbf{P}_{k+1} = \mathbf{P}_k - \mathbf{k}_k \mathbf{u}_k^T \mathbf{P}_k, \tag{2.24}$$

where the diagonal elements of $\mathbf{P}_0$ are initialized to large positive values, and $\mathbf{w}_0$ to a vector of small random values. Also, $\mathbf{w}_f = \mathbf{w}_m$ after a single presentation of all training data (i.e., after a single epoch).

We recover a batch, least-squares solution to this single-node training problem via an extreme application of the multistream concept, where we associate m unique streams with each of the m training instances. In this case, we arrange the input vectors into a matrix $\mathbf{U}$ of size $d \times m$, where each column corresponds to a unique training pattern. Similarly, we arrange the target values into a single m-dimensional column vector $\mathbf{y}$, where elements of $\mathbf{y}$ are ordered identically with the matrix $\mathbf{U}$. As before, we select the initial weight vector $\mathbf{w}_0$ to consist of randomly chosen values, and we select $\mathbf{P}_0 = \epsilon^{-1}\mathbf{I}$, with ϵ small. Given the choice of initial weight vector, we can compute the network output for each training pattern, and arrange all the results using the matrix notation

$$\hat{\mathbf{y}}_0 = \mathbf{U}^T\mathbf{w}_0. \tag{2.25}$$

A single weight update step of the Kalman filter recursion applied to this m-dimensional output problem at the beginning of training can be written as

$$\mathbf{A}_0 = [\mathbf{I} + \mathbf{U}^T\mathbf{P}_0\mathbf{U}]^{-1}, \tag{2.26}$$

$$\mathbf{K}_0 = \mathbf{P}_0\mathbf{U}\mathbf{A}_0, \tag{2.27}$$

$$\mathbf{w}_1 = \mathbf{w}_0 + \mathbf{K}_0(\mathbf{y} - \hat{\mathbf{y}}_0), \tag{2.28}$$

where we have chosen not to include the error covariance update here for reasons that will soon become clear. At the beginning of training, we recognize that $\mathbf{P}_0$ is large, and we assume that the training data set is scaled so that $\mathbf{U}^T\mathbf{P}_0\mathbf{U} \gg \mathbf{I}$. This allows $\mathbf{A}_0$ to be approximated by

$$\mathbf{A}_0 \approx \epsilon[\epsilon\mathbf{I} + \mathbf{U}^T\mathbf{U}]^{-1}, \tag{2.29}$$

since $\mathbf{P}_0$ is diagonal. Given this approximation, we can write the Kalman gain matrix as

$$\mathbf{K}_0 = \mathbf{U}[\epsilon\mathbf{I} + \mathbf{U}^T\mathbf{U}]^{-1}. \tag{2.30}$$

We now substitute Eqs. (2.25) and (2.30) into Eq. (2.28) to derive the weight vector after one time step of this m-stream Kalman filter procedure:

$$\begin{aligned} \mathbf{w}_1 &= \mathbf{w}_0 + \mathbf{U}[\epsilon\mathbf{I} + \mathbf{U}^T\mathbf{U}]^{-1}[\mathbf{y} - \mathbf{U}^T\mathbf{w}_0] \\ &= \mathbf{w}_0 - \mathbf{U}[\epsilon\mathbf{I} + \mathbf{U}^T\mathbf{U}]^{-1}\mathbf{U}^T\mathbf{w}_0 + \mathbf{U}[\epsilon\mathbf{I} + \mathbf{U}^T\mathbf{U}]^{-1}\mathbf{y}. \end{aligned} \tag{2.31}$$

If we apply the matrix equality $\lim_{\epsilon\to 0}\mathbf{U}[\epsilon\mathbf{I} + \mathbf{U}^T\mathbf{U}]^{-1}\mathbf{U}^T = \mathbf{I}$, we obtain the pseudoinverse solution:

$$\mathbf{w}_f = \mathbf{w}_1 = [\mathbf{U}\mathbf{U}^T]^{-1}\mathbf{U}\mathbf{y}, \tag{2.32}$$

where we have made use of

$$\lim_{\epsilon \to 0} \mathbf{U}[\epsilon\mathbf{I} + \mathbf{U}^T\mathbf{U}]^{-1}\mathbf{U}^T = \mathbf{I}, \tag{2.33}$$

$$\lim_{\epsilon \to 0} \mathbf{U}[\epsilon\mathbf{I} + \mathbf{U}^T\mathbf{U}]^{-1}\mathbf{U}^T = [\mathbf{U}\mathbf{U}^T]^{-1}\mathbf{U}\mathbf{U}^T, \tag{2.34}$$

$$\lim_{\epsilon \to 0} \mathbf{U}[\epsilon\mathbf{I} + \mathbf{U}^T\mathbf{U}]^{-1} = [\mathbf{U}\mathbf{U}^T]^{-1}\mathbf{U}. \tag{2.35}$$

Thus, one step of the multistream Kalman recursion recovers very closely the least-squares solution. If m is too large to make the inversion operation practical, we could instead divide the problem into subsets and perform the procedure sequentially for each subset, arriving eventually at nearly the same result (in this case, however, the covariance update needs to be performed).

As illustrated in this one-node example, the multistream EKF update is not an average of the individual updates, but rather is coordinated through the global scaling matrix **A**. It is intuitively clear that this coordination is most valuable when the various streams place contrasting demands on the network.

2.5.2 Advantages and Extensions of Multistream Training

Discussions of the training of networks with external recurrence often distinguish between series–parallel and parallel configurations. In the former, target values are substituted for the corresponding network outputs during the training process. This scheme, which is also known as *teacher forcing*, helps the network to get "on track" and stay there during training. Unfortunately, it may also compromise the performance of the network when, in use, it must depend on its own output. Hence, it is not uncommon to begin with the series–parallel configuration, then switch to the parallel configuration as the network learns the task. Multistream training seems to lessen the need for the series–parallel scheme; the response of the training process to the demands of multiple streams tends to keep the network from getting too far off-track. In this respect, multistream training seems particularly well suited for training networks with internal recurrence (e.g., recurrent multilayered perceptrons), where the opportunity to use teacher forcing is limited, because *correct* values for most if not all outputs of recurrent nodes are unknown.

Though our presentation has concentrated on multistreaming simply as an enhanced training technique, one can also exploit the fact that the streams used to provide input–output data need not arise homogeneously, that is, from the same training task. Indeed, we have demonstrated that a single fixed-weight, recurrent neural network, trained by multistream EKF, can carry out multiple tasks in a control context, namely, to act as a stabilizing controller for multiple distinct and unrelated systems, without explicit knowledge of system identity [14]. This work demonstrated that the trained network was capable of exhibiting what could be considered to be adaptive behavior: the network, acting as a controller, observed the behavior of the system (through the system's output), implicitly identified which system the network was being subjected to, and then took actions to stabilize the system. We view this somewhat unexpected behavior as being the direct result of combining an effective training procedure with enabling representational capabilities that recurrent networks provide.

2.6 COMPUTATIONAL CONSIDERATIONS

We discuss here a number of topics related to implementation of the various EKF training procedures from a computational perspective. In particular, we consider issues related to computation of derivatives that are critical to the EKF methods, followed by discussions of computationally efficient formulations, methods for avoiding matrix inversions, and the use of square-root filtering as an alternative means of insuring stable performance.

2.6.1 Derivative Calculations

We discussed above both the global and decoupled versions of the EKF algorithm, where we consider the global EKF to be a limiting form of decoupled EKF (i.e., DEKF with a single weight group). In addition, we have described the multistream EKF procedure as a means of batching training instances, and have noted that multistreaming can be used with any form of decoupled EKF training, for both feedforward and recurrent networks. The various EKF procedures can all be compactly described by the DEKF recursion of Eqs. (2.12)–(2.15), where we have assumed that the derivative matrices $\mathbf{H}_k^i$ are given. However, the implications for computationally efficient and clear implementations of the various forms of EKF training depend upon the derivative calculations, which are dictated by whether a network architecture is static or dynamic (i.e., feedforward or recurrent), and whether or not multistreaming is used. Here we provide insight into the nature of derivative calculations for training of both static and dynamic networks with EKF methods (see [12] for implementation details).

We assume the convention that a network's weights are organized by node, regardless of the degree of decoupling, which allows us to naturally partition the matrix of derivatives of network outputs with respect to weight parameters, $\mathbf{H}_k$, into a set of G submatrices $\mathbf{H}_k^i$, where G is the number of nodes of the network. Then, each matrix $\mathbf{H}_k^i$ denotes the matrix of derivatives of network outputs with respect to the weights associated with the ith node of the network. For feedforward networks, these submatrices can be written as the outer product of two vectors [3],

$$\mathbf{H}_k^i = \mathbf{u}_k^i(\boldsymbol{\psi}_k^i)^T,$$

where $\mathbf{u}_k^i$ is the ith node's input vector and $\boldsymbol{\psi}_k^i$ is a vector of partial derivatives of the network's outputs with respect to the ith node's net input, defined as the dot product of the weight vector $\mathbf{w}_k^i$ with the corresponding input vector $\mathbf{u}_k^i$. Note that the vectors $\boldsymbol{\psi}_k^i$ are computed via the backpropagation process, where the dimension of each of these vectors is determined by the number of network outputs. In contrast to the standard backpropagation algorithm, which begins the derivative calculation process (i.e., backpropagation) with error signals for each of the network's outputs, and effectively combines these error signals (for multiple-output problems) during the backpropagation process, the EKF methods begin the process with signals of unity for each network output and backpropagate a separate signal for each unique network output.

In the case of recurrent networks, we assume the use of truncated backpropagation through time for calculation of derivatives, with a truncation depth of h steps; this process is denoted by BPTT(h). Now, each submatrix $\mathbf{H}_k^i$ can no longer be expressed as a simple outer product of two vectors; rather, each of these submatrices is expressed as the sum of a series of outer products:

$$\mathbf{H}_k^i = \sum_{j=1}^{h} \mathbf{H}_k^{i,j} = \sum_{j=1}^{h} \mathbf{u}_k^{i,j}(\boldsymbol{\psi}_k^{i,j})^T,$$

where the matrix $\mathbf{H}_k^{i,j}$ is the contribution from the jth step of backpropagation to the computation of the total derivative matrix for the ith node; the vector $\mathbf{u}_k^{i,j}$ is the vector of inputs to the ith node at the jth step of backpropagation; and $\boldsymbol{\psi}_k^{i,j}$ is the vector of backpropagated derivatives of network outputs with respect to the ith node's net input at the jth step of backpropagation. Here, we have chosen arbitrarily to have j increase as we step back in time.

Finally, consider multi-stream training, where we assume that the training problem involves a recurrent network architecture, with derivatives computed by BPTT(h) (feedforward networks are subsumed by the case of $h = 1$). We again assume an N_o-component cost function with N_s streams, and define $l = k + N_s - 1$. Then, each submatrix $\mathbf{H}^i_{k:l}$ becomes a concatenation of a series of submatrices, each of which is expressed as the sum of a series of outer products:

$$\mathbf{H}^i_{k:l} = \left[\sum_{j=1}^{h} \mathbf{H}_k^{i,j,1} \sum_{j=1}^{h} \mathbf{H}_k^{i,j,2} \cdots \sum_{j=1}^{h} \mathbf{H}_k^{i,j,N_s} \right] \tag{2.36}$$

$$= \left[\sum_{j=1}^{h} \mathbf{u}_k^{i,j,1} (\boldsymbol{\psi}_k^{i,j,1})^T \sum_{j=1}^{h} \mathbf{u}_k^{i,j,2} (\boldsymbol{\psi}_k^{i,j,2})^T \cdots \sum_{j=1}^{h} \mathbf{u}_k^{i,j,N_s} (\boldsymbol{\psi}_k^{i,j,N_s})^T \right]. \tag{2.37}$$

Here we have expressed each submatrix $\mathbf{H}^i_{k:l}$ as an $M_i \times (N_o N_s)$ matrix, where M_i is the number of weights corresponding to the networks ith node. The submatrices $\mathbf{H}_k^{i,j,m}$ are of size $M_i \times N_o$, corresponding to a single training stream. For purposes of a compact representation, we express each matrix $\mathbf{H}^i_{k:l}$ as a sum of matrices (as opposed to a concatenation) by forming vectors $\boldsymbol{\Psi}_k^{i,j,m}$ from the vectors $\boldsymbol{\psi}_k^{i,j,m}$ in the following fashion. The vector $\boldsymbol{\Psi}_k^{i,j,m}$ is of length $N_o N_s$ with components set to zero everywhere except for in the mth (out of N_s) block of length N_o, where this subvector is set equal to the vector $\boldsymbol{\psi}_k^{i,j,m}$. Then,

$$\mathbf{H}^i_{k:l} = \sum_{j=1}^{h} \sum_{m=1}^{N_s} \mathbf{u}_k^{i,j,m} (\boldsymbol{\Psi}_k^{i,j,m})^T.$$

Note that the matrix is expressed in this fashion for notational convenience and consistency, and that we would make use of the sparse nature of the vector $\boldsymbol{\Psi}_k^{i,j,m}$ in implementation.

2.6.2 Computationally Efficient Formulations for Multiple-Output Problems

We now consider implications for the computational complexity of EKF training due to expressing the derivative calculations as a series of vector outer products as shown above. We consider the simple case of feedforward networks trained by node-decoupled EKF (NDEKF) in which each node's weights comprise a unique group for purposes of the error covariance update. The NDEKF recursion can then be written as

$$\mathbf{A}_k = \left[\mathbf{R}_k + \sum_{j=1}^{G} \alpha_k^j \boldsymbol{\psi}_k^j (\boldsymbol{\psi}_k^j)^T\right]^{-1}, \tag{2.38}$$

$$\mathbf{K}_k^i = \mathbf{v}_k^i (\boldsymbol{\gamma}_k^i)^T, \tag{2.39}$$

$$\hat{\mathbf{w}}_{k+1}^i = \hat{\mathbf{w}}_k^i + [(\boldsymbol{\psi}_k^i)^T (\mathbf{A}_k \boldsymbol{\xi}_k)]\mathbf{v}_k^i, \tag{2.40}$$

$$\mathbf{P}_{k+1}^i = \mathbf{P}_k^i - \beta_k^i \mathbf{v}_k^i (\mathbf{v}_k^i)^T + \mathbf{Q}_k^i, \tag{2.41}$$

where we have used the following equations in intermediate steps:

$$\mathbf{v}_k^i = \mathbf{P}_k^i \mathbf{u}_k^i, \tag{2.42}$$

$$\boldsymbol{\gamma}_k^i = \mathbf{A}_k \boldsymbol{\psi}_k^i, \tag{2.43}$$

$$\alpha_k^i = (\mathbf{u}_k^i)^T \mathbf{v}_k^i, \tag{2.44}$$

$$\beta_k^i = (\boldsymbol{\gamma}_k^i)^T \boldsymbol{\psi}_k^i. \tag{2.45}$$

Based upon this partitioning of the derivative matrix $\mathbf{H}_k$, we find that the computational complexity of NDEKF is reduced from $\mathcal{O}(N_o^2 M + N_o \sum_{i=1}^{G} M_i^2)$ to $\mathcal{O}(N_o^2 G + \sum_{i=1}^{G} M_i^2)$, indicating a distinct advantage for feedforward networks with multiple output nodes. On the other hand, the partitioning of the derivative matrix does not provide any computational advantage for GEKF training of feedforward networks.

2.6.3 Avoiding Matrix Inversions

A complicating factor for effective implementation of EKF training schemes is the need to perform matrix inversions for those problems with multiple cost function components. We typically perform these types of calculations with matrix inversion routines based on singular-value decomposition [15]. Although these techniques have served us well over the years, we recognize that this often discourages "quick-and-dirty" implementations and may pose a large obstacle to hardware implementation.

Two classes of methods have been developed that allow EKF training to be performed for multiple-output problems without explicitly resorting to matrix inversion routines. The first class [16] depends on the partitioning of the derivative matrices described above. This method computes the global scaling matrix $\mathbf{A}_k$ by recursively applying the matrix inversion lemma. This procedure provides results that are mathematically identical to conventional matrix inversion procedures, regardless of the degree of decoupling employed. In addition, it can be employed for training of any form of network, static or dynamic, as well as for the multistream procedure. On the other hand, we have found that this method often requires the use of double-precision arithmetic to produce results that are statistically identical to EKF implementations based on explicit matrix inversion methods.

The second class, developed by Plumer [17], treats each output component individually in an iterative procedure. This sequential update procedure accumulates the weight vector update as each output component is processed, and only applies the weight vector update after all output signals have been processed. The error covariance matrix is updated in a sequential fashion. Plumer's sequential-update form of EKF turns out to be exactly equivalent to the batch form of GFKF given above in which all output signals are processed simultaneously. However, for decoupled EKF training, it turns out that sequential updates only approximate the updates obtained via the simultaneous DEKF recursion of Eqs. (2.12)–(2.15), though this has been reported to not pose any problems during training.

The sequential DEKF method is compactly given by a set of equations that are similar to the simultaneous DEKF equations. We again assume a decoupling with g mutually exclusive groups of weights, with a limit of $g = 1$ reducing to the global version, and use the superscript i to refer to the individual weight groups. We handle the multistream case by labeling each cost function component from $l = 1$ to $N_o N_s$, where N_o and N_s refer to the number of network outputs and number of processing streams, respectively. A single weight vector update with the sequential multistream DEKF procedure requires an initialization step of $\Delta\hat{\mathbf{w}}^i_{k,0} = \mathbf{0}$ and $\mathbf{P}^i_{k,0} = \mathbf{P}^i_k$, where $\Delta\hat{\mathbf{w}}^i_{k,l}$ is used to accumulate the update to the weight vector. Then, the sequential multistream DEKF procedure is compactly represented by the following equations:

$$a_{k,l} = \left[r_{k,l} + \sum_{j=1}^{g} (\mathbf{h}^i_{k,l})^T \mathbf{P}^i_{k,l-1} \mathbf{h}^i_{k,l} \right]^{-1}, \tag{2.46}$$

$$\mathbf{k}^i_{k,l} = \mathbf{P}^i_{k,l-1} \mathbf{h}^i_{k,l} a_{k,l}, \tag{2.47}$$

$$\Delta\hat{\mathbf{w}}^i_{k,l} = \Delta\hat{\mathbf{w}}^i_{k,l-1} + \mathbf{k}^i_{k,l} \xi_{k,l} - \mathbf{k}^i_{k,l} [(\mathbf{h}^i_{k,l})^T \Delta\hat{\mathbf{w}}^i_{k,l-1}], \tag{2.48}$$

$$\mathbf{P}^i_{k,l} = \mathbf{P}^i_{k,l-1} - \mathbf{k}^i_{k,l} (\mathbf{h}^i_{k,l})^T \mathbf{P}^i_{k,l-1}. \tag{2.49}$$

Note that the scalar $r_{k,l}$ is the lth diagonal element of the measurement covariance matrix $\mathbf{R}_k$ in the simultaneous form of DEKF, that the scalar $\xi_{k,l}$ is the lth error signal, and that the vector $\mathbf{h}^i_{k,l}$ is the lth column of the augmented derivative matrix $\mathbf{H}^i_k$. After all output signals of all training streams have been processed, the weight vectors and error covariance matrices for all weight groups are updated by

$$\hat{\mathbf{w}}^i_{k+1} = \hat{\mathbf{w}}^i_k + \Delta\hat{\mathbf{w}}^i_{k,N_oN_s}, \tag{2.50}$$

$$\mathbf{P}^i_{k+1} = \mathbf{P}^i_{k,N_oN_s} + \mathbf{Q}^i_k. \tag{2.51}$$

Structurally, these equations for sequential updates are nearly identical to those of the simultaneous update, with the exception of an additional correction term in the delta-weight update equation; this term is necessary to account for the fact that the weight estimate changes at each step of this sequential multiple-output recursion.

2.6.4 Square-Root Filtering

2.6.4.1 Without Artificial Process Noise Sun and Marko [18] have described the use of square-root filtering as a numerically stable, alternative method to performing the approximate error covariance matrix update given by the Riccati equation (2.6). The square-root filter methods are well known in the signal processing community [19], and were developed so as to guarantee that the positive-definiteness of the matrix is maintained throughout training. However, this insurance is accompanied by increased computational complexity. Below, we summarize the square-root formulation for the case of no artificial process noise, with proper treatment of the EKF learning rate as given in Eq. (2.7) (we again assume $\mathbf{S}_k = \mathbf{I}$).

The square-root covariance filter update is based on the matrix factorization lemma, which states that for any pair of $J \times K$ matrices $\mathbf{B}_1$ and $\mathbf{B}_2$, with $J \leq K$, the relation $\mathbf{B}_1\mathbf{B}_1{}^T = \mathbf{B}_2\mathbf{B}_2{}^T$ holds if and only if there exists a unitary matrix Θ such that $\mathbf{B}_2 = \mathbf{B}_1\Theta$. With this in mind, the covariance update equations (2.3) and (2.6) can be written in matrix form as

$$\begin{bmatrix} \mathbf{R}_k^{1/2} & \mathbf{H}_k^T\mathbf{P}_k^{1/2} \\ \mathbf{0} & \mathbf{P}_k^{1/2} \end{bmatrix} \begin{bmatrix} \mathbf{R}_k^{1/2} & \mathbf{0} \\ \mathbf{P}_k^{1/2}\mathbf{H}_k & \mathbf{P}_k^{1/2} \end{bmatrix} = \begin{bmatrix} \mathbf{A}_k^{-1/2} & \mathbf{0} \\ \mathbf{P}_k\mathbf{H}_k\mathbf{A}_k^{1/2} & \mathbf{P}_{k+1}^{1/2} \end{bmatrix} \begin{bmatrix} \mathbf{A}_k^{-1/2} & \mathbf{A}_k^{1/2}\mathbf{H}_k^T\mathbf{P}_k \\ \mathbf{0} & \mathbf{P}_{k+1}^{1/2} \end{bmatrix}. \tag{2.52}$$

Now, the idea is to find a unitary transformation Θ such that

$$\begin{bmatrix} \mathbf{A}_k^{-1/2} & \mathbf{0} \\ \mathbf{P}_k\mathbf{H}_k\mathbf{A}_k^{1/2} & \mathbf{P}_{k+1}^{1/2} \end{bmatrix} = \begin{bmatrix} \mathbf{R}_k^{1/2} & \mathbf{H}_k^T\mathbf{P}_k^{1/2} \\ \mathbf{0} & \mathbf{P}_k^{1/2} \end{bmatrix}\Theta. \qquad (2.53)$$

This is easily accomplished by applying a series of 2×2 Givens rotations to annihilate the elements of the submatrix $\mathbf{H}_k^T\mathbf{P}_k^{1/2}$, thereby yielding the left-hand-side matrix. Given this result of the square-root filtering procedure, we can perform the network weight update via the following additional steps: (1) compute $\mathbf{A}_k^{1/2}$ by inverting $\mathbf{A}_k^{-1/2}$; (2) compute the Kalman gain matrix by $\mathbf{K}_k = (\mathbf{P}_k\mathbf{H}_k\mathbf{A}_k^{1/2})\mathbf{A}_k^{1/2}$; (3) perform the weight update via Eq. (2.5).

2.6.4.2 *With Artificial Noise*

In our original work [3] on EKF-based training, we introduced the use of artificial process noise as a simple and easily controlled mechanism to help assure that the approximate error covariance matrix $\mathbf{P}_k$ would retain the necessary property of nonnegative-definiteness, thereby allowing us to avoid the more computationally complicated square-root formulations. In addition to controlling the proper evolution of $\mathbf{P}_k$, we have also found that artificial process noise, when carefully applied, helps to accelerate the training process and, more importantly, leads to solutions superior to those found without artificial process noise. We emphasize that the use of artificial process noise is not ad hoc, but appears due to the process noise term in Eq. (2.1) (i.e., the covariance matrix $\mathbf{Q}_k$ in Eq. (2.6) disappears only when $\boldsymbol{\omega}_k = 0$ for all k). We have continued to use this feature in our implementations as an effective means for escaping poor local minima, and have not experienced problems with divergence. Other researchers with independent implementations of various forms of EKF [13, 17, 20] for neural network training have also found the use of artificial process noise to be beneficial. Furthermore, other gradient-based training algorithms have effectively exploited weight noise (e.g., see [21]).

We now demonstrate that the square-root filtering formulation of Eq. (2.53) is easily extended to include artificial process noise, and that the use of artificial process noise and square-root filtering are not mutually exclusive. Again, we wish to express the error covariance update in a factorized form, but we now augment the left-hand-side of Eq. (2.52) to include the square root of the process-noise covariance matrix:

$$\begin{bmatrix} \mathbf{R}_k + \mathbf{H}_k^T\mathbf{P}_k\mathbf{H}_k & \mathbf{H}_k^T\mathbf{P}_k \\ \mathbf{P}_k\mathbf{H}_k & \mathbf{P}_k + \mathbf{Q}_k \end{bmatrix} = \begin{bmatrix} \mathbf{R}_k^{1/2} & \mathbf{H}_k^T\mathbf{P}_k^{1/2} & \mathbf{0} \\ \mathbf{0} & \mathbf{P}_k^{1/2} & \mathbf{Q}_k^{1/2} \end{bmatrix} \begin{bmatrix} \mathbf{R}_k^{1/2} & \mathbf{0} \\ \mathbf{P}_k^{1/2}\mathbf{H}_k & \mathbf{P}_k^{1/2} \\ \mathbf{0} & \mathbf{Q}_k^{1/2} \end{bmatrix}. \tag{2.54}$$

Similarly, the right-hand side of Eq. (2.52) is augmented by blocks of zeroes, so that the matrices are of the same size as those of the right-hand side of Eq. (2.54):

$$\begin{bmatrix} \mathbf{A}_k^{-1} & \mathbf{H}_k^T\mathbf{P}_k \\ \mathbf{P}_k\mathbf{H}_k & \mathbf{P}_{k+1} + \mathbf{P}_k\mathbf{H}_k\mathbf{A}_k\mathbf{H}_k^T\mathbf{P}_k \end{bmatrix} = \begin{bmatrix} \mathbf{A}_k^{-1/2} & \mathbf{0} & \mathbf{0} \\ \mathbf{P}_k\mathbf{H}_k\mathbf{A}_k^{1/2} & \mathbf{P}_{k+1}^{1/2} & \mathbf{0} \end{bmatrix} \begin{bmatrix} \mathbf{A}_k^{1/2} & \mathbf{A}_k^{1/2}\mathbf{H}_k^T\mathbf{P}_k \\ \mathbf{0} & \mathbf{P}_{k+1}^{1/2} \\ \mathbf{0} & \mathbf{0} \end{bmatrix}. \tag{2.55}$$

Here, Eqs. (2.54) and (2.55) are equivalent to one another, and Eq. (2.53) is appropriately modified to

$$\begin{bmatrix} \mathbf{A}_k^{1/2} & \mathbf{0} & \mathbf{0} \\ \mathbf{P}_k\mathbf{H}_k\mathbf{A}_k^{1/2} & \mathbf{P}_{k+1}^{1/2} & \mathbf{0} \end{bmatrix} = \begin{bmatrix} \mathbf{R}_k^{1/2} & \mathbf{H}_k^T\mathbf{P}_k^{1/2} & \mathbf{0} \\ \mathbf{0} & \mathbf{P}_k^{1/2} & \mathbf{Q}_k^{1/2} \end{bmatrix} \Theta. \tag{2.56}$$

Thus, square-root filtering can easily accommodate the artificial process-noise extension, where the matrices $\mathbf{H}_k^T\mathbf{P}_k^{1/2}$ and $\mathbf{Q}_k^{1/2}$ are both annihilated via a sequence of Givens rotations. Note that this extension involves substantial additional computational costs beyond those incurred when $\mathbf{Q}_k = \mathbf{0}$. For a network with M weights and N_o outputs, the use of artificial process noise introduces $O(M^3)$ additional computations for the annihilation of $\mathbf{Q}_k^{1/2}$, whereas the annihilation of the matrix $\mathbf{H}_k^T\mathbf{P}_k^{1/2}$ only involves $\mathcal{O}(M^2N_o)$ computations (here we assume $M \gg M_o$).

2.7 OTHER EXTENSIONS AND ENHANCEMENTS

2.7.1 EKF Training with Constrained Weights

Due to the second-order properties of the EKF training procedure, we have observed that, for certain problems, networks trained by EKF tend to develop large weight values (e.g., between 10 and 100 in magnitude). We view this capability as a double-edged sword: on the one hand, some problems may require that large weight values be developed, and the EKF procedures are effective at finding solutions for these problems. On the other hand, trained networks may need to be deployed with execution performed in fixed-point arithmetic, which requires that limits be imposed on the range of values of network inputs, outputs and weights. For nonlinear sigmoidal nodes, the node outputs are usually limited to values between -1 and $+1$, and input signals can usually be linearly transformed so that they fall within this range. On the other hand, the EKF procedures as described above place no limit on the weight values. We describe here a natural mechanism, imposed during training, that limits the range of weight values. In addition to allowing for fixed-point deployment of trained networks, this weight-limiting mechanism may also promote better network generalization.

We wish to set constraints on weight values during the training process while maintaining rigorous consistency with the EKF recursion. We can accomplish this by converting the unconstrained nonlinear optimization problem into one of optimization with constraints. The general idea is to treat each of the network's weight values as the output of a monotonically increasing function $\phi(\cdot)$ with saturating limits at the function's extremes (e.g., a sigmoid function). Thus, the EKF recursion is performed in an unconstrained space, while the network's weight values are nonlinear transformations of the corresponding unconstrained values that evolve during the training process. This transformation requires that the EKF recursion be modified to take into account the function $\phi(\cdot)$ as applied to the parameters (i.e., unconstrained weight values) that evolve during training.

Assume that the vector of network's weights $\mathbf{w}_k$ is constrained to take on values in the range $-\alpha$ to $+\alpha$ and that each component $w_k^{i,j}$ (the jth weight of the ith node) of the constrained weight vector is related to an unconstrained value $\tilde{w}_k^{i,j}$ via a function $w_k^{i,j} = \phi(\tilde{w}_k^{i,j}, \alpha)$. We formulate the EKF recursion so that weight updates are performed in the unconstrained weight space, while the steps of forward propagation and backpropagation of derivatives are performed in the constrained weight space.

The training steps are carried out as follows. At time step k, an input vector is propagated through the network, and the network outputs are computed and stored in the vector $\hat{\mathbf{y}}_k$. The error vector $\boldsymbol{\xi}_k$ is also formed as defined above. Subsequently, the derivatives of each component of $\hat{\mathbf{y}}_k$ with respect to each node's weight vector $\mathbf{w}_k^i$ are computed and stored into the matrices $\mathbf{H}_k^i$, where the component $\mathbf{H}_k^{i,j,l}$ contains the derivative of the lth component of $\hat{\mathbf{y}}_k$ with respect to the jth weight of the ith node. In order to perform the EKF recursion in the unconstrained space, we must perform three steps in addition to those that are normally carried out. First, we transform weight values from the constrained space via $\tilde{w}_k^{i,j} = \phi^{-1}(w_k^{i,j}, \alpha)$ for all trainable weights of all nodes of the network, which yields the vectors $\tilde{\mathbf{w}}_k^i$. Second, the derivatives that have been previously computed with respect to weights in the constrained space must be transformed to derivatives with respect to weights in the unconstrained space. This is easily performed by the following transformation for each derivative component:

$$\tilde{H}_k^{i,j,l} = H_k^{i,j,l} \frac{\partial \omega_k^{i,j}}{\partial \tilde{w}_k^{i,j}} = H_k^{i,j,l} \frac{\partial \phi(\tilde{w}_k^{i,j}, \alpha)}{\partial \tilde{w}_k^{i,j}}. \tag{2.57}$$

The EKF weight update procedure of Eqs. (2.8)–(2.11) is then applied using the unconstrained weights and derivatives with respect to unconstrained weights. Note that no transformation is applied to either the scaling matrix $\mathbf{S}_k$ or the error vector $\boldsymbol{\xi}_k$ before they are used in the update. Finally, after the weight updates are performed, the unconstrained weights are transformed back to the constrained space by $w_k^{i,j} = \phi(\tilde{w}_k^{i,j}, \alpha)$ for all weights of all nodes in the network.

We now consider specific forms for the function $\phi(\tilde{w}_k^{i,j}, \alpha)$ that transforms weight values from an unconstrained space to a constrained space. We require that the function obey the following properties:

1. $\phi(\tilde{w}_k^{i,j}, \alpha)$ is monotonically increasing.
2. $\phi(0, \alpha) = 0$.
3. $\phi(-\infty, \alpha) = -\alpha$.
4. $\phi(+\infty, \alpha) = +\alpha$.
5. $\dfrac{\partial w_k^{i,j}}{\partial \tilde{w}_k^{i,j}} |_{\tilde{w}_k^{i,j}=0} = 1$.
6. $\lim_{\alpha \to \infty} \phi(\tilde{w}_k^{i,j}, \alpha) = \tilde{w}_k^{i,j} = w_k^{i,j}$.

Property 5 imposes a constraint on the gain of the transformation, while property 6 imposes the constraint that in the limit of large α, the constrained optimization problem operates identically to the unconstrained problem. This last constraint also implies that $\lim_{\alpha\to\infty}(\partial w_k^{i,j}/\partial \tilde{w}_k^{i,j}) = 1$.

One candidate function is a symmetric saturating linear transformation where $w_k^{i,j} = \tilde{w}_k^{i,j}$ when $-\alpha \le \tilde{w}_k^{i,j} \le \alpha$, and is otherwise equal to either saturating value. The major disadvantage with this constraint function is that its inverse is multivalued outside the linear range. Thus, once the training process pushes the constrained weight value into the saturated region, the derivative of constrained weight with respect to unconstrained weight becomes zero, and no further training of that particular weight value will occur due to the zero-valued derivative.

Alternatively, we may consider various forms of symmetric sigmoid functions that are everywhere differentiable and have well-defined inverses. The Elliott sigmoid [22] conveniently does not involve transcendental functions. We choose to consider a generalization of this monotonic and symmetric saturating function given by

$$w_k^{i,j} = \phi(\tilde{w}_k^{i,j}, \alpha) = \frac{\alpha \tilde{w}_k^{i,j}}{\beta + |\tilde{w}_k^{i,j}|} = \frac{\tilde{w}_k^{i,j}}{\beta/\alpha + |\tilde{w}_k^{i,j}|/\alpha}, \tag{2.58}$$

where β is a positive quantity that determines the function's gain. We must choose the value of β so that the derivative of $\phi(\tilde{w}_k^{i,j}, \alpha)$ with respect to $\tilde{w}_k^{i,j}$, evaluated at $\tilde{w}_k^{i,j} = 0$, is equal to 1. This condition can be shown to be satisfied by the choice $\beta = \alpha$. Thus, the constraint function we choose is given by

$$w_k^{i,j} = \phi(\tilde{w}_k^{i,j}, \alpha) = \frac{\alpha \tilde{w}_k^{i,j}}{\alpha + |\tilde{w}_k^{i,j}|} = \frac{\tilde{w}_k^{i,j}}{1 + |\tilde{w}_k^{i,j}|/\alpha}. \tag{2.59}$$

By inspection, we see that this function satisfies the various requirements. For example, for large α (i.e., as $\alpha \to \infty$), $w_k^{i,j} \to \tilde{w}_k^{i,j}$; for smaller values of α, when $|\tilde{w}_k^{i,j}| \gg \alpha$, $w_k^{i,j} \to \alpha\ \mathrm{sgn}(\tilde{w}_k^{i,j})$. The inverse of this function is easily found to be given by

$$\tilde{w}_k^{i,j} = \phi^{-1}(w_k^{i,j}, \alpha) = \frac{\alpha w_k^{i,j}}{\alpha - |w_k^{i,j}|} = \frac{w_k^{i,j}}{1 - |w_k^{i,j}|/\alpha}. \tag{2.60}$$

In this case, for $\alpha \gg w_k^{i,j}$, $\tilde{w}_k^{i,j} \to w_k^{i,j}$; similarly, as $w_k^{i,j} \to \alpha$, $\tilde{w}_k^{i,j} \to \infty$. As a final note, the derivative of constrained weight with respect to unconstrained weight, which is needed for computing the proper derivatives in the EKF recursion, can be expressed in many different ways, some of which are given by

$$\frac{\partial w_k^{i,j}}{\partial \tilde{w}_k^{i,j}} = \left(\frac{\alpha}{\alpha + |\tilde{w}_k^{i,j}|} \right)^2 = \left(\frac{\alpha - |w_k^{i,j}|}{\alpha} \right)^2 = \left(1 - \frac{|w_k^{i,j}|}{\alpha} \right)^2 = \left(\frac{w_k^{i,j}}{\tilde{w}_k^{i,j}} \right)^2. \quad (2.61)$$

2.7.2 EKF Training with an Entropic Cost Function

As defined above, the EKF training algorithm assumes that a quadratic function of some error signal is being minimized over all network outputs and all training patterns. However, other cost functions are often useful or necessary. One such function that has been found to be particularly appropriate for pattern classification problems, and for which a sound statistical basis exists, is a cost function based on minimizing cross-entropy [23]. We consider a prototypical problem in which a network is trained to act as a pattern classifier; here network outputs encode binary pattern classifications. We assume that target values of ± 1 are provided for each training pattern. Then the contribution to the total entropic cost function at time step n is given by

$$\mathscr{E}_k = \sum_{l=1}^{N_o} \mathscr{E}_k^l = \sum_{l=1}^{N_o} \left[(1 + y_k^l) \log \frac{1 + y_k^l}{1 + \hat{y}_k^l} + (1 - y_k^l) \log \frac{1 - y_k^l}{1 - \hat{y}_k^l} \right]. \quad (2.62)$$

Since the components of the vector $\mathbf{y}_k$ are constrained to be either $+1$ or -1, we note that only one of the two components for each output l will be nonzero. This allows the cost function to be expressed as

$$\mathscr{E}_k|_{y_k^l = \pm 1} = \sum_{l=1}^{N_o} \mathscr{E}_k^l = \sum_{l=1}^{N_o} 2 \log \frac{2}{1 + y_k^l \hat{y}_k^l}. \quad (2.63)$$

The EKF training procedure assumes that at each time step k a quadratic cost function is being minimized, which we write as $C_k = \sum_{l=1}^{N_o} (z_k^l - \hat{z}_k^l)^2 = \sum_{l=1}^{N_o} (\xi_k^l)^2$, where z_k^l and $\hat{z}_k^l$ are target and output values, respectively. (We assume here the case of $\mathbf{S}_k = \mathbf{I}$; this procedure is easily extended to nonuniform weighting matrices.) At this point, we would like to find appropriate transformations between the $\{z_k^l, \hat{z}_k^l\}$ and $\{y_k^l, \hat{y}_k^l\}$ so that C_k and $\mathscr{E}_k$ are equivalent, thereby allowing us to use the EKF procedure to minimize the entropic cost function. We first note that both the quadratic and entropic cost functions are calculated by summing individual cost function components from all targets and output nodes. We immediately see that this leads to the equality $(\xi_k^l)^2 = \varepsilon_k^l$ for all N_o outputs, which implies $z_k^l - \hat{z}_k^l = (\varepsilon_k^l)^{1/2}$. At this point, we assume that we can assign all target values $z_k^l = 0$,[3] so that $\xi_k^l = -\hat{z}_k^l = (\varepsilon_k^l)^{1/2}$. Now the EKF recursion can be applied to minimize the entropic cost function, since $C^k = \Sigma_{l=1}^{N_o} (\xi_k^l)^2 = \Sigma_{l=1}^{N_o} [(\varepsilon_k^l)^{1/2}]^2 = \Sigma_{l=1}^{N_o} \varepsilon_k^l = \mathscr{E}_k$.

The remainder of the derivation is straightforward. The EKF recursion in the case of the entropic cost function requires that derivatives of $\hat{z}_k^l = (-\varepsilon_k^l)^{1/2}$ be computed for all N_o outputs and all weight parameters, which are subsequently stored in the matrices $\mathbf{H}_k^i$. Applying the chain rule, these derivatives are expressed as a function of the derivatives of network outputs with respect to weight parameters:

$$H_k^{i,j,l} = \left.\frac{\partial \hat{z}_k^l}{\partial w_k^{i,j}}\right|_{y_k^l = \pm 1} = -\frac{\partial (\varepsilon_k^l)^{1/2}}{\partial w_k^{i,j}} = \frac{1}{(\varepsilon_k^l)^{1/2} (y_k^l + \hat{y}_k^l)} \frac{\partial \hat{y}_k^l}{\partial w_k^{i,j}}. \tag{2.64}$$

Note that the effect of the relative entropy cost function on the calculation of derivatives is handled entirely in the initialization of the backpropagation process, where the term $1/[(\varepsilon_k^l)^{1/2}(y_k^l + \hat{y}_k^l)]$ is used for each of the N_o output nodes to start the backpropagation process, rather than starting with a value of unity for each output node as in the nominal formulation.

In general, the EKF procedure can be modified in the manner just described for a wide range of cost functions, provided that they meet at least three simple requirements. First, the cost function must be a differentiable function of network outputs. Second, the cost function should be expressed as a sum of contributions, where there is a separate target value for each individual component. Third, each component of the cost function must be non-negative.

2.7.3 EKF Training with Scalar Errors

When applied to a multiple-output training problem, the EKF formulation in Eqs. (2.3)–(2.6) requires a separate backpropagation for each output and a matrix inversion. In this section, we describe an approximation to

[3]The idea of using a modified target value of zero with the actual targets appearing in expressions for system outputs can be applied to the EKF formulation of Eqs. (2.3)–(2.6) without any change in its underlying behavior.

the EKF neural network training procedure that allows us to treat such problems with single-output training complexity. In this approximation, we require only the computation of derivatives of a scalar quantity with respect to trainable weights, thereby reducing the backpropagation computation and eliminating the need for a matrix inversion in the multiple-output EKF recursion.

For the sake of simplicity, we consider here the prototypical network training problem for which network outputs directly encode signals for which targets are defined. The square root of the contribution to the total cost function at time step k is given by

$$\tilde{y}_k = C_k^{1/2} = \left(\sum_{l=1}^{N_o} |y_k^l - \hat{y}_k^l|^2 \right)^{1/2}, \tag{2.65}$$

where we are again treating the simple case of uniform scaling of network errors (i.e., $\mathbf{S}_k = \mathbf{I}$). The goal here is to train a network so that the sum of squares of this scalar error measure is minimized over time. As in the case of the entropic cost function, we consider the target for training to be zero for all training instances, and the scalar error signal used in the Kalman recursion to be given by $\xi_k = 0 - \tilde{y}_k$. The EKF recursion requires that the derivatives of the scalar observation $\tilde{y}_k$ be computed with respect to all weight parameters. The derivative of the scalar error with respect to the jth weight of the ith node is given by

$$H_k^{i,j,1} = \frac{\partial \tilde{y}_k}{\partial w_k^{i,j}} = \sum_{l=1}^{N_o} \frac{\partial \tilde{y}_k}{\partial \hat{y}_k^l} \frac{\partial \hat{y}_k^l}{\partial w_k^{i,j}} = \sum_{l=1}^{N_o} \frac{y_k^l - \hat{y}_k^l}{\xi_k} \frac{\partial \hat{y}_k^l}{\partial w_k^{i,j}}. \tag{2.66}$$

In this scalar formulation, the derivative calculations via backpropagation are initialized with the terms $(y_k^l - \hat{y}_k^l)/\xi_k$ for all N_o network output nodes (as opposed to initializing the backpropagation calculations with values of unity for the nominal EKF recursion of Eqs. (2.3)–(2.6)). Furthermore, only one quantity is backpropagated, rather than N_o quantities for the nominal formulation. Note that this scalar approximation reduces exactly to the nominal EKF algorithm in the limit of a single-output problem:

$$\tilde{y}_k = (|y_k^1 - \hat{y}_k^1|^2)^{1/2} = |y_k^1 - \hat{y}_k^1|, \tag{2.67}$$

$$\xi_k = 0 - |y_k^1 - \hat{y}_k^1|, \tag{2.68}$$

$$H_k^{i,j,1} = \frac{\partial \tilde{y}_k}{\partial w_k^{i,j}} = -\mathrm{sgn}(y_k^1 - \hat{y}_k^1) \frac{\partial \hat{y}_k^1}{\partial w_k^{i,j}}. \tag{2.69}$$

Consider the case $y_k^1 \leq \hat{y}_k^1$. Then, the error signal is given by $\xi_k = y_k^1 - \hat{y}_k^1$. Similarly, $\partial\tilde{y}_k/\partial w = \partial\hat{y}_k^1/\partial w$, since $\partial\tilde{y}_k/\partial y_k^1 = 1$. Otherwise, when $y_k^1 > \hat{y}_k^1$, the error signal is given by $\xi_k = -(y_k^1 - \hat{y}_k^1)$, and $\partial\tilde{y}_k/\partial w = -\partial\hat{y}_k^1/\partial w$, since $\partial\tilde{y}_k/\partial\hat{y}_k^1 = -1$. Since both the error and the derivatives are the negatives of what the nominal EKF recursion provides, the effects of negation cancel one another. Thus, in either case, the scalar formulation for a single-output problem is exactly equivalent to that of the EKF procedure of Eqs. (2.3)–(2.6).

Because the procedure described here is an approximation to the base procedure, we suspect that classes of problems exist for which it is not as effective; further work will be required to clarify this question. In this regard, we note that once criteria are available to guide the decision of whether to scalarize or not, one may also consider a hybrid approach to problems with many outputs. In this approach, selected outputs would be combined as described above to produce scalar error variables; the latter would then be treated with the original procedure.

2.8 AUTOMOTIVE APPLICATIONS OF EKF TRAINING

The general area of automotive powertrain control, diagnosis, and modeling has offered substantial opportunity for the application of neural network methods. These opportunities are driven by the steadily increasing demands that are placed on the performance of vehicle control and diagnostic systems as a consequence of global competition and government mandates. Modern automotive powertrain control systems involve several interacting subsystems, any one of which can involve significant engineering challenges. We summarize the application of EKF training to three signal processing problems related to automotive diagnostics and emissions modeling, as well as its application to two automotive control problems. In all five cases, we have found EKF training of recurrent neural networks to be an enabler for developing effective solutions to these problems.

Figure 2.7 provides a diagrammatic representation of these five neural network applications and how they potentially interact with one another. We observe that the neural network controllers for engine idle speed and air/fuel (A/F) ratio control produce signals that affect the operation of the engine, while the remaining neural network models are used to describe various aspects of engine operation as a function of measurable engine outputs.

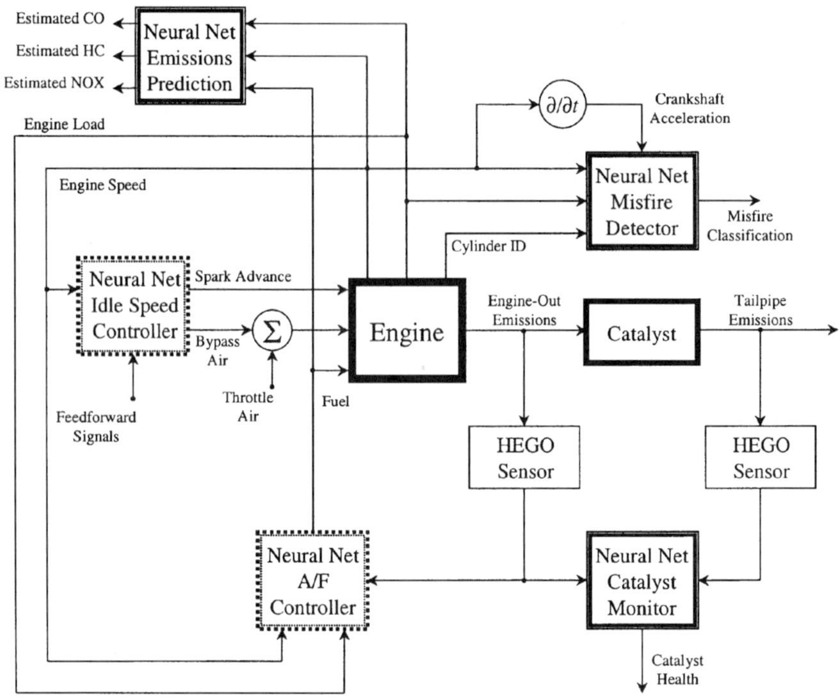

Figure 2.7 Block-diagram representation of neural network applications for automotive engine control and diagnosis. Solid boxes represent physical components of the engine system, double-lined solid boxes represent neural network models or diagnostic processes, and double-lined dashed boxes represent neural network controllers. For the sake of simplicity, we have not shown all relevant sensors and their corresponding signals (e.g., engine coolant temperature).

2.8.1 Air/Fuel Ratio Control

At a very basic level, the role of the A/F controller is to supply fuel to the engine such that it matches the amount of air pumped into the engine via the throttle and idle speed bypass valve. This is accomplished with an electronic feedback control system that utilizes a heated exhaust gas oxygen (HEGO) sensor whose role is to indicate whether the engine-out exhaust is rich (i.e., too much fuel) or lean (too much air). Depending on the measured state of the exhaust gases, as well as engine operating conditions such as engine speed and load, the A/F control is changed so as to drive the system toward stoichiometry. Since the HEGO sensor is largely considered to be a binary sensor (i.e., it produces high/low voltage levels for rich/lean operations, respectively), and since there are time-varying transport delays, the closed-loop A/F control strategy often takes the form of a jump/ramp strategy, which effectively causes the HEGO output to oscillate between the two voltage levels. We have demonstrated that an open-loop recurrent neural network controller can be trained to provide a correction signal to the closed-loop A/F control in the face of transient conditions (i.e., dynamic changes in engine speed and load), thereby eliminating large deviations from stoichiometry. This is accomplished by using an auxiliary universal EGO (UEGO) sensor, which provides a continuous measure of A/F ratio (as opposed to the rich/lean indication provided by the HEGO), during the in-vehicle training process. Deviations of measured A/F ratio from stoichiometric A/F ratio provide the error signal for the EKF training process; however, the measured A/F ratio is not used as an input, and since the A/F control does not have a major effect on engine operating conditions when operated near stoichiometry, then this can be viewed as a problem of training an open-loop controller. Nevertheless, we use recurrent network controllers to provide the capability of representing the condition-dependent dynamics associated with the operation of the engine system under A/F control, and must take care to properly compute derivatives with BPTT(h).

2.8.2 Idle Speed Control

A second engine control task is that of maintaining smooth engine operation at idle conditions. In this case, no air is provided to the intake manifold of the engine via the throttle; in order to keep the engine running, a bypass air valve is used to regulate the flow of air into the engine. The role of the idle speed control system is to maintain a relatively low (for purposes of fuel economy) and constant engine speed, in the face of disturbances that place and remove additional loads on the engine (e.g., shifting from neutral to drive, activating the air conditioning system, and locking up the power steering); feedforward signals encoding these events are provided as input to the idle speed controller. The control range of the bypass air signal is large (more than 1000 rpm under idle conditions), but its effect is delayed by a time inversely proportional to engine speed. The spark advance command, which regulates the timing of ignition, has an immediate effect on engine speed, but over a small range (on order of 100 rpm). Thus, an effective engine idle speed controller coordinates the two controls to maintain a constant engine speed. The error signals for the EKF training process are a weighted sum of squared deviations of engine speed from a desired speed, combined with constraints on the controls expressed as squared error signals. We have used recurrent neural networks, trained by on-line EKF methods, to develop effective idle speed control strategies, and have documented this work in [5]. Note that unlike the case of the A/F controller, this is an example of a closed-loop controller, since the bypass air and spark advance controls affect engine speed, which is used as a controller input.

2.8.3 Sensor-Catalyst Modeling

A particularly critical component of a vehicle's emissions control system is the catalytic converter. The role of the catalytic converter is to chemically transform noxious and environmentally damaging engine-out emissions, which are the byproduct of the engine's combustion process, to environmentally benign chemical compounds. An ideal three-way catalytic converter should completely perform the following three tasks during continuous vehicle operation: (1) oxidation of hydrocarbon (HC) exhaust gases to carbon dioxide (CO_2) and water (H_2O); (2) oxidation of carbon monoxide (CO) to CO_2; and (3) reduction of nitrogen oxides (NO_x) to nitrogen (N_2) and oxygen (O_2). In practice, it is possible to achieve high conversion efficiencies for all three types of exhaust gases only when the engine is operating near stoichiometry. An effective A/F control strategy enables such conversion.

However, even in the presence of effective A/F control, vehicle-out (i.e., tailpipe) emissions may be unreasonably high if the catalytic converter has been damaged. Government regulations require that the performance of a vehicle's catalytic converter be continuously monitored to detect when conversion efficiencies have dropped below some threshold. Unfortunately, it is currently infeasible to equip vehicles with sensors that can measure the various exhaust gas species directly. Instead, catalytic converter monitors are based on comparing the output of a HEGO sensor that is exposed to engine-out emissions with the output of a second sensor that is mounted downstream of the catalytic converter and is exposed to the tailpipe emissions. This approach is based on the observation that the postcatalyst HEGO sensor switches infrequently, relative to the precatalyst HEGO sensor, when the catalyst is operating efficiently. Similarly, the average rate of switching of the postcatalyst sensor increases as catalyst efficiency decreases (due to decreasing oxygen storage capability).

A catalyst monitor can be developed based on a neural network model of the dynamic operation of the postcatalyst HEGO sensor as a function of the precatalyst HEGO sensor and engine operating conditions [12] for a catalyst of nominal conversion efficiency. This is a difficult task, especially given the nonlinear responses of the various components and the condition-dependent time delays, which can range from less than 0.1 s at high engine speeds to more than 1 s at low speeds. We employed a RMLP network with structure 15-20R-15R-10R-1 and a sparse tapped delay line representation to directly capture the long-term temporal characteristics of the precatalyst HEGO sensor. Because of the size of the network (over 1,500 weights) and the number of training samples (63,000), we chose to employ decoupled EKF training. The trained network effectively represented the condition-dependent time delays and nonlinearities of the system, as shown in [12].

2.8.4 Engine Misfire Detection

Engine misfire is broadly defined as the condition in which a substantial fraction of a cylinder's air–fuel mixture fails to ignite. Frequent misfire will lead to a deterioration of the catalytic converter, ultimately resulting in unacceptable levels of emitted pollutants. Consequently, government mandates require that onboard misfire detection capability be provided for nearly all engine operating conditions.

While there are many ways of detecting engine misfire, all currently practical methods rely on observing engine crankshaft dynamics with a position sensor located at one end of the shaft. Briefly stated, one looks for a crankshaft acceleration deficit following a cylinder firing and attempts to determine whether such a deficit is attributable to a lack of power provided on the most recent firing stroke.

Since every engine firing must be evaluated, the natural "clock" for misfire detection is based on crankshaft rotation, rather than on time. For an n-cylinder engine, there are n engine firings, or events, per engine cycle, which requires two engine revolutions. The actual time interval between events varies considerably, from 20 ms at 750 rpm to 2.5 ms at 6000 rpm for an eight-cylinder engine. Engine speed, as required for control, is typically derived from measured intervals between marks on a timing wheel. As used in misfire detection, an acceleration value is calculated from the difference between successive intervals.

A serious problem associated with measuring crankshaft acceleration is the presence of complex torsional dynamics of the crankshaft, even in the absence of misfire. This is due to the finite stiffness of the crankshaft. The magnitude of acceleration induced by such torsional vibrations may be large enough to dwarf acceleration deficits from misfire. Further, the torsional vibrations are themselves altered by misfire, so that normal engine firings followed by misfire may be misinterpreted.

We have approached the misfire detection problem with recurrent neural networks trained by GEKF [12] to act as dynamic pattern classifiers. We use as inputs engine speed, engine load, crankshaft acceleration, and a binary flag to identify the beginning of the cylinder firing sequence. The training target is a binary signal, according to whether a misfire had been artificially induced for the current cylinder during the previous engine cycle. This phasing enables the network to make use of information contained in measured accelerations that follow the engine event being classified. We find that trained networks make remarkably few classification errors, most of which occur during moments of rapid acceleration or deceleration.

2.8.5 Vehicle Emissions Estimation

Increasing levels of pollutants in the atmosphere – observed despite the imposition of stricter emission standards and technological improvements in emissions control systems – have led to models being developed to predict emissions inventories. These are typically based on the emissions levels that are mandated by the government for a particular driving schedule and a given model year. It has been found that the emissions inventories based on these mandated levels do not accurately reflect those that are actually found to exist. That is, actual emission rates depend heavily upon driving patterns, and real-world driving patterns are not comprehensively represented by the mandated driving schedules. To better assess the emissions that occur in practice and to predict emissions inventories, experiments have been conducted using instrumented vehicles that are driven in actual traffic. Unfortunately, such vehicles are costly and are difficult to operate and maintain.

We have found that recurrent neural networks can be trained to estimate instantaneous engine-out emissions from a small number of easily measured engine variables. Under the assumption of a properly operating fuel control system and catalytic converter, this leads to estimates of tailpipe emissions as well. This capability then allows one to estimate the sensitivity of emissions to driving style (e.g., aggressive versus conservative). Once trained, the network requires only information already available to the powertrain processor. Because of engine dynamics, we have found the use of recurrent networks trained by EKF methods to enable accurate estimation of instantaneous emissions levels. We provide a detailed description of this application in [24].

2.9 DISCUSSION

We have presented in this chapter an overview of neural network training methods based on the principles of extended Kalman filtering. We summarize our findings by considering the virtues and limitations of these methods, and provide guidelines for implementation.

2.9.1 Virtues of EKF Training

The EKF family of training algorithms develops and employs second-order information during the training process using only first-order approximations. The use of second-order information, as embedded in the approximate error covariance matrix, which co-evolves with the weight vector during training, provides enhanced capabilities relative to first-order methods, both in terms of training speed and quality of solution. The amount of second-order information utilized is controlled by the level of decoupling, which is chosen on the basis of computational considerations. Thus, the computational complexity of the EKF methods can be scaled to meet the needs of specific applications.

We have found that EKF methods have enabled the training of recurrent neural networks, for both modeling and control of nonlinear dynamical systems. The sequential nature of the EKF provides advantages relative to batch second-order methods, since weight updates can be performed on an instance-by-instance basis with EKF training. On the other hand, the ability to batch multiple training instances with multistream EKF training provides a level of scalability in addition to that provided by decoupling. The sequential nature of the EKF, in both single- and multistream operation, provides a stochastic component that allows for more effective search of the weight space, especially when used in combination with artificial process noise.

The EKF methods are easily implemented in software, and there is substantial promise for hardware implementation as well. Methods for avoiding matrix inversions in the EKF have been developed, thereby enabling easy implementations. Finally, we believe that the greatest virtue of EKF training of neural networks is its established and proven applicability to a wide range of difficult modeling and control problems.

2.9.2 Limitations of EKF Training

Perhaps the most significant limitation of EKF training is its limited applicability to cost functions other then minimizing sum of squared error. Although we have shown that other cost functions can be used (e.g. entropic measures), we are nevertheless restricted to those optimization problems that can be converted to minimizing a sum of squared error criterion. On the other hand, many problems, particularly in control, require other optimization criteria. For example, in a portfolio optimization problem, we should like to *maximize* the total return over time. Converting such an optimization criterion to a sum of squared errors criterion is usually not straightforward. However, we do not view the sum of squared-error optimization criterion as a limitation for most problems that can be viewed as belonging to the class of traditional supervised training problems.

The EKF procedures described in this chapter are derived on the basis of a first-order linearization of the nonlinear system; this may provide a limitation in the form of large errors in the weight estimates and covariance matrix, since the second-order information is effectively developed by taking outer products of the gradients. Chapter 7 introduces the unscented Kalman filter (UKF) as an alternative to the EKF. The UKF is expected to provide a more accurate means of developing the required second-order information than the EKF, without increasing the computational complexity.

2.9.3 Guidelines for Implementation and Use

1. Decoupling should be used when computation is a concern (e.g., for on-line applications). Node and layer decoupling are the two most appropriate choices. Otherwise, we recommend the use of global EKF, regardless of network architecture, as it should be expected to find better solutions than any of the decoupled versions because of the use of full second-order information.
2. Effectively, two parameter values need to be chosen for training of networks with EKF methods. We assume that the approximate error covariance matrices are always initialized with diagonal value of 100 and 1,000 for weights corresponding to nonlinear and linear nodes, respectively. Then, the user of these methods must set values for the learning rate and process-noise term according to characteristics of the training problem.

3. Training of recurrent networks, either as supervised training tasks or for controller training, can often be improved by multistreaming. The choice of the number of streams is dictated by problem characteristics.
4. Matrix inversions can be avoided by use of sequential EKF update procedures. In the case of decoupling, the order in which outputs are processed can affect training performance in detail. We recommend that outputs be processed in random order when these methods are used.
5. Square-root filtering can be employed to insure computational stability for the error covariance update equation. However, the use of square-root filtering with artificial process noise for covariance updates results in a substantial increase in computational complexity. We have noted that nonzero artificial process noise benefits training, by providing a mechanism to escape poor local minima and a mechanism that maintains stable covariance updates when using the Riccati update equation. We recommend that square-root filtering only be employed when no artificial process noise is used (and only for GEKF).
6. The EKF procedures can be modified to allow for alternative cost functions (e.g., entropic cost functions) and for weight constraints to be imposed during training, which thereby allow networks to be deployed in fixed-point arithmetic.

REFERENCES

[1] D.E. Rumelhart, G.E. Hinton, and R.J. Williams, "Learning representations of back-propagation errors," *Nature* **323**, 533–536 (1986).

[2] S. Singhal and L. Wu, "Training multilayer perceptrons with the extended Kalman algorithm," in D.S. Touretzky, Eds., *Advances in Neural Information Processing Systems 1*, San Mateo, CA: Morgan Kaufmann, 1989, pp. 133–140.

[3] G.V. Puskorius and L. A. Feldkamp, "Decoupled extended Kalman filter training of feedforward layered networks," in *Proceedings of International Joint Conference of Neural Networks*, Seattle, WA, 1991, Vol. 1, pp. 771–777.

[4] G.V. Puskorius and L.A. Feldkamp, "Automotive engine idle speed control with recurrent neural networks," in *Proceedings of the 1993 American Control Conference*, San Francisco, CA, pp. 311–316.

[5] G.V. Puskorius, L.A. Feldkamp, and L.I. Davis, Jr., "Dynamic neural network methods applied to on-vehicle idle speed control," *Proceedings of the IEEE*, **84**, 1407–1420 (1996).

[6] R.J. Williams and D. Zipser, "A learning algorithm for continually running fully recurrent neural networks," *Neural Computation*, **1**, 270–280 (1989).

[7] G.V. Puskorius and L.A. Feldkamp, "Neurocontrol of nonlinear dynamical systems with Kalman filter-trained recurrent networks," *IEEE Transactions on Neural Networks*, **5**, 279–297 (1994).

[8] P.J. Werbos, "Backpropagation through time: What it does and how to do it," *Proceedings of the IEEE*, **78**, 1550–1560 (1990).

[9] G.V. Puskorius and L.A. Feldkamp, "Extensions and enhancements of decoupled extended Kalman filter training," in *Proceedings of the 1997 International Conference on Neural Networks*, Houston, TX, Vol. 3, pp. 1879–1883.

[10] L.A. Feldkamp and G.V. Puskorius, "Training controllers for robustness: multi-stream DEKF," in *Proceedings of the IEEE International Conference on Neural Networks*, Orlando, FL, 1994, Vol. IV, pp. 2377–2382.

[11] L.A. Feldkamp and G.V. Puskorius, "Training of robust neurocontrollers," in *Proceedings of the 33rd IEEE International Conference on Decision and Control*, Orlando, FL, 1994, Vol. III, pp. 2754–2760.

[12] L.A. Feldkamp and G.V. Puskorius, "A signal processing framework based on dynamic neural networks with application to problems in adaptation, filtering and classification," *Proceedings of the IEEE*, **86**, 2259–2277 (1998).

[13] F. Heimes, "Extended Kalman filter neural network training: experimental results and algorithm improvements," in *Proceedings of the 1998 IEEE Conference on Systems, Man and Cybernetics*, Orlando, FL., pp. 1639–1644.

[14] L.A. Feldkamp and G.V. Puskorius, "Fixed weight controller for multiple systems," in *Proceedings of the 1997 IEEE International Conference on Neural Networks*, Houston, TX, Vol. 2, pp 773–778.

[15] G.H. Golub and C.F. Van Loan, *Matrix Computations*, 2nd ed. Baltimore, MD: The John Hopkins University Press, 1989.

[16] G.V. Puskorius and L.A. Feldkamp, "Avoiding matrix inversions for the decoupled extended Kalman filter training algorithm," in *Proceedings of the World Congress on Neutral Networks*, Washington, DC, 1995, pp. I-704–I-709.

[17] E.S. Plumer, "Training neural networks using sequential extended Kalman filtering," in *Proceedings of the World Congress on Neural Networks*, Washington DC, 1995 pp. I-764–I-769.

[18] P. Sun and K. Marko, "The square root Kalman filter training of recurrent neural networks," in *Proceedings of the 1998 IEEE Conference on Systems, Man and Cybernetics*, Orlando, FL, pp. 1645–1651.

[19] S. Haykin, *Adaptive Filter Theory*, 3rd ed. Englewood Cliffs, NJ: Prentice-Hall.

[20] E.W. Saad, D.V. Prokhorov, and D.C. Wunsch III, "Comparative study of stock trend prediction using time delay, recurrent and probabilistic neural networks," *IEEE Transactions on Neural Networks*, **9**, 1456–1470 (1998).

[21] K.-C. Jim, C.L. Giles, and B.G. Horne, "An analysis of noise in recurrent neural networks: convergence and generalization," *IEEE Transactions on Neural Networks*, **7**, 1424–1438 (1996).

[22] D.L. Elliot, "A better activation function for artificial neural networks," Institute for Systems Research, University of Maryland, Technical Report TR93-8, 1993.

[23] J. Hertz, A. Krogh, and R.G. Palmer, *Introduction to the Theory of Neural Computation*, Redwood City, CA: Addison-Wesley, 1991.

[24] G. Jesion, C.A. Gierczak, G.V. Puskorius, L.A. Feldkamp, and J.W. Butler, "The application of dynamic neural networks to the estimation of feedgas vehicle emissions," in *Proceedings of the 1998 International Joint Conference on Neural Networks*, Anchorage, AK, pp. 69–73.

3

LEARNING SHAPE AND MOTION FROM IMAGE SEQUENCES

Gaurav S. Patel

Department of Electrical and Computer Engineering, McMaster University, Hamilton, Ontario, Canada

Sue Becker and Ron Racine

Department of Psychology, McMaster University, Hamilton, Ontario, Canada

(beckers@mcmaster.ca)

3.1 INTRODUCTION

In Chapter 2, Puskorius and Feldkamp described a procedure for the supervised training of a recurrent multilayer perceptron – the node-decoupled extended Kalman filter (NDEKF) algorithm. We now use this model to deal with high-dimensional signals: moving visual images. Many complexities arise in visual processing that are not present in one-dimensional prediction problems: the scene may be cluttered with back-

Kalman Filtering and Neural Networks, Edited by Simon Haykin
ISBN 0-471-36998-5

ground objects, the object of interest may be occluded, and the system may have to deal with tracking differently shaped objects at different times. The problem we have dealt with initially is tracking objects that vary in both shape and location. Tracking differently shaped objects is challenging for a system that begins by performing local feature extraction, because the features of two different objects may appear identical locally even though the objects differ in global shape (e.g., squares versus rectangles). However, adequate tracking may still be achievable without a perfect three-dimensional model of the object, using locally extracted features as a starting point, provided there is continuity between image frames.

Our neural network model is able to make use of short-term continuity to track a range of different geometric shapes (circles, squares, and triangles). We evaluate the model's abilities in three experiments. In the first experiment, the model was trained on images of two different moving shapes, where each shape had its own characteristic movement trajectory. In the second experiment, the training set was made more difficult by adding a third object, which also had a unique motion trajectory. In the third and final experiment, the restriction of one direction of motion per shape was lifted. Thus, the model experienced the same shape traveling in different trajectories, as well as different shapes traveling in the same trajectory. Even under these conditions, the model was able to learn to track a given shape for many time steps and anticipate both its shape and location many time steps into the future.

3.2 NEUROBIOLOGICAL AND PERCEPTUAL FOUNDATIONS

The architecture of our model is motivated by two key anatomical features of the mammalian neocortex, the extensive use of feedback connections, and the hierarchical multiscale structure. We discuss briefly the evidence for, and benefits of, each of these in turn.

Feedback is a ubiquitous feature of the brain, both between and within cortical areas. Whenever two cortical areas are interconnected, the connections tend to be bidirectional [1]. Additionally, within every neocortical area, neurons within the superficial layers are richly interconnected laterally via a network of horizontal connections [2]. The dense web of feedback connections within the visual system has been shown to be important in suppressing background stimuli and amplifying salient or foreground stimuli [3]. Feedback is also likely to play an important role in processing sequences. Clearly, we view the world as a continuously varying sequence rather than as a disconnected collection of snapshots. Seeing the world in this way allows recent experience to play a role in the anticipation or prediction of what will come next. The generation of predictions in a perceptual system may serve at least two important functions: (1) To the extent that an incoming sensory signal is consistent with expectations, intelligent filtering may be done to increase the signal-to-noise ratio and resolve ambiguities using context. (2) When the signal violates expectations, an organism can react quickly to such changing or salient conditions by de-emphasizing the expected part of the signal and devoting more processing capacity to the unexpected information. Top-down connections between processing layers, or lateral connections within layers, or both, might be used to accomplish this. Lateral connections allow for local constraints about moving contours to guide one's expectations, and this is the basis for our model.

Prediction in a high-dimensional space is computationally complex in a fully connected network architecture. The problem requires a more constrained network architecture that will reduce the number of free parameters. The visual system has done just that. In the earliest stages of processing, cells' receptive fields span only a few degrees of visual angle, while in higher visual areas, cells' receptive fields span almost the entire visual field (for a review, see [4]). Therefore, we designed our model network with a similar hierarchical architecture, in which the first layer of units were connected to relatively small, local regions of the image and a subsequent layer spanned the entire visual field (see Figure 3.1).

3.3 NETWORK DESCRIPTION

Prediction in a high-dimensional space such as a 50×50 pixel image, using a fully connected recurrent network is not feasible, because the number of connections is typically one or more orders of magnitude larger than the dimensionality of the input, and the NDEKF training procedure requires adapting these parameters for typically hundreds to thousands of iterations. The problem requires a more constrained network architecture that will reduce the number of free parameters. Motivated by the hierarchical architecture of real visual systems, we designed our model network with a similar hierarchical architecture in which the first layer of units were connected to relatively small, local 5×5 pixel regions of the image and a subsequent layer spanned the entire visual field (see Figure 3.1).

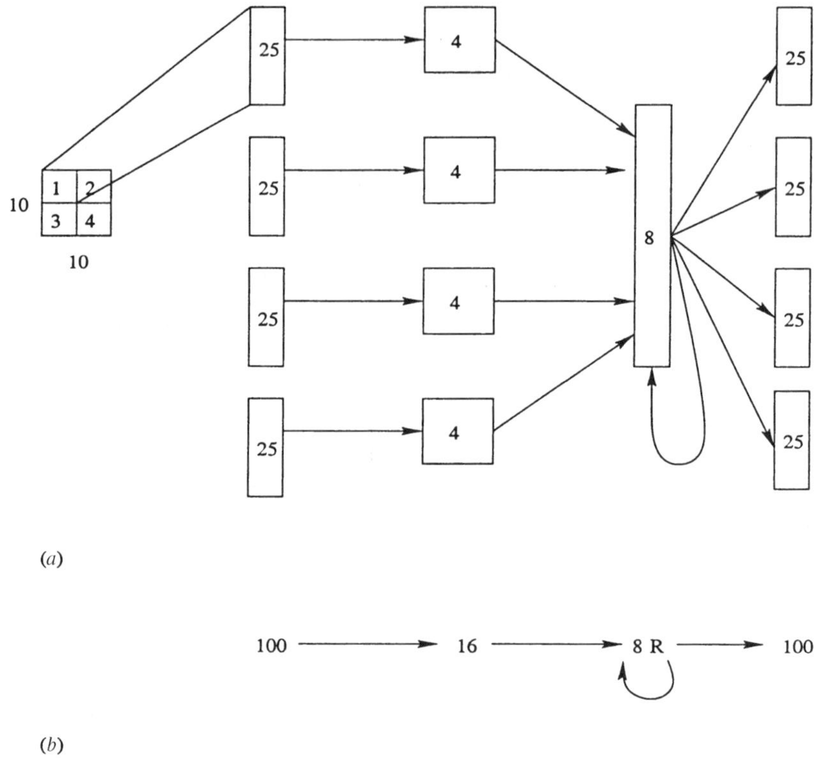

Figure 3.1 A diagram of the network used. The numbers in the boxes indicate the number of units in each layer or module, except in the input layer, where the receptive fields are numbered 1, ..., 4. Local receptive fields of size 5×5 at the input are fed to the four banks of four units in the first hidden layer. The second layer of eight units then combines these local features learned by the first hidden layer. Note the recurrence in the second hidden layer.

A four-layer network of size 100-16-8R-100, as depicted in Figures 3.1*a* and 3.1*b*, was used in the following experiments. Training images of size 10×10, which are arranged in a vector format of size 100×1, were used to form the input to the networks. As depicted in Figure 3.1*a*, the input image is divided into four non-overlapping receptive fields of size 5×5. Further, the 16 units in the first hidden layer are divided into four banks of four units each. Each of the four units within a bank receive inputs from one of the four receptive fields. This describes how the 10×10 image is connected to the 16 units in the first hidden layer. Each of these 16 units feed into a second hidden layer of 8 units. The second hidden layer has recurrent connections (note that recurrence is only within the layer and not between layers).

Thus, the input layer of the network is connected to small and local regions of the image. The first layer processes these local receptive fields separately, in an effort to extract relevant local features. These features are then combined by the second hidden layer to predict the next image in the sequence. The predicted image is represented at the output layer. The prediction error is then used in the EKF equations to update the weights. This process is repeated over several epochs through the training image sequences until a sufficiently small incremental mean-squared error is obtained.

3.4 EXPERIMENT 1

In the first experiment, the model is trained on images of two different moving shapes, where each shape has its own characteristic movement, that is, shape and direction of movement are perfectly correlated. The sequence of eight 10×10 pixel images in Figure 3.2*a* is used to train a four-layered (100-16-8R-100) network to make one-step predictions of the image sequence. In the first four time steps, a circle moves upward within the image; and in the last four time steps, a triangle moves downward

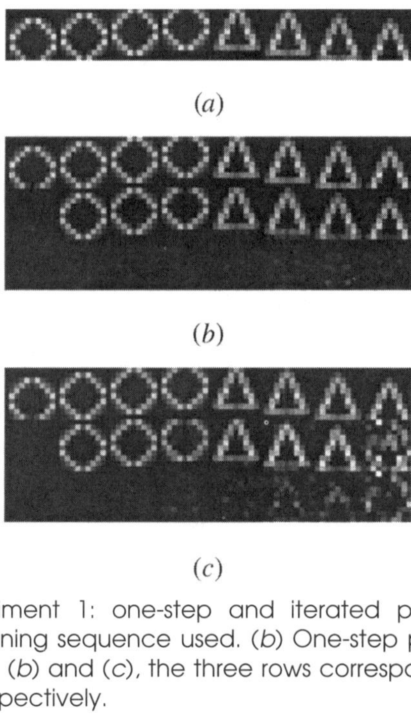

(*a*)

(*b*)

(*c*)

Figure 3.2 Experiment 1: one-step and iterated prediction of image sequence. (*a*) Training sequence used. (*b*) One-step prediction. (*c*) multi-step prediction. In (*b*) and (*c*), the three rows correspond to input, prediction, and error, respectively.

within the image. At each time step, the network is presented with one of the eight 10×10 images as input (divided into four 5×5 receptive fields as described above), and generates in its output layer a prediction of the input at the next time step, but it is always given the correct input at the next time step. Training was stopped after 20 epochs through the training sequence. Figure 3.2*b* shows the network operating in one-step prediction mode on the training sequence after training. It makes excellent predictions of the object shape and also its motion. Figure 3.2*c* shows the network operating in an autonomous mode after being shown only the first image of the sequence. In this multistep prediction case, the network is only given external input at the first time step in the sequence. Beyond the first time step, the network is given its prediction from time $t - 1$ as its input at time t, which could potentially lead to a buildup of prediction errors over many time steps. This shows that the network has reconstructed the entire dynamics, to which it was exposed during training, when provided with only the first image. This is indeed a difficult task. It is seen that as the iterative prediction proceeds, the residual errors (the third row in Figure 3.2*c*) are amplified at each step.

3.5 EXPERIMENT 2

Next, an ND-EKF network with the same 100-16-8R-100 architecture used in Experiment 1 was trained with three sequences, each consisting of four images, in the following order:

- circle moving right and up;
- triangle moving right and down;
- square moving right and up.

During training, at the beginning of each sequence, the network states were initialized to zero, so that the network would not learn the order of presentation of the sequences. The network was therefore expected to learn the motions associated with each of the three shapes, and not the order of presentation of the shapes.

During testing, the order of presentation of the three sequences varied, as shown in Figure 3.3*a*. The trained network does well at the task of one-step prediction, only failing momentarily at transition points where we switch between sequences. It is important to note that one-step prediction, in this case, is a difficult and challenging task because the network has to determine (1) what shape is present and (2) which direction it is moving in, without direct knowledge of inputs some time in the past. In order to

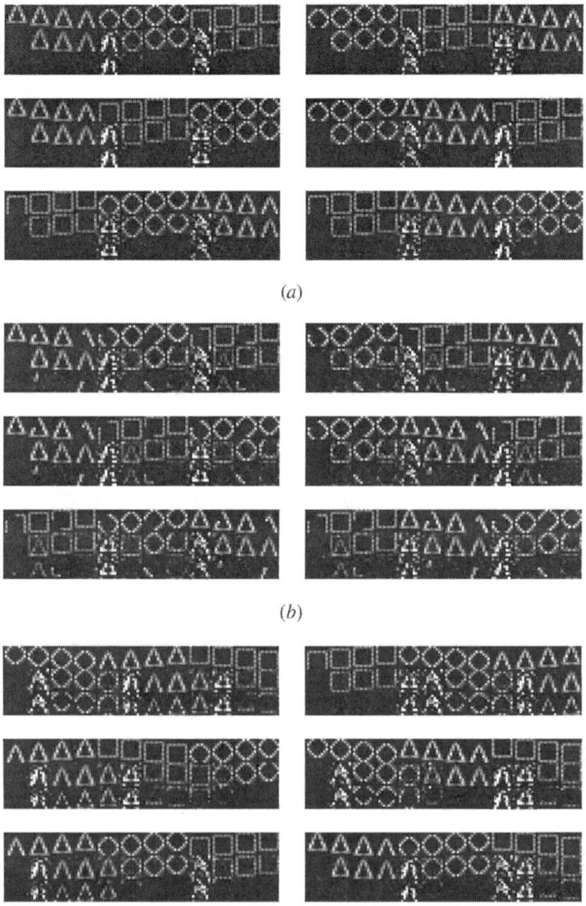

(*a*) (*b*) (*c*)

Figure 3.3 Experiment 2: one-step prediction of image sequences using the trained network. (*a*) Various combinations of sequences used in training. (*b*) Same sequences as in (*a*), but with occlusions. (*c*) Prediction on some sequences not seen during training. The three rows in each image correspond to input, prediction, and error, respectively.

make good predictions, it must rely on its recurrent or feedback connections, which play a crucial role in the present model.

We also tested the model on a set of occluded images – images with regions that are intentionally zeroed. Remarkably, the network makes correct one-step predictions, even in the presence of occlusions as shown in Figure 3.3*b*. In addition, the predictions do not contain occlusions; that is, they are correctly filled in, demonstrating the robustness of the model to occlusions. In Figure 3.3*c*, when the network is presented with sequences that it had not been exposed to during training, a larger residual error is obtained, as expected. However, the network is still capable of identifying the shape and motion, although not as accurately as before.

3.6 EXPERIMENT 3

In Experiment 1, the network was presented with short sequences (four images) of only two shapes (circle and triangle), and in experiment 2 an extra shape (square) was added. In Experiment 3, to make the learning task even more challenging, the length of the sequences was increased to 10 and the restriction of one direction of motion per shape was lifted. Specifically, each shape was permitted to move right and either up or down. Thus, the network was exposed to different shapes traveling in similar directions and also the same shape traveling in different directions, increasing the total number of images presented to the network from 8 images in Experiment 1 and 12 images in Experiment 2 to 100 images in this experiment. In effect, there is a substantial increase in the number of learning patterns, and thus a substantial increase in the complexity of the learning task. However, since the number of weights in the network is limited and remains the same as in the other experiments, the network cannot simply memorize the sequences.

We trained a network of the same 100-16-8R-100 architecture on six sequences, each consisting of 10 images (see Fig. 3.4) in the following order:

- circle moving right and up;
- square moving right and down;
- triangle moving right and up;
- circle moving right and down;
- square moving right and up;
- triangle moving right and down.

Training was performed in a similar manner as Experiment 2. During testing, the order of presentation of the six sequences was varied; several examples are shown in Figure 3.5. As in the previous experiments, even with the larger number of training patterns, the network is able to predict the correct motion of the shapes, only failing during transitions between shapes. It is able to distinguish between the same shapes moving in different directions as well as different shapes moving in the same direction, using context available via the recurrent connections.

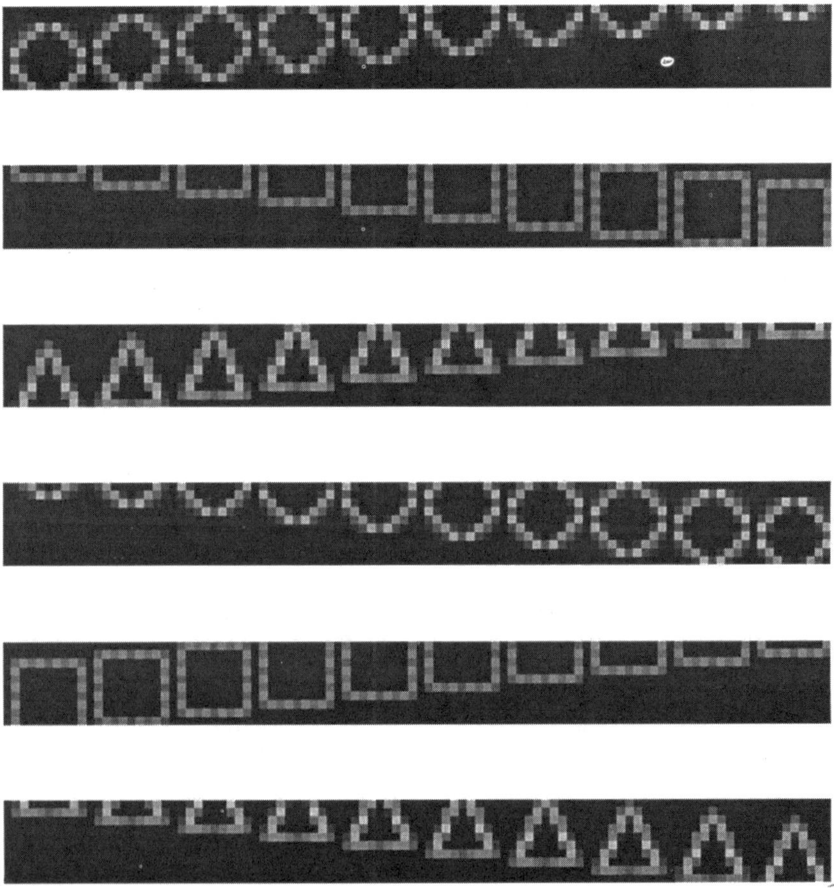

Figure 3.4 Experiment 3: six image sequences used for training.

The failure of the model to make accurate predictions at transitions between shapes can also be seen in the residual error that is obtained during prediction. The residual error in the predicted image is quantified by calculating the mean-squared prediction error, as shown in Figure 3.6. The figure shows how the mean-squared prediction error varies as the prediction continues. Note the transient increase in error at transitions between shapes.

3.7 DISCUSSION

In this chapter, we have dealt with time-series prediction of high-dimensional signals: moving visual images. This situation is much more

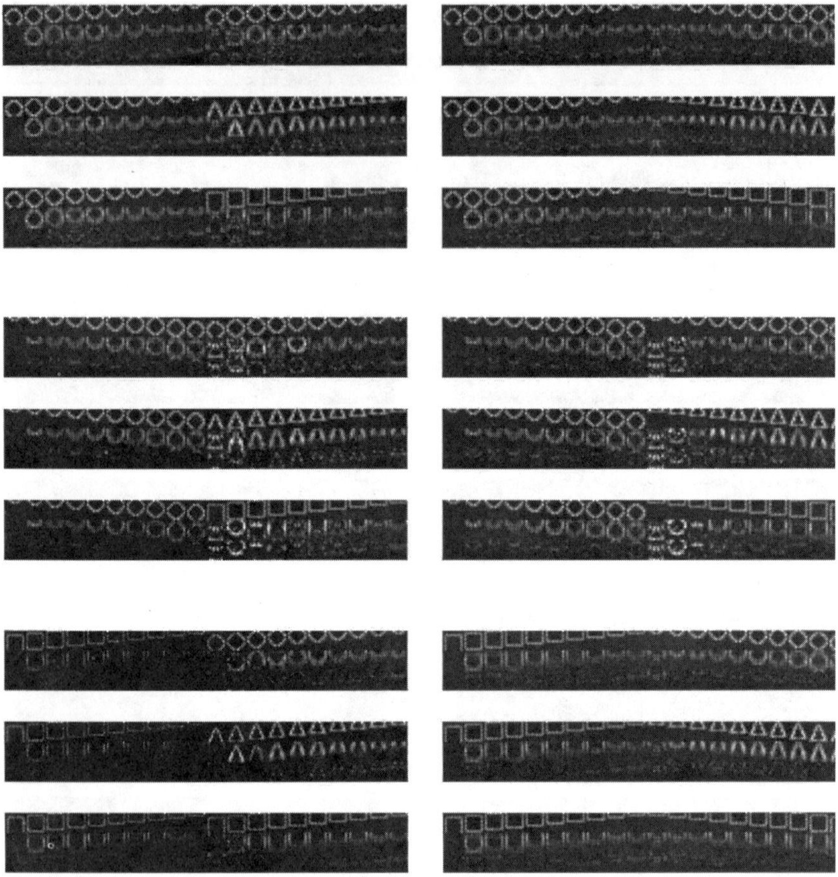

Figure 3.5 Experiment 3: one-step prediction of image sequences using the trained network. The three rows in each image correspond to input, prediction, and error, respectively.

complicated than a one-dimensional case, in that the system has to deal with simultaneous shape and motion prediction. The network was trained by the EKF method to perform one-step prediction of image sequences in a specific order. Then, during testing, the order of the sequences was varied and the network was asked to predict the correct shape and location of the next image in the sequence. The complexity of the problem was increased from Experiment 1 to 3 as we introduced occlusions, increased both the length of the training sequences and the number of shapes presented, and allowed shape and motion to vary independently. In all

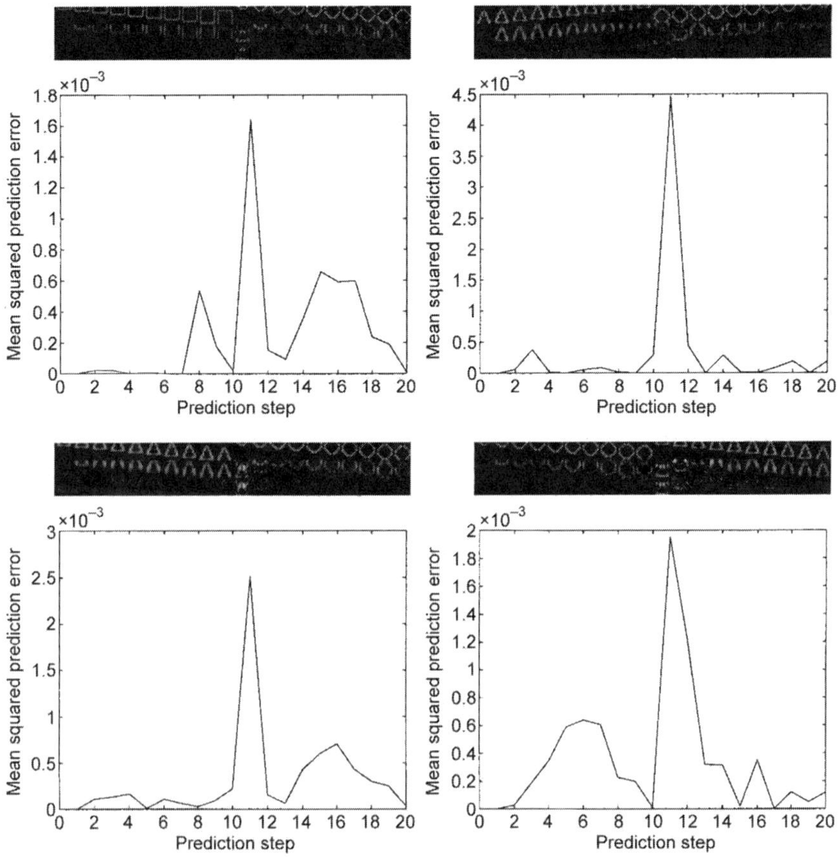

Figure 3.6 Mean-squared prediction error in one-step prediction of image sequences using the trained network. The three rows in each image correspond to input, prediction, and error, respectively. The graphs show how the mean-squared prediction error varies as the prediction progresses. Notice the increase in error at transitions between shapes.

cases, the network was able to predict the correct motion of the shapes, failing only momentarily at transitions between shapes.

The network described here is a first step toward modeling the mechanisms by which the human brain might simultaneously recognize and track moving stimuli. Any attempt to model both shape and motion processing simultaneously within a single network may seem to be at odds with the well-established finding that shape and spatial information are processed in separate pathways of the visual system [5]. An extreme version of this view posits that form-related features are processed strictly by the ventral "what" pathway and motion features are processed strictly by the dorsal "where" pathway. Anatomically, however, there are cross-connections between the two pathways at several points [6]. Furthermore, there is ample behavioral evidence that the processes of shape and motion perception are not completely separate. For example, it has long been established that we are able to infer shape from motion (see e.g., [7]). Conversely, under certain conditions, object recognition can be shown to drive motion perception [8]. In addition, Stone [9] has shown that viewers are much better at recognizing objects when they are moving in characteristic, familiar trajectories as compared with unfamiliar trajectories. These data suggest that when shape and motion are tightly correlated, viewers will learn to use them together to recognize objects. This is exactly what happens in our model.

To accomplish temporal processing in our model, we have incorporated within-layer recurrent connections in the architecture used here. Another possibility would be to incorporate top-down recurrent connections. A key anatomical feature of the visual system is top-down feedback between visual areas [3]. Top-down connections could allow global expectations about the three-dimensional shape of a moving object to guide predictions. Thus, an important direction for future work is to extend the model to allow top-down feedback. Rao and Ballard [10] have proposed an alternative neural network implementation of the EKF that employs top-down feedback between layers, and have applied their model to both static images and time-varying image sequences [10, 11]. Other models of cortical feedback for modeling the generation of expectations have also been proposed (see, e.g., [12, 13]).

Natural visual systems can deal with an enormous space of possible images, under widely varying viewing conditions. Another important direction for future work is to extend our model to deal with more realistic images. Many additional complexities arise in natural images that were not present in the artificial image sequences used here. For example, the simultaneous presence of both foreground and background objects may hinder the prediction accuracy. Natural visual systems likely use attentional filtering and binding strategies to alleviate this problem; for example, Moran and Desimone [14] have observed cells that show a suppressed neural response to a preferred stimulus if unattended and in the presence of an attended stimulus. Another simplification of our images is that shape remained constant for many time frames, whereas for real three-dimensional objects, the shape projected onto a two-dimensional image may change dramatically over time, because of rotations as well as non-rigid motions (e.g. bending). Humans are able to infer three-dimensional shape from non-rigid motion, even from highly impoverished stimuli such as moving light displays [7]. It is likely that the architecture described here could handle changes in shape, provided shape changes predictably and gradually over time.

REFERENCES

[1] D.J. Felleman and D.C. Van Essen, "Distributed hierarchical processing in the primate cerebral cortex", *Cerebral Cortex*, **1**, 1–47 (1991).

[2] J.S. Lund, Q. Wu and J.B. Levitt, "Visual cortex cell types and connections", in M.A. Arbib, Ed., *Handbook of Brain Theory and Neural Networks*, Cambridge, MA: MIT Press, 1995.

[3] J.M. Hupé, A.C. James, B.R. Payne, S.G. Lomber, P. Girard and J. Bullier, "Cortical feedback improves discrimination between figure and background by V1, V2 and V3 neurons", *Nature*, **394**, 784–787 (1998).

[4] M.W. Oram and D.I. Perrett, "Modeling visual recognition from neurobiological constraints", *Neural Networks*, **7**, 945–972 (1994).

[5] M. Mishkin, L.G. Ungerleider and K.A. Macko, "Object vision and spatial vision: Two cortical pathways", *Trends in Neurosciences*, **6**, 414–417 (1983).

[6] E.A. De Yoe and D.C. Van Essen, "Concurrent processing streams in monkey visual cortex", *Trends in Neurosciences*, **11**, 219–226, (1988).

[7] G. Johanssen, "Visual perception of biological motion and a model for its analysis", *Perception and Psychophysics*, **14**, 201–211 (1973).

[8] V.S. Ramachandran, C. Armel, C. Foster and R. Stoddard, "Object recognition can drive motion perception", *Nature*, **395**, 852–853 (1998).

[9] J.V. Stone, "Object recognition: View-specificity and motion-specificity", *Vision Research*, **39**, 4032–4044, (1999).

[10] R.P.N. Rao and D.H. Ballard, "Dynamic model of visual recognition predicts neural response properties in the visual cortex", *Neural Computation*, **9(4)**, 721–763 (1997)

[11] R.P.N. Rao, "Correlates of attention in a model of dynamic visual recognition", in M.I. Jordan, M.J. Kearns and S.A. Solla, Eds., *Advances in Neural Information Processing Systems*, Vol. 10. Cambridge, MA: MIT Press, 1998.

[12] E. Harth, K.P. Unnikrishnan and A.S. Panday, "The inversion of sensory processing by feedback pathways: A model of visual cognitive functions", *Science*, **237**, 184–187 (1987).

[13] D. Mumford, "On the computational architecture of the neocortex", *Biological Cybernetics*, **65**, 135–145 (1991).

[14] J. Moran and R. Desimone, "Selective attention gates visual processing in the extrastriate cortex", *Science*, **229**, 782–784, (1985).

4

CHAOTIC DYNAMICS

Gaurav S. Patel
Department of Electrical and Computer Engineering, McMaster University, Hamilton, Ontario, Canada

Simon Haykin
Communications Research Laboratory, McMaster University, Hamilton, Ontario, Canada
(haykin@mcmaster.ca)

4.1 INTRODUCTION

In this chapter, we consider another application of the extended Kalman filter recurrent multilayer perceptron (EKF-RMLP) scheme: the modeling of a chaotic time series or one that could be potentially chaotic.

The generation of a *chaotic process* is governed by a coupled set of nonlinear differential or difference equations. The hallmark of a chaotic process is sensitivity to *initial conditions*, which means that if the starting point of motion is perturbed by a very small increment, the deviation in

Kalman Filtering and Neural Networks, Edited by Simon Haykin
ISBN 0-471-36998-5

Table 4.1 Summary of data sets used in the study

	Network size	Training length	Testing length	Sampling frequency f_s(Hz)	Largest Lyapunov exponent λ_{max} (nats/sample)	Correlation dimension D_{ML}
Logistic	6-4R-2R-1	5,000	25,000	1	0.69	1.04
Ikeda	6-6R-5R-1	5,000	25,000	1	0.354	1.51
Lorenz	3-8R-7R-1	5,000	25,000	40	0.040	2.09
NH_3 laser	9-10R-8R-1	1,000	9,000	1[a]	0.147	2.01
Sea clutter	6-8R-7R-1	40,000	10,000	1000	0.228	4.69

[a]The sampling frequency for the laser data was not known. It was assumed to be 1 Hz for the Lyapunov exponent calculations.

the resulting waveform, compared to the original waveform, increases exponentially with time. Consequently, unlike an ordinary deterministic process, a chaotic process is predictable only in the short term.

Specifically, we consider five data sets categorized as follows:

- The logistic map, Ikeda map, and Lorenz attractor, whose dynamics are governed by known equations; the corresponding time series can therefore be numerically generated by using the known equations of motion.
- Laser intensity pulsations and sea clutter (i.e., radar backscatter from an ocean surface) whose underlying equations of motion are unknown; in this second case, the data are obtained from real-life experiments.

Table 4.1 shows a summary of the data sets used for model validation. The table also shows the lengths of the data sets used, and their division into the training and test sets, respectively. Also shown is a partial summary of the dynamic invariants for each of the data sets used and the size of network used for modeling the dynamics for each set.

4.2 CHAOTIC (DYNAMIC) INVARIANTS

The *correlation dimension* is a measure of the complexity of a chaotic process [1]. This chaotic invariant is always a fractal number, which is one reason for referring to a chaotic process as a "strange" attractor. The other chaotic invariants, the *Lyapunov exponents*, are, in part, responsible for sensitivity of the process to initial conditions, the occurrence of which requires having at least one positive Lyapunov exponent. The *horizon of predictability* (HOP) of the process is determined essentially by the largest positive Lyapunov exponent [1]. Another useful parameter of a chaotic process is the *Kaplan–York dimension* or *Lyapunov dimension*, which is defined in terms of a Lyapunov spectrum by

$$D_{KY} = K + \frac{\sum_{i=1}^{K} \lambda_i}{|\lambda_{K+1}|}, \tag{4.1}$$

where the λ_i are the Lyapunov exponents arranged in decreasing order and K is the largest integer for which the following inequalities hold

$$\sum_{i=1}^{K} \lambda_i \geq 0 \quad \text{and} \quad \sum_{i=1}^{K+1} \lambda_i < 0.$$

Typically, the Kaplan–Yorke dimension is close in numerical value to the correlation dimension. Yet another byproduct of the Lyapunov spectrum is the *Kolmogorov entropy*, which provides a measure of information generated due to sensitivity to initial conditions. It can be calculated as the sum of all the positive Lyapunov exponents of the process. The chaotic invariants were estimated as follows:

1. The correlation dimension was estimated using an algorithm based on the method of maximum likelihood [2] – hence the notation D_{ML} for the correlation dimension.
2. The Lyapunov exponents were estimated using an algorithm, involving the QR - decomposition applied to a Jacobian that pertains to the underlying dynamics of the time series.
3. The Kolmogorov entropy was estimated directly from the time series using an algorithm based on the method of maximum likelihood [2] – hence the notation KE_{ML} for the Kolmogorov entropy so estimated. The indirect estimate of the Kolmogorov entropy from the Lyapunov spectrum is denoted by KE_{LE}.

4.3 DYNAMIC RECONSTRUCTION

The attractor of a dynamical system is constructed by plotting the evolution of the state vector in state space. This construction is possible when we have access to every state variable of the system. In practical situations dealing with dynamical systems of unknown state-space equations, however, all that we have available is a set of measurements taken from the system. Given such a situation, we may raise the following question: Is it possible to reconstruct the attractor of a system (with many state variables) using a single time series of measurements? The answer to this question is an emphatic yes; it was first illustrated by Packard et al. [3], and then given a firm mathematical foundation by Takens [4] and Mañé [5]. In essence, the celebrated *Takens embedding* theorem guarantees that by applying the *delay coordinate method* to the measurement time series, the original dynamics could be reconstructed, under certain assumptions. In the delay coordinate method (sometimes referred to as the method of delays), delay coordinate vectors are formed using time-delayed values of the measurements, as shown here:

$$\mathbf{s}(n) = [s(n), s(n-\tau), \ldots, s(n-(d_E-2)\tau), s(n-(d_E-1)\tau)]^T,$$

where d_E is called the *embedding dimension* and τ is known as the *embedding delay*, taken to be some suitable multiple of the sampling time τ_s. By means of such an embedding, it is possible to reconstruct the true dynamics using only one measurement. Takens' theorem assumes the existence of d_E and τ such that mapping from $s(n)$ to $s(n+\tau)$ is possible. The concept of dynamic reconstruction using delay coordinate embedding is very elegant, because we can use it to build a model of a nonlinear dynamical system, given a set of measured data on the system. We can use it to "reverse-engineer" the dynamics, i.e., use the time series to deduce characteristics of the physical system that was responsible for its generation. Put it another way, the reconstruction of the dynamics from a time series is in reality an *ill-posed inverse problem*. The direct problem is: given the dynamics, describe the iterates; and the inverse problem is: given the iterates, describe the dynamics. The inverse problem is ill-posed because, depending on the quality of the data, a solution may not be stable, may not be unique, or may not even exist. One way to make the problem well-posed is to include prior knowledge about the input–output mapping. In effect, the use of delay coordinate embedding inserts some prior knowledge into the model, since the embedding parameters are determined from the data.

To estimate the embedding delay τ, we used the method of mutual information proposed by Fraser [6]. According to this method, the embedding delay is determined by finding the particular delay for which the mutual information between the observable time series and its delayed version is minimized for the first time. Given such an embedding delay, we can construct a delay coordinate vector whose adjacent samples are as statistically independent as possible.

To estimate the embedding dimension d_E, we use the method of false nearest neighbors [1]; the embedding dimension is the smallest integer dimension that unfolds the attractor.

4.4 MODELING NUMERICALLY GENERATED CHAOTIC TIME SERIES

4.4.1 Logistic Map

In this experiment, the EKF-RMLP-based modeling scheme is applied to the logistic map (also known as the quadratic map), which was first used as a model of population growth. The logistic map is described by the difference equation:

$$x(k+1) = \alpha x(k)[1 - x(k)], \tag{4.2}$$

where the nonlinearity parameter α is chosen to be 4.0 so that the produced behavior is chaotic. The logistic map exhibits deterministic chaos in the interval $\alpha \in (3.5699, 4]$. An initial value of $x(0) = 0.5$ was used, and 35,000 points were generated, of which the first 5000 points were discarded, leaving a data set of 30,000 samples. A training set, consisting of the first 5000 samples, was used to train an RMLP on a one-step prediction task by means of the EKF method. An RMLP configuration of 6-4R-2R-1, which has a total of 61 weights including the bias terms, was selected for this modeling problem. The training converged after only 5 epochs and a sufficiently low MSE was achieved as shown in Figure 4.1.

Open-Loop Evaluation A test set, consisting of the unexposed 25,000 samples, was used to evaluate the performance of the network at the task of one-step prediction as well as recursive prediction. Figure 4.2*a* shows the one-step prediction performance of the network on a short portion of the test data. It is visually observed that the two curves are

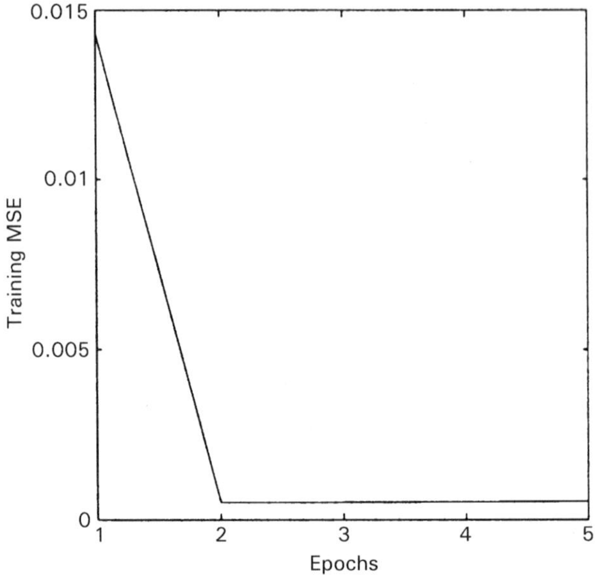

Figure 4.1 Training MSE versus epochs for the logistic map.

almost identical. Also, for numerical one-step performance evaluation, signal-to-error ratio (SER) is used. This measure, expressed in decibels, is defined by

$$\mathrm{SER} = 10\log_{10}\frac{\mathrm{MSS}}{\mathrm{MSE}}, \tag{4.3}$$

where MSS is the mean-squared value of the actual test data and MSE is the mean-squared value of the prediction error at the output. MSS is found to be 0.374 for the 25,000-testing sequence and MSE is found to be 1.09×10^{-5} for the trained RMLP network prediction error. This gives an SER of 45.36 dB, which is certainly impressive because it means that the power of the one-step prediction error over 25,000 test samples is many times smaller than the power of the signal.

Closed-Loop Evaluation To evaluate the autonomous behavior of the network, its node outputs are first initialized to zero, it is then seeded with points selected from the test data, and then passed through a *priming phase* where it operates in the one-step mode for $p_l = 30$ steps. At the end of priming, the network's output is fed back to its input, and autonomous

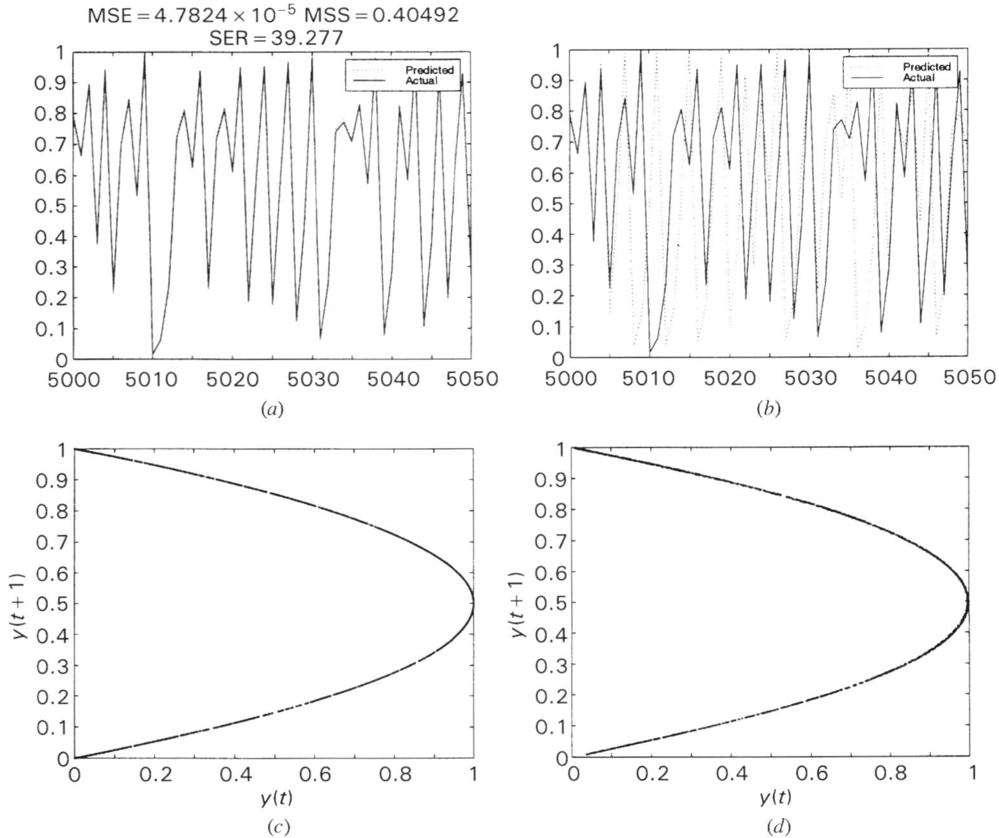

Figure 4.2 Results for the dynamic reconstruction of the logistic map. (*a*) One-step prediction. (*b*) Iterated prediction. (*c*) Attractor of original signal. (*d*) Attractor of iteratively reconstructed signal.

operation begins. At this point, the network is operating on its own without further inputs, and the task that is asked of the network is indeed challenging. The autonomous behavior of the network, which begins after priming, is shown in Figure 4.2*b*, and it is observed that the predictions closely follow the actual data for about 5 steps on average [which is close to the theoretical horizon of predictability (HOP) of 5 calculated from the Lyapunov spectrum], after which they start to deviate significantly. Figure 4.3 plots the one-step prediction of the logistic map for three different starting points

The overall trajectory of the predicted signal, in the long term, has a structure that is very similar to the actual logistic map. The similarity is clearly seen by observing their attractors, which are shown in Figures 4.2*c* and 4.2*d*. For numerical autonomous performance evaluation, the dynamical invariants of both the actual data and the model-generated data are compared in Table 4.2. For the logistic map, $d_L = 1$; it therefore has only one Lyapunov exponent, which happens to be 0.69 nats/sample. This means that the sum of Lyapunov exponents is not negative, thus violating one of the conditions in the Kaplan–Yorke method, and it is for this reason that the Kaplan–Yorke dimension D_{KY} could not be calculated. However, by comparing the other calculated invariants, it is seen that the Lyapunov exponent and the correlation dimension of the two signals are in close agreement with each other. In addition, the Kolmogorov entropy values for the two signals also match very closely. The theoretical horizons of predictability of the two signals are also in agreement with each other. These results demonstrate very convincingly that the original dynamics have been accurately modeled by the trained RMLP. Furthermore, the robustness of the model is tested by starting the predictions from various locations on the test data, corresponding to indices of $N_0 = 3060$, 5060, and 10,060. The results, shown in Figure 4.4, clearly indicate that the RMLP network is able to reconstruct the logistic series beginning from any location, chosen at random.

Table 4.2 Comparison of chaotic invariants of logistic map

Time series	d_E	τ	d_L	D_{ML}	D_{KY}	λ_1 (nats/sample)	KE_{LE} (nats/sample)	KE_{ML} (nats/sample)	HOP (samples)
Actual logistic	6	5	1	1.04	—[a]	0.69	0.69	0.64	5
Reconstructed	6	12	1	1.00	—[a]	0.61	0.61	0.65	6

[a]Since the sum of Lyapunov exponents is not negative, D_{KY} could not be calculated.

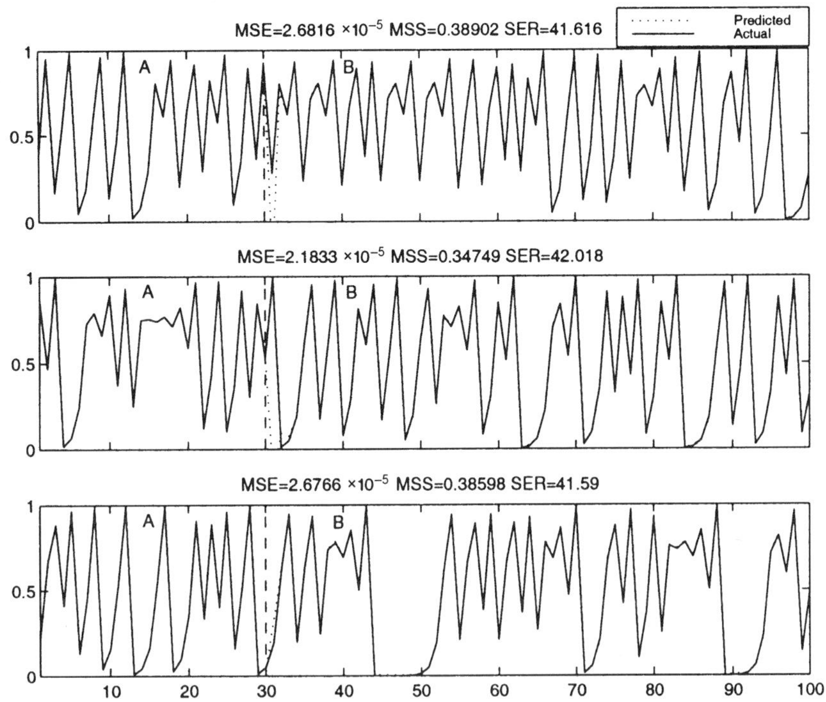

Figure 4.3 One-step prediction of logistic map from different starting points. Note that A = initialization and B = one-step phase.

4.4.2 Ikeda Map

This second experiment uses the Ikeda map (which is substantially more complicated than the logistic map) to test the performance of the EKF-RMLP modeling scheme. The Ikeda map is a complex-valued map and is generated using the following difference equations:

$$m(k) = 0.4 - \frac{6.0}{1 + x_1^2(k) + x_2^2(k)}, \tag{4.4}$$

$$x_1(k+1) = 1.0 + \mu\{x_1(k)\cos[m(k)] - x_2(k)\sin[m(k)]\}, \tag{4.5}$$

$$x_2(k+1) = 1.0 + \mu\{x_1(k) + x_2(k)\cos[m(k)]\}, \tag{4.6}$$

where x_1 and x_2 are the real and imaginary components, respectively, of x and the parameter μ is carefully chosen to be 0.7 so that the produced behavior is chaotic. The initial values of $x_1(0) = 0.5$ and $x_2(0) = 0.5$ were selected and, as pointed out earlier, a data set of 30,000 samples was

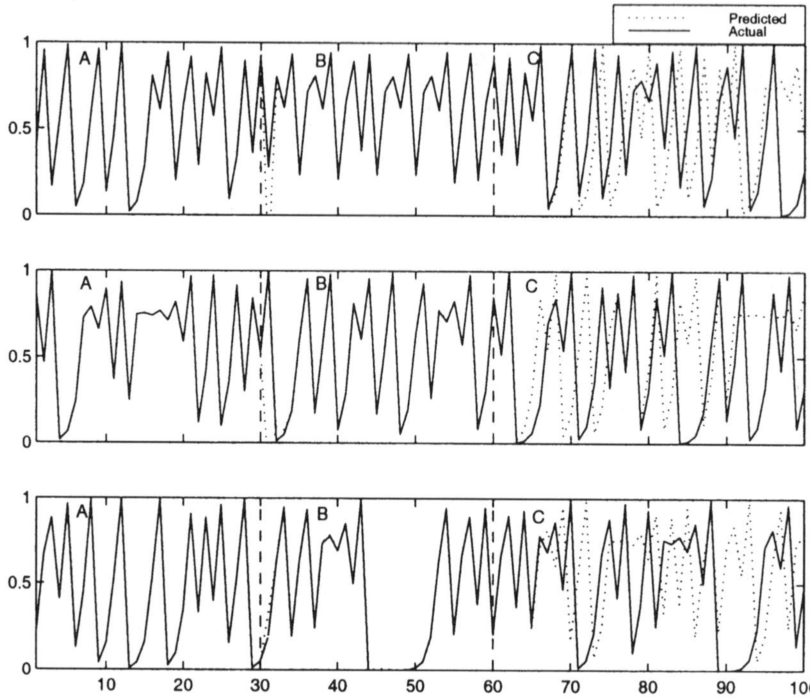

Figure 4.4 Iterative prediction of logistic map from different starting points, corresponding to $N_0 = 3060$, 5060, and 10,060, respectively, with $p_l = 30$. Note that A = initialization, B = priming phase, and C = autonomous phase.

generated. In this experiment, only the x_1 component of the Ikeda map is used, for which the embedding parameters of $d_E = 6$ and $\tau = 10$ were determined. The first 5000 samples of this data set were used to train an RMLP with the EKF algorithm at one-step prediction. During training, a truncation depth $t_d = 10$ was used for the backpropagation through-time (BPTT) derivative calculations. The RMLP configuration of 6-6R-5R-1, which has a total of 144 weights including the bias terms, was chosen to model the Ikeda series. The training converged after only 15 epochs, and a sufficiently low incremental training mean-squared error was achieved, as shown in Figure 4.5.

Open-Loop Evaluation The test set, consisting of the unexposed 25,000 samples of data, is used for performance evaluation, and Figure 4.6*a* shows one-step performance of the network on a short portion of the test data. It is indeed difficult to distinguish between the actual and predicted signals, thus visually verifying the goodness of the predictions.

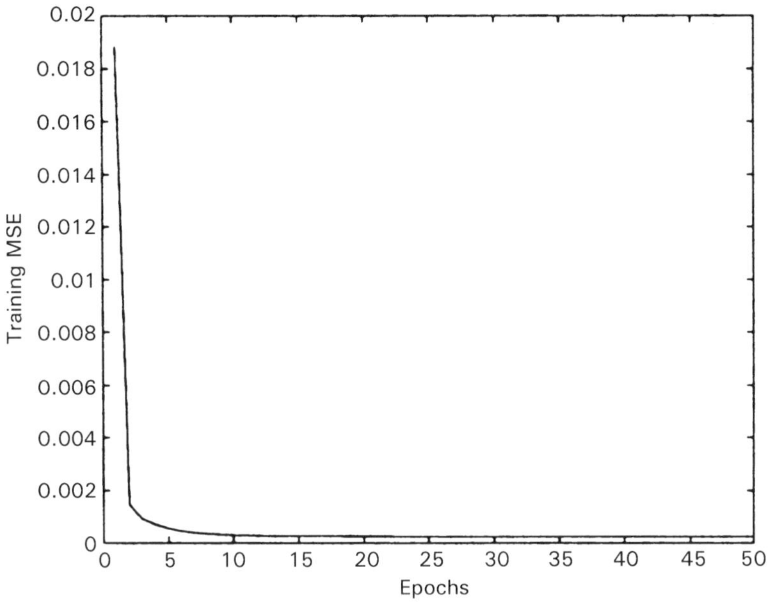

Figure 4.5 Training MSE versus epochs for the Ikeda map.

For a numerical measure, the mean-squared value of the 25,000 sample Ikeda test series was calculated to be MSS $= 0.564$ and the mean-squared prediction error of MSE $= 1.4 \times 10^{-4}$ was produced by the trained RMLP network, thus giving an SER of 36.02 dB.

Closed-Loop Evaluation To begin autonomous prediction, a delay vector consisting of 6 taps spaced by 10 samples apart is constructed as dictated by the embedding parameters d_E and τ. The RMLP is initialized with a delay vector, constructed from the test samples, and passed through a priming phase with $p_l = 60$, after which the network operates in closed-loop mode. The autonomous continuation from where the training data end is shown in Figure 4.6*b*. Note that the predictions follow closely for about 10 steps on average, which is in agreement with the theoretical horizon of predictability of 11 calculated by from the Lyapunov spectrum. A length of 25,000 autonomous samples were generated using the trained EKF-RMLP model, and the reconstructed attractor is plotted in Figure 4.6*d*. The reconstructed attractor has exactly the same form as the original attractor, which is plotted in Figure 4.6*c* using the actual Ikeda samples. These figures clearly demonstrate that the RMLP network has captured the underlying dynamics of the Ikeda map series. For numerical performance

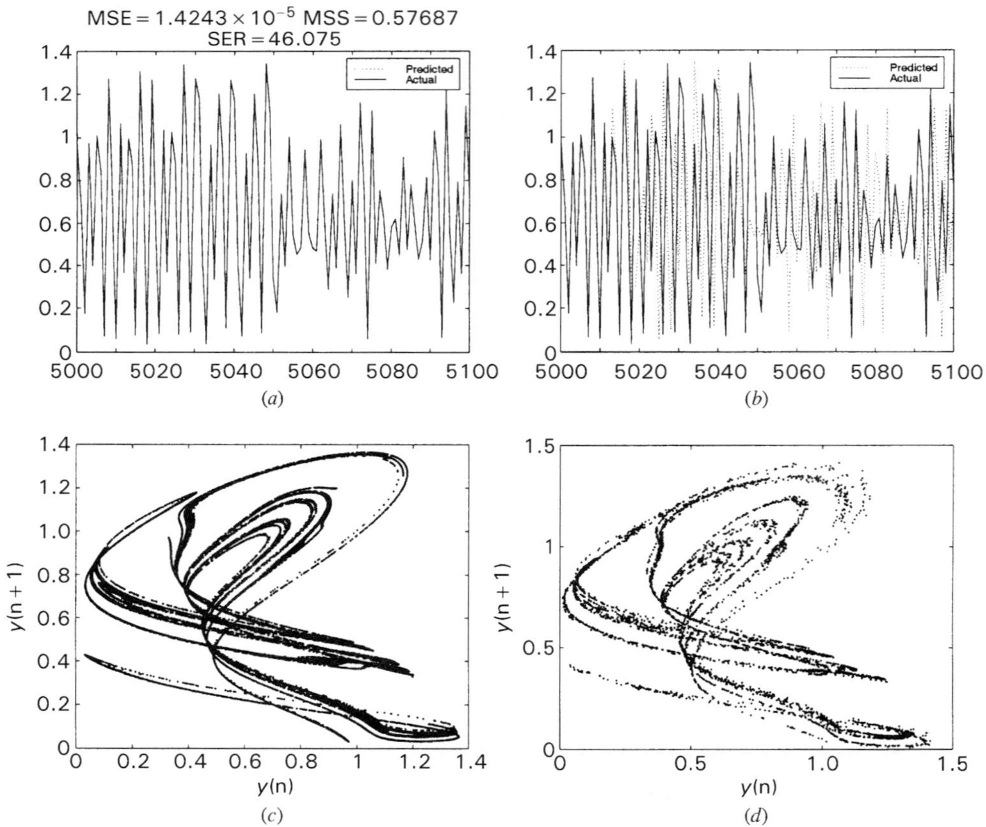

Figure 4.6 Results for the dynamic reconstruction of the Ikeda map. (*a*) One-step prediction. (*b*) Iterated prediction. (*c*) Attractor of original signal. (*d*) Attractor of iteratively reconstructed signal.

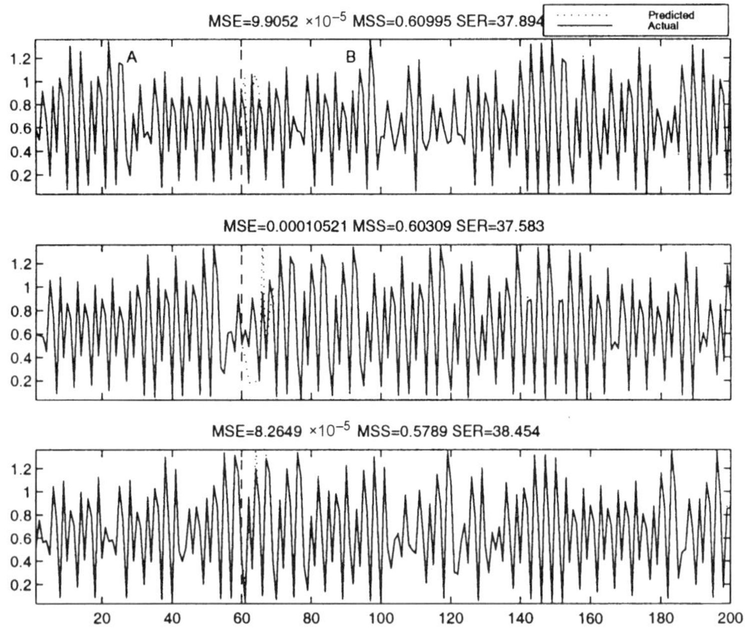

Figure 4.7 One-step prediction of Ikeda series from different starting points. Note that A = initialization and B = one-step phase.

evaluation, the correlation dimension, Lyapunov exponents and Kolmogorov entropy of both the actual Ikeda series and the autonomously generated samples are calculated. Table 4.3, which summarizes the results, shows that the dynamic invariants of both the actual and reconstructed signals are in very close agreement with each other. This illustrates that the true dynamics of the data were captured by the trained network. Figure 4.7 plots the one-step prediction of the Ikeda map for three different starting points. The reconstruction produced here is robust and stable, regardless of the position of the initializing delay vector on the test data, as demonstrated in Figure 4.8, which shows autonomous operation starting at indices of $N_0 = 3120$, 10,120, and 17,120, respectively.

Noisy Ikeda Series It was shown above that the noise-free Ikeda series can be modeled by the RMLP scheme. In a real environment, observables signals are usually corrupted by additive noise, which makes the problem more difficult. Thus, to make the modeling task more challenging than it already is, computer-generated noise is added to the Ikeda series such that the resulting signal-to-noise ratios (SNRs) of two sets of the noisy observables signals are 25 dB and 10 dB, respectively.

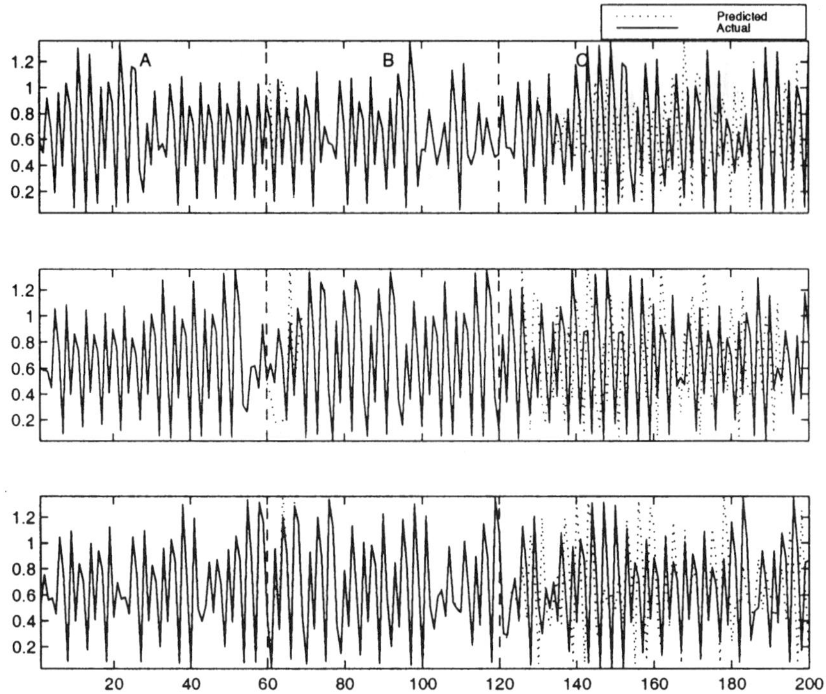

Figure 4.8 Iterative prediction of Ikeda series from different starting points, corresponding to indices of $N_0 = 3120$, 10,120 and 17,120, respectively, with $p_l = 60$. Note that A = initialization, B = priming phase and C = autonomous phase.

The attractors of the noisy signals are shown in the left-hand parts of Figures 4.9*a* and 4.9*b*, respectively. The increase in noise level was more substantial for the 10 dB case, and this corrupted the signal very significantly. It is apparent from Figure 4.9*b* that the intricate details of the attractor trajectories are lost owing to the high level of noise.

The noisy signals were used to train two distinct 6-6R-5R-1 networks using the first 5000 samples in the same fashion as in the noise-free case. The right-hand plots of Figures 4.9*a* and 4.9*b* show the attractors of the autonomously generated Ikeda series produced by the two trained RMLP networks. Whereas the network trained with a 25 dB SNR was able to capture the Ikeda dynamics, the network trained with a 10 dB SNR was unable to do so. This shows that because of the substantial amount of noise in the 10 dB case, the network was unable to capture dynamics of the Ikeda series. However, for the 25 dB case, the network was not only able to capture the predictable part of the Ikeda series but also filtered the noise. Table 4.3 displays the chaotic invariants of the original and

Table 4.3 Comparison of chaotic invariants of Ikeda map under noiseless and noisy conditions

Time series	d_E	τ	d_L	D_{ML}	D_{KY}	λ_1 (nats/sample)	λ_2 (nats/sample)	λ_3 (nats/sample)	λ_4 (nats/sample)	λ_5 (nats/sample)	λ_6 (nats/sample)	KE_{LE} (nats/sample)	KE_{ML} (nats/sample)	HOP (samples)
Actual Ikeda	6	10	2	1.51	1.44	0.354	−0.802	—	—	—	—	0.354	0.316	11
Reconstructed	6	19	2	1.68	1.33	0.248	−0.736	—	—	—	—	0.248	0.347	16
25 dB SNR Ikeda	6	1	5	1.92	3.9	0.354	0.076	−0.081	−0.296	−0.765	—	0.3548	0.314	14
Reconstructed	6	15	3	1.54	2.15	0.248	−0.14	−0.792	—	—	—	0.262	0.304	15
10 dB SNR Ikeda	6	6	6	3.46	—[a]	1.145	0.948	0.778	0.543	0.215	0.495	0.363	0.662	3
Reconstructed	6	2	3	0.26	—[a]	−3.47	−6.01	−3.47	—	—	—	—[a]	—[a]	—[a]

[a]Quantities that could not be calculated.

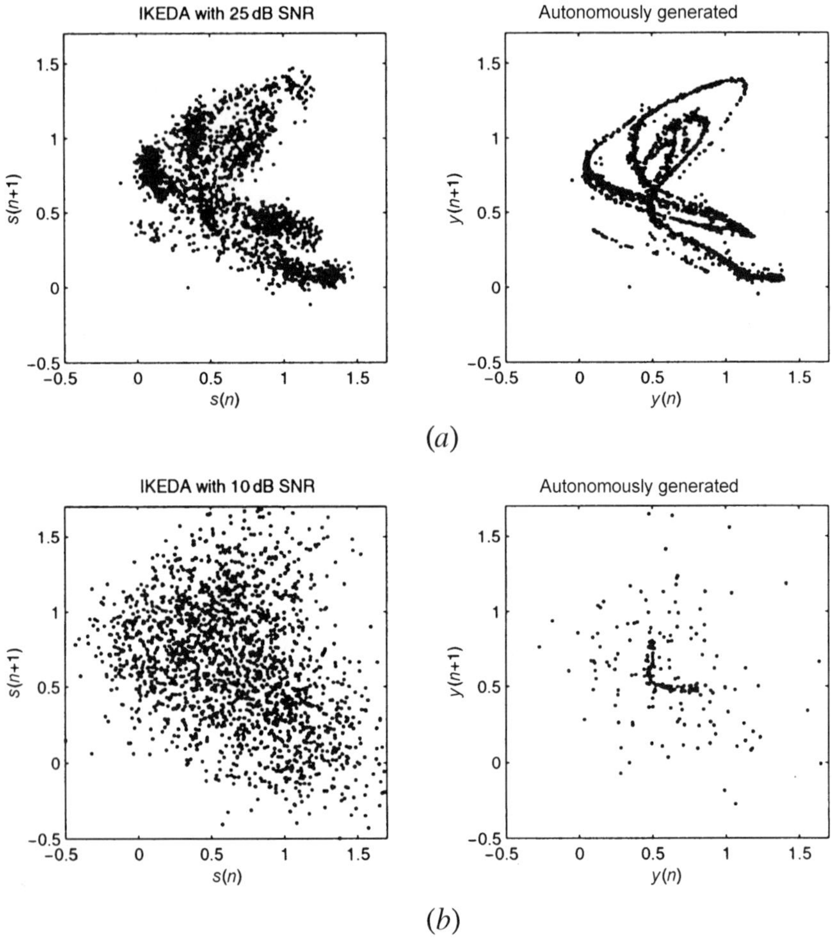

Figure 4.9 Dynamic reconstruction of the noisy Ikeda map. (*a*) Autonomous performance for Ikeda map with 25 dB SNR. (*b*) Autonomous performance for Ikeda map with 10 dB SNR. Plots on the left to noisy original signals, and those on the right to reconstructed signals.

reconstructed signals for both levels of noise. The addition of noise has the effect of increasing the number of active degrees of freedom, and thus the number of Lyapunov exponents increases in a corresponding way. The invariants of the reconstructed signal corresponding to the 25 dB SNR case match more closely with the original noise-free Ikeda invariants as compared with the noisy invariants. However, for the failed reconstruction in the 10 dB case, there is a large disagreement between the reconstructed invariants and the actual invariants, which is to be expected. In fact, some of the invariants could not even be calculated.

Table 4.4 SER results for several test cases for the Ikeda series

	Training noise		
Testing noise	Clean	25 dB SNR	10 dB SNR
Clean	36.0	27.1	16.3
25 dB SNR	14.1	18.1	14.9
10 dB SNR	1.7	5.8	6.6

Evaluation of one-step prediction for the more challenging noisy cases was also done. Table 4.4 summarizes the one-step SER results collected over a number of distinct test cases. For example, the first column of the table shows how a network trained with clean training data performs on test data with various levels of noise. It is important to note that it is difficult to achieve an SER larger than the SNR of the test data. The best overall generalization results are obtained when the network is trained with 25 dB SNR.

4.4.3 Lorenz Attractor

The Lorenz attractor is more challenging than the Ikeda or logistic map; it is described by a coupled set of three nonlinear differential equations:

$$\dot{x} = \sigma[y(n) - x(n)], \tag{4.7a}$$

$$\dot{y} = x(n)[r - z(n)] - y(n), \tag{4.7b}$$

$$\dot{z} = x(n)y(n) - bz(n), \tag{4.7c}$$

where the fixed parameters $r = 45.92$, $\sigma = 16$, and $b = 4$ are used, and $\dot{x}$ means the derivative of x with respect to time t. As before, a data set of 30,000 samples was generated at 40 Hz, of which the first 5000 samples were used to train the RMLP model and the remaining 25,000 were used for testing. For the experimental Lorenz series, an embedding dimension of $d_E = 3$ and a delay of $\tau = 4$ were calculated. An RMLP network configuration of 3-8R-7R-1, consisting of 216 weights including the biases, was trained with the EKF algorithm, and the convergence of the training MSE is shown in Figure 4.10.

Open-Loop Evaluation The results shown in Figure 4.11 were arrived at in only 10 epochs of working through the training set. The

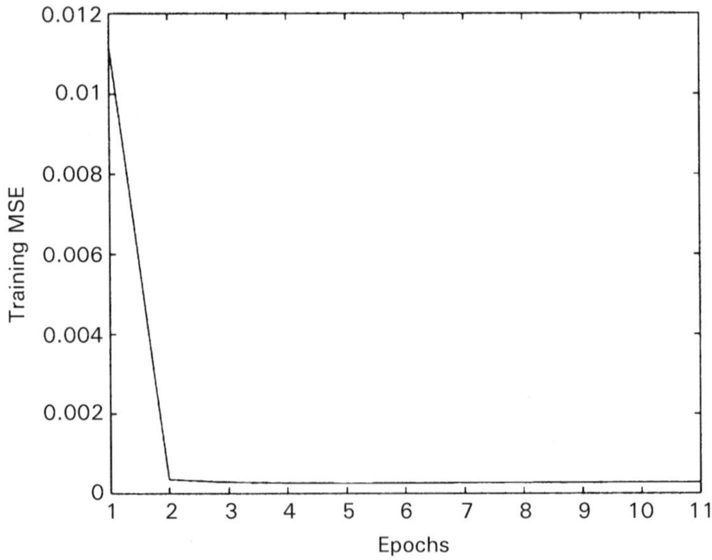

Figure 4.10 Training MSE versus epochs for the Lorenz series.

plot of one-step prediction over a portion of the test data is depicted in Figure 4.11*a*. The one-step MSE over 25,000 samples of test data was calculated to be 1.57×10^{-5}, which corresponds to an SER value of 40.2 dB.

Closed-Loop Evaluation The autonomous continuation, from where the test data end, is shown in Figure 4.11*b*, which demonstrates that the network has learned the dynamics of the Lorenz attractor very well. The iterated predictions follow the trajectory very closely for about 80 time steps on average, and then demonstrate chaotic divergence, as expected. This is in close agreement with the theoretical horizon of predictability of 97 calculated from the Lyapunov spectrum. A further testament to the success of the EKF-RMLP model is that the reconstructed attractor, shown in Figure 4.11*d*, is similar in shape to the attractor of the original series, which is shown in Figure 4.11*c*, demonstrating that the network has indeed captured the dynamics well. In addition, the dynamic invariants of the original and reconstructed series are compared in Table 4.5, which shows close agreement between their respective correlation dimension, Lyapunov spectrum, and Kolmogorov entropy, thus indicating the strong presence of the original dynamics in the reconstructed signal. Figure 4.12

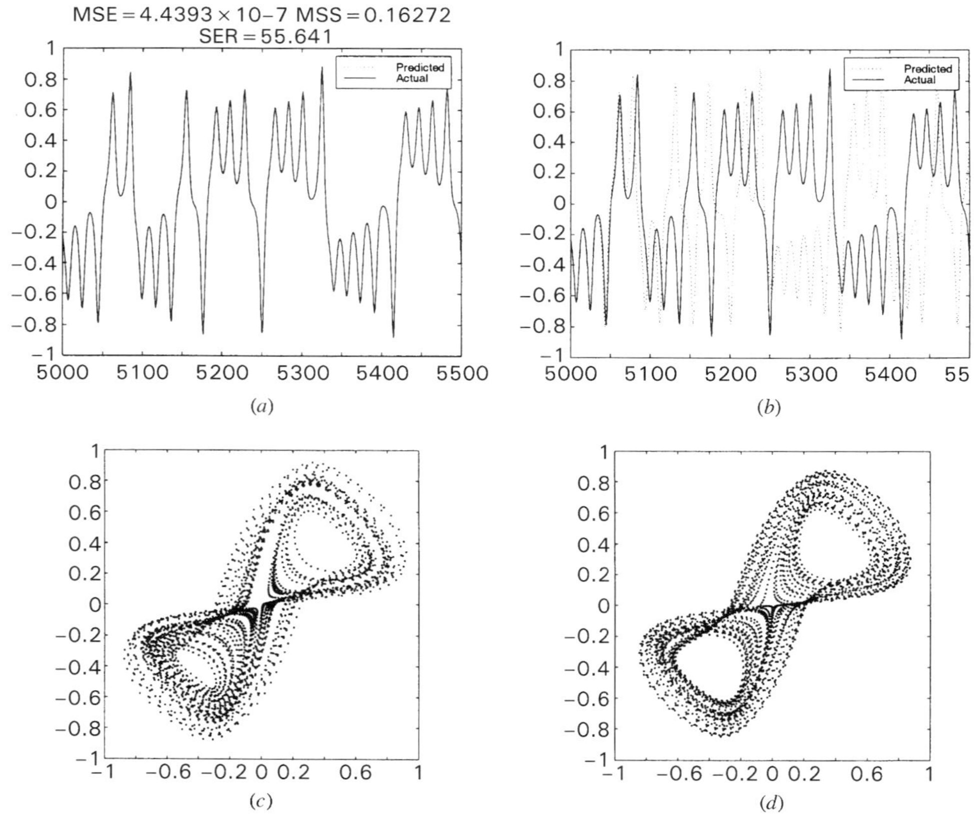

Figure 4.11 Results for the dynamic reconstruction of the Lorenz series. (*a*) One-step prediction. (*b*) Iterated prediction. (*c*) Attractor of original signal. (*d*) Attractor of iteratively reconstructed signal.

Table 4.5 Comparison of chaotic invariants of Lorenz series

Time series	d_E	τ	d_L	D_{ML}	D_{KY}	λ_1	λ_2	λ_3	λ_4	λ_5	λ_6	λ_7	KE_{LE}	KE_{ML}	HOP
						(nats/sample)							(nats/sample)		(samples)
Actual Lorenz	3	4	3	2.09	2.03	0.040	−0.0005	−1.286	—	—	—	—	0.040	0.042	97
Reconstructed	3	4	3	1.99	2.26	0.050	−0.0009	−0.1828	—	—	—	—	0.050	0.037	78
25 dB SNR Lorenz	5	4	4	2.84	3.75	0.489	0.249	−0.092	−0.850	—	—	—	0.738	0.151	8
Reconstructed	3	4	3	2.00	2.11	0.039	−0.006	−0.303	—	—	—	—	0.039	0.022	99
10 dB SNR Lorenz	7	14	7	7.18	6.71	1.554	0.419	0.269	0.109	−0.10	−0.442	−1.128	1.35	1.06	7
Reconstructed	3	5	3	1.84	2.18	0.062	0.003	−0.360	—	—	—	—	0.065	0.024	63

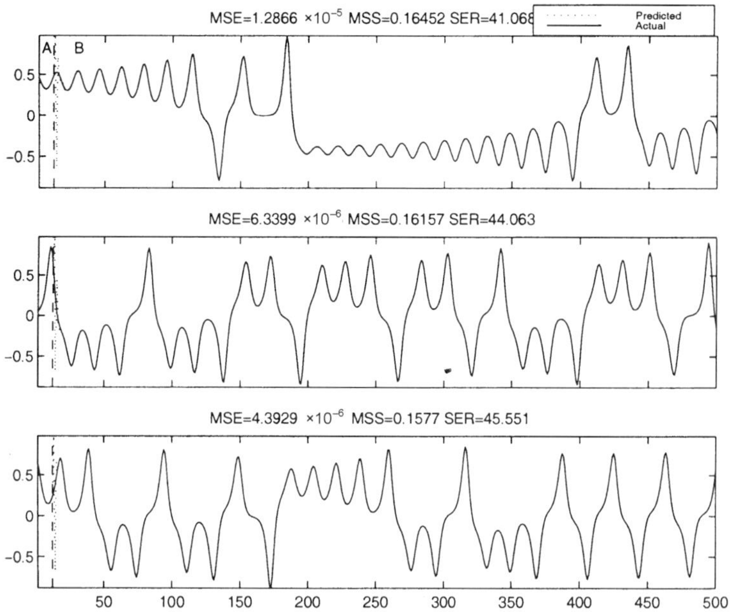

Figure 4.12 One-step prediction of Lorenz series from different starting points. Note that A = initialization and B = one-step phase.

plots one-step predictions of the Lorenz time series for three different starting points. The reconstruction of the Lorenz series was indeed very robust and stable, regardless of the starting position on the test series, as demonstrated in Figure 4.13, which shows autonomous operation starting at indices of $N_0 = 3042$, 10,042, and 17,042, respectively.

Noisy Lorenz Series To make the problem even more challenging than it already is, a significant level of noise was added to the clean Lorenz series just as was done for the Ikeda map. An RMLP network architecture was selected similar to the noise-free case, and two distinct networks were trained using the noisy Lorenz signals with 25 dB SNR and 10 dB SNR, respectively. The networks were trained with a learning rate of $p_r = 0.001$ for 15 epochs through the first 5000 samples, as was done for the noise-free case. Then, both one-step prediction and autonomous prediction results were obtained with the remaining unexposed 25,000 samples. Figures 4.14*a* and 4.15*b* show that excellent dynamic reconstruction of the Lorenz series was possible even in the presence of 25 dB and 10 dB SNR,

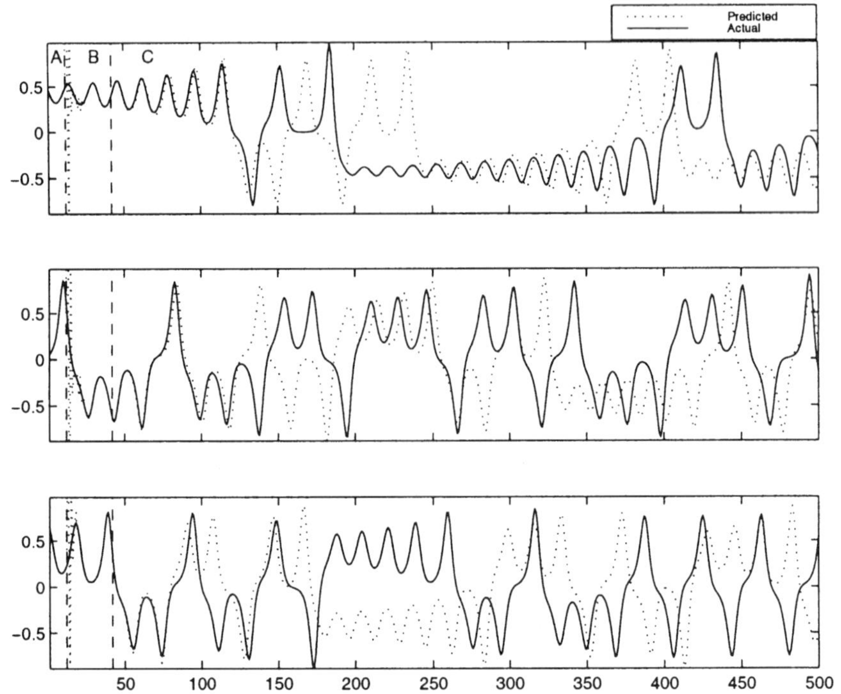

Figure 4.13 Iterative prediction of Lorenz series from different starting points, corresponding to indices of $N_0 = 3042$, 10,042, and 17,042, respectively with $p_l = 30$. Note that A = initialization, B = priming phase, and C = autonomous phase.

respectively, despite the fact that the additive noise corrupts the attractors considerably. Moreover, comparison of the dynamic invariants of the noisy and reconstructed signals in Table 4.5 reveals that the reconstructed signals and their invariants are reasonably close to the noise-free signal and the iterated predictions are smoother in comparison to the noisy signals, as shown in Figs. 4.15*a* and 4.15*b*.

It was shown earlier that dynamic reconstruction of the Ikeda map was not possible for the case of 10 dB SNR. In order to explain this, the normalized frequency spectra of both the Ikeda map and Lorenz series are plotted in Figure 4.16. This figure illustrates how much more significantly the Ikeda spectrum is corrupted by 10 dB noise as compared with the Lorenz spectrum. Whereas the Ikeda spectrum is completely drowned at 10 dB SNR, the Lorenz spectrum is still discernible, and as such its dynamic reconstruction was possible. Another way of explaining the

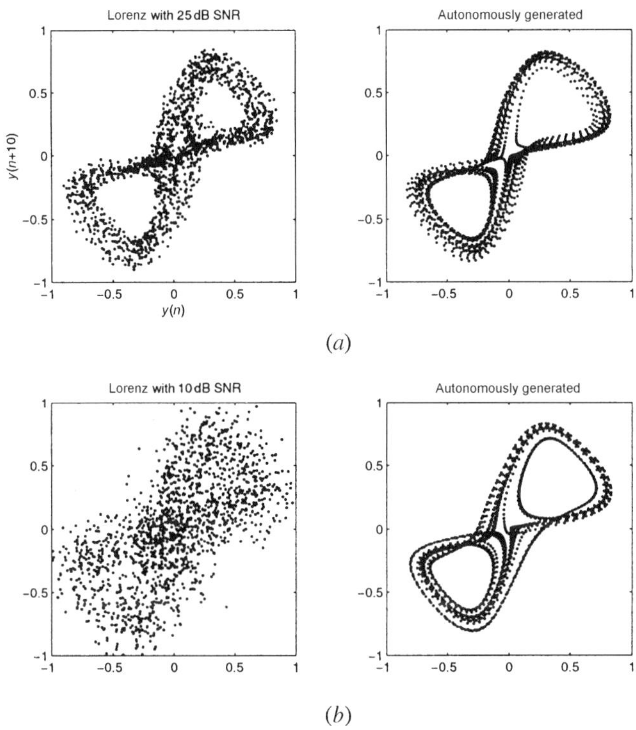

Figure 4.14 Dynamic reconstruction of the noisy Lorenz series. (*a*) Autonomous performance for Lorenz series with 25 dB SNR. (*b*) Autonomous performance for Lorenz series with 10 dB SNR. Plots on the left pertain to the original data, and those on the right to reconstructed data.

reason for the robust behavior of the Lorenz attractor to the presence of additive noise is to recall that the correlation dimension of the Lorenz attractor is greater than that of the Ikeda map; see Table 4.1. Accordingly, the Lorenz attractor is more complex than the Ikeda map – hence the greater robustness of the Lorenz attractor to noise.

Evaluation of one-step performance for the more challenging noisy cases was also done. Table 4.6 summarizes the one-step SER results collected over a number of distinct test cases. For example, the first column of the table shows how a network trained with clean training data performs on test data with various levels of noise. The best overall generalization results are obtained when the network is trained with 25 dB SNR.

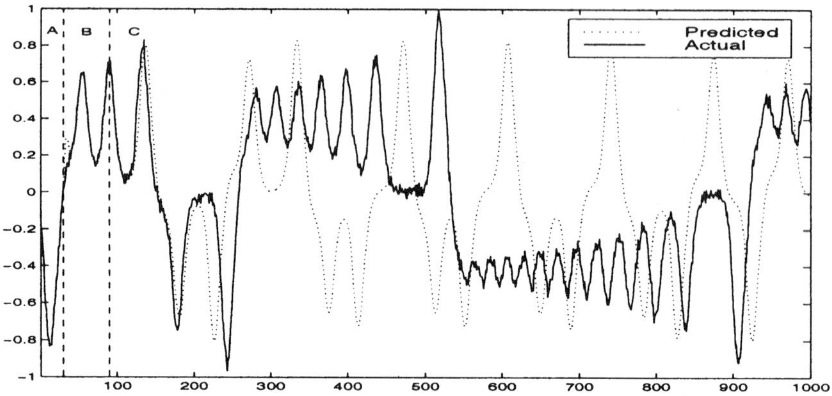

(*a*)

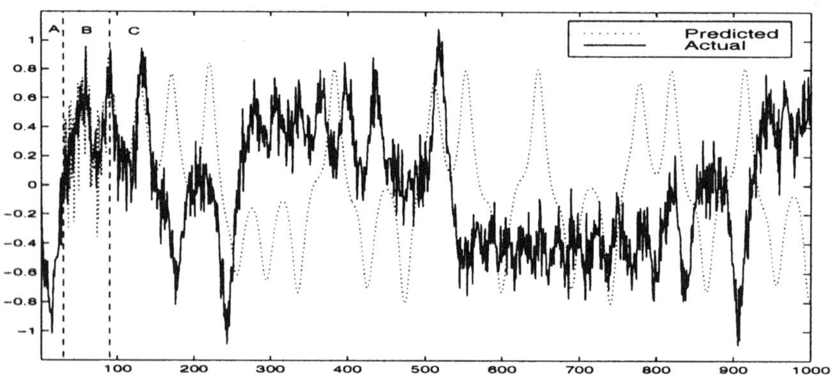

(*b*)

Figure 4.15 Iterative prediction of noisy Lorenz series with (*a*) 25 dB and (*b*) 10 dB SNR. Note that A = initialization, B = priming phase and C = autonomous phase.

4.5 NONLINEAR DYNAMIC MODELING OF REAL-WORLD TIME SERIES

4.5.1 Laser Intensity Pulsations

In the first experiment with real-world data, a series recorded from a far-infrared laser in a chaotic state is used. This data set was provided as part of a data package used in the Santa Fe Institute Time Series Prediction and Analysis Competition [7]. During the time of the competition, only the first 1000 samples were provided, and the goal was to predict the following

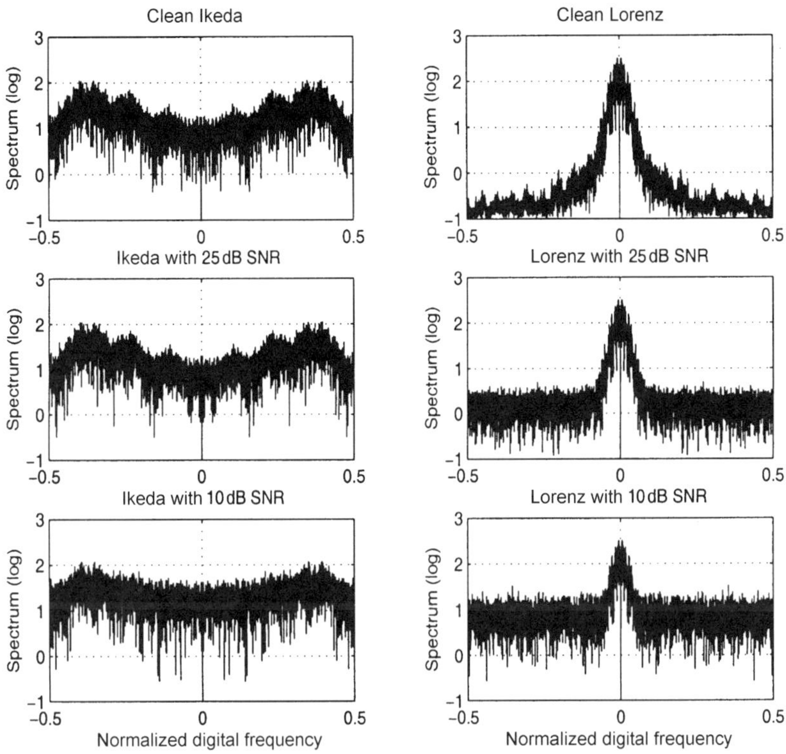

Figure 4.16 Lorenz and Ikeda series: normalized frequency spectra.

Table 4.6 SER results for several test cases for the Lorenz series

	Training noise		
Testing noise	Clean	25 dB SNR	10 dB SNR
Clean	40.2	30.5	36.8
25 dB SNR	17.8	23.1	17.2
10 dB SNR	3.5	9.3	3.8

100 samples as accurately as possible. Now, however, the full data set consisting of 10,000 samples is available. The first 1100 samples are plotted in Figure 4.17. The data consist of high-frequency pulsations, which gradually rise in amplitude and suddenly collapse from a high to low amplitude. The rapid decays of oscillations in this data set occur with no periodicity, and are a challenge to model. During these collapses, the

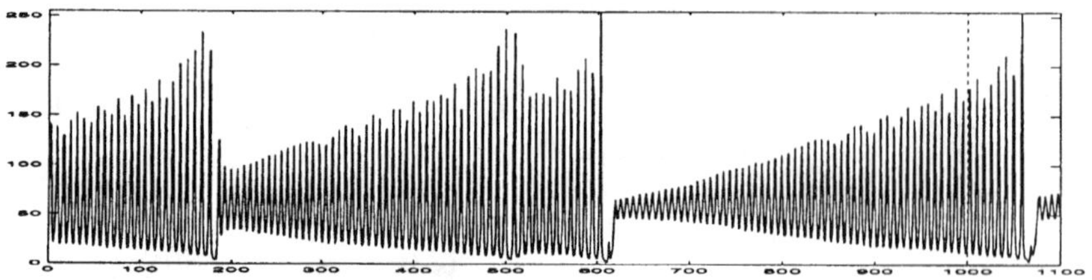

Figure 4.17 First 1100 samples of original laser series.

signal abruptly jumps from one region of the attractor to another distinct region, and these transitions are the most difficult parts of this series to model.

To remain consistent with what was done in the SFI competition, only the first 1000 samples were presented to the network, while the remaining 9000 samples of data were reserved for testing purposes and not shown to the network. The optimal embedding delay was found to be $\tau = 2$ and the embedding dimension to be $d_E = 9$. A network architecture of 9-10R-8R-1, consisting of 361 weights including the biases, was used to model this data. The network was trained through the first 1000 points for 50 epochs.

Open-Loop Evaluation The network makes excellent one-step predictions as shown in Figure 4.18*a*, with very low prediction errors in most regions. The accuracy of prediction drops slightly during the sudden signal collapses, but then increases quickly as the network adapts. The SER value for the one-step prediction performance on the unexposed 9000 samples of the test set was calculated to be 26.05 dB.

Closed-Loop Evaluation The iterated predictions of the network are also very remarkable. The autonomous continuation from sample 1000, where the training data end, is shown in Figure 4.18*b*. This was achieved by first initializing the network with a delay vector, consisting of 9 taps, each spaced 2 samples apart, from the beginning of the training set and then priming for $p_l = 982$ steps, after which the network operates in an autonomous mode. Since the network was initialized with the first samples of the actual data, autonomous operation does not begin until the point $(9 \times 2) + 982 = 1000$, as shown in Figure 4.19. The autonomous output of the network closely follows the original series for about 60 steps (which is higher than the average horizon of predictability of 26 calculated from the Lyapunov spectrum) after which it begins to diverge. Figure 4.20 shows the one-step prediction of the laser time series with three different starting points.

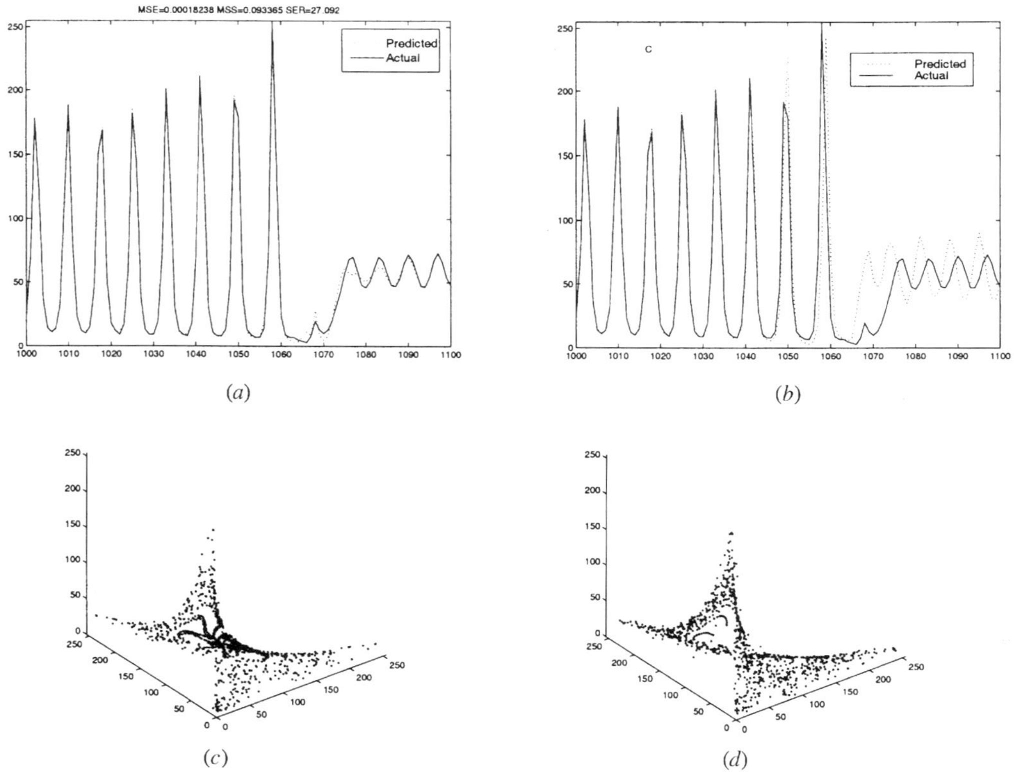

Figure 4.18 Results of the dynamic reconstruction of the laser series. (*a*) One-step prediction. (*b*) Iterated prediction. (*c*) Attractor of original signal. (*d*) Attractor of iteratively reconstructed signal.

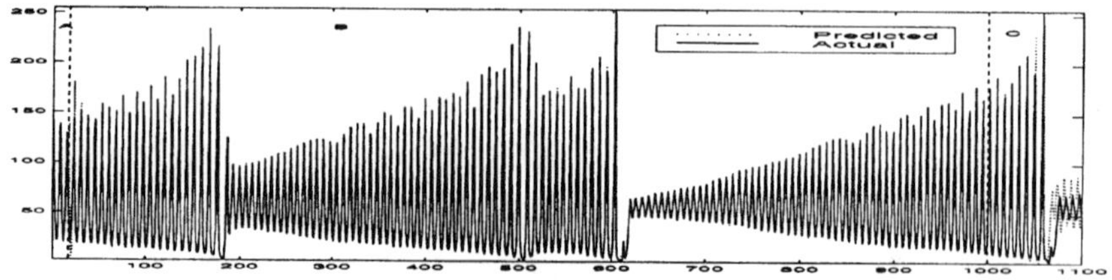

Figure 4.19 Iterated prediction of the laser data beginning at sample 1000, using $p_l = 982$. Note that A = initialization, B = priming phase, and C = autonomous phase. Phase C is enlarged in Figure 4.18*b*.

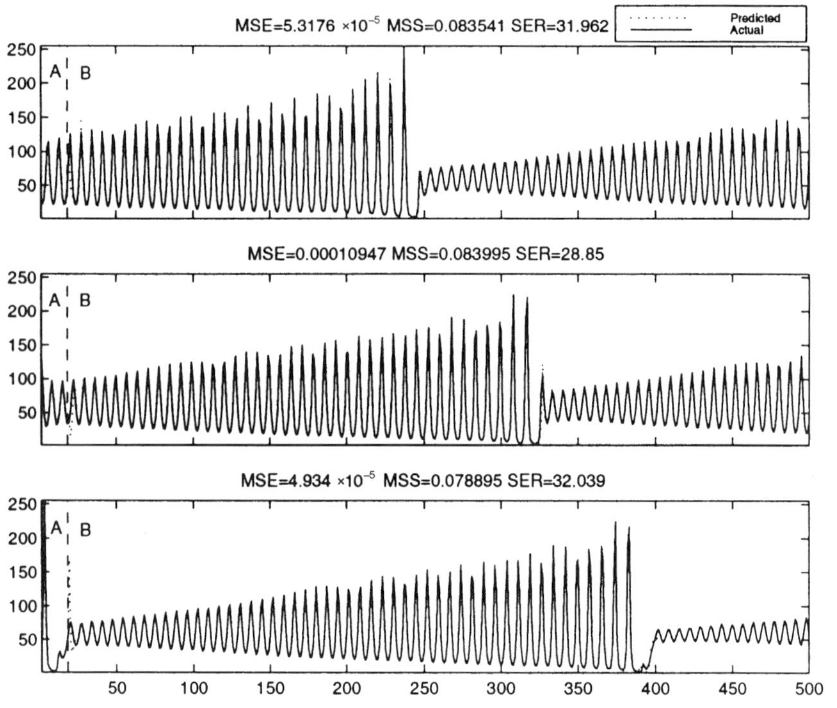

Figure 4.20 One-step prediction of laser series with different starting points. Note that A = initialization and B = one-step phase.

Figure 4.21 shows that the network is capable of stable and robust reconstruction even when it is initialized from different starting points on the test set, corresponding to indices of $N_0 = 48$, 1048, and 4048, respectively. Note that the network has not seen this data set before, and is still able to predict the signal collapses up to an impressive degree of accuracy. Figure 4.22, which shows the 9000-sample iterated output of the network, demonstrates that the chaotic pulsations have been learned by the

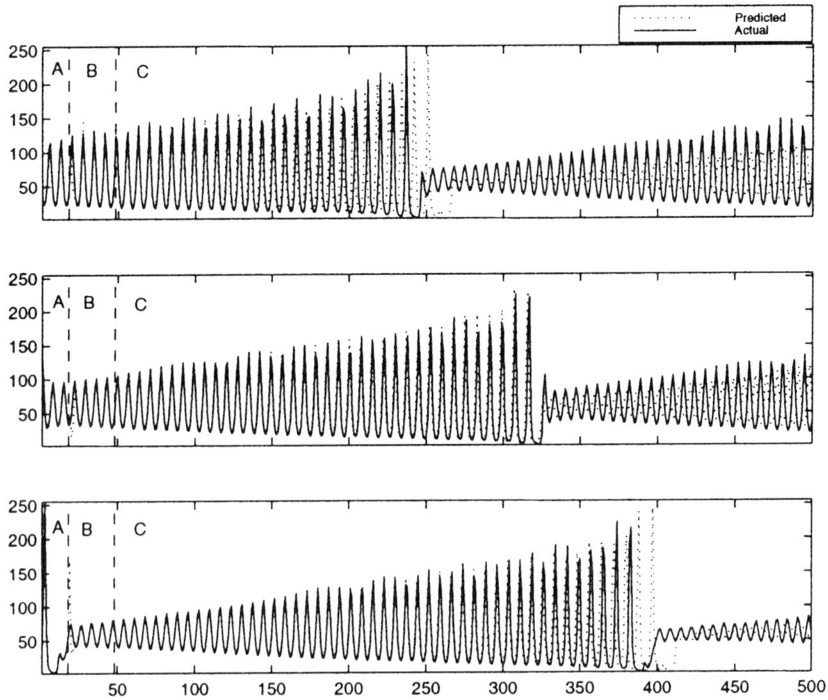

Figure 4.21 Iterative prediction of laser series from different starting points, corresponding to indices of $N_0 = 48$, 1048, and 4048, respectively, with $p_l = 30$. Note that A = initialization, B = printing phase, and C = autonomous phase.

network from just the first 1000 samples. The attractor of the original series is plotted in Figure 4.18*c* using the actual laser data. The attractor in Figure 4.18*d* is plotted using the reconstructed points as produced by the trained RMLP network. It is clear from the attractor of the reconstructed signal that the dynamics of the laser data has been faithfully captured by the EKF trained RMLP network. Table 4.7 shows that the correlation dimension, Lyapunov exponents, and Kolmogorov entropy of the reconstructed signal are in close agreement with those obtained from the original data, thus indicating a faithful reconstruction of the original dynamics. The correlation dimension of NH_3 laser pulsations calculated here is consistent with what is reported by Hubner et al. [8].

Longer Training Length In another experiment, the same-sized network was trained on the first 5000 points, as compared with the first 1000 used above. The iterated output produced by the network is shown in Figure 4.23. Since the network was exposed to more of the signal

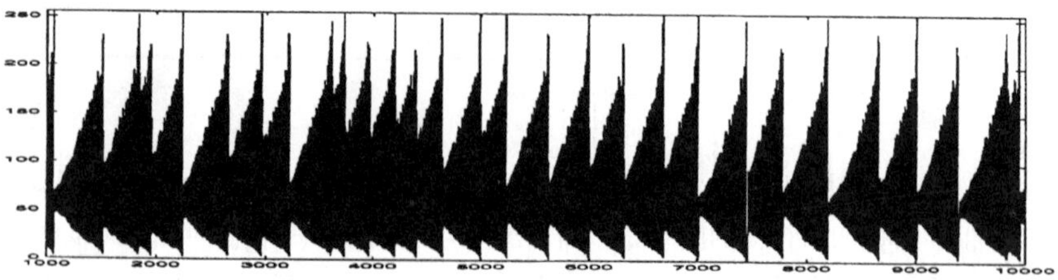

Figure 4.22 9000 samples of iterated output of the RMLP trained on the first 1000 samples.

Table 4.7 Comparison of chaotic invariants of laser series

Time series	d_E	τ	d_L	D_{ML}	D_{KY}	λ_1 (nats/sample)	λ_2 (nats/sample)	λ_3 (nats/sample)	KE_{LE} (nats/sample)	KE_{ML} (nats/sample)	HOP (samples)
Actual laser	9	2	3	2.01	2.19	0.147	−0.039	−0.547	0.147	0.195	26
Reconstructed	9	2	3	1.90	2.13	0.115	−0.037	−0.602	0.115	0.133	34

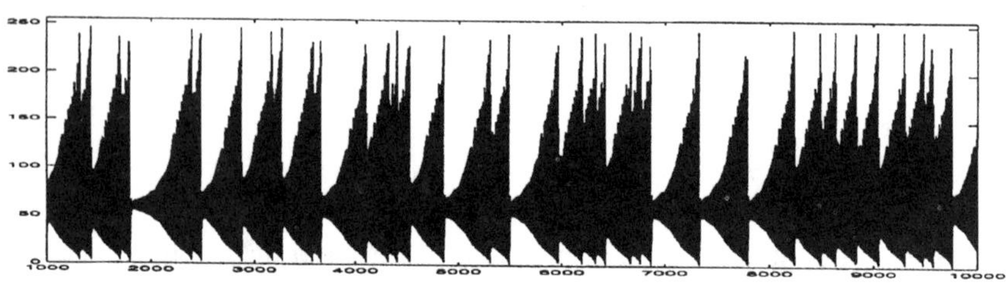

Figure 4.23 9000 samples of iterated output of the RMLP trained on the first 5000 samples.

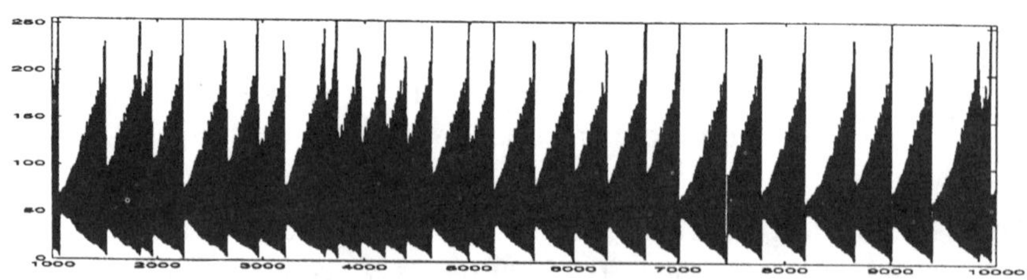

Figure 4.24 Last 9000 samples of the actual laser series.

dynamics, in this case, the iterated output of the network demonstrates more frequent chaotic collapses and is closer in appearance to the actual signal shown in Figure 4.24.

4.5.2 Sea Clutter Data

Sea clutter, or *sea echo*, refers to the radar backscatter from an ocean surface. Radars operating in a maritime environment have a serious limitation imposed on their performance by unwanted sea clutter. Therefore, a problem of fundamental interest in the radar literature is the modeling of sea clutter to design optimum target detectors.

For many years, clutter echoes in radar systems were modeled as a stochastic process. However, in recent years, some researchers have concluded that the generation of sea clutter is governed by deterministic chaos [9–13]. These claims, in varying degrees, have been largely based on the estimation of chaotic invariants of sea clutter, namely, the correlation dimension, Lyapunov exponents, Kaplan-Yorke dimension, and the Kolmogorov entropy, the latter two are derived from the Lyapunov exponents. But, recently, serious concerns have been raised on the *discriminative power* of the state-of-the-art algorithms currently available for distinguishing between sea clutter and different forms of surrogate data derived from the original data; the surrogate data are known to be stochastic by purposely introducing randomization into their generation. By estimating the correlation dimension and Lyapunov exponents, it is found that there is little difference between the chaotic invariants of sea clutter and those of their surrogate counterparts [14, 15], which casts doubt on deterministic chaos being the nonlinear dynamical model foe sea clutter.

In [14], it is pointed out physical realities dictate the presence of noise not just in the measurement equation (owing to instrument errors) but also in the process equation (owing to the rate of variability of the forces affecting the ocean dynamics). According to Sugihara [16], there is unavoidable practical difficulty in disentangling the process noise from the measurement noise when we try to reconstruct an invariant measure such as the Lyapunov exponents. This may explain the reason for the inability of chaotic-invariant estimation algorithms to distinguish between sea clutter and its surrogates.

Our primary interest in this chapter is in dynamic reconstruction. In this context, we may raise the following question: Can the EKF-RMLP be configured to perform dynamic reconstruction on sea clutter in a robust manner? In what follows, an attempt to answer this question is presented;

Table 4.8 Comparison of chaotic invariants of sea clutter

						λ_1	λ_2	λ_3	λ_4	λ_5	KE_{LE}	KE_{ML}	
Time series	d_E	τ	d_L	D_{ML}	D_{KY}	(nats/sample)					(nats/sample)		HOP (samples)
Actual sea clutter	6	10	5	4.69	4.42	0.228	0.102	0.005	−0.118	−0.448	0.335	0.369	17
Reconstructed from $N_2 = 0$	6	9	5	3.12	4.51	0.214	0.078	0.010	−0.083	−0.427	0.302	0.292	18
Reconstructed from $N_0 = 5160$	7	6	5	3.58	4.75	0.216	0.129	0.049	−0.112	−0.373	0.394	0.303	18
Reconstructed from $N_0 = 6160$	6	8	5	4.43	4.62	0.232	0.136	0.023	−0.113	−0.434	0.391	0.317	17
Reconstructed from $N_0 = 7160$	6	6	5	3.67	4.23	0.191	0.088	0.029	−0.142	−0.713	0.308	0.220	20

the answer has bearing on whether sea clutter is chaotic or not. Figure 4.25 shows the in-phase component of typical sea clutter, recorded by an instrument-quality radar on the East Coast of Canada, with the radar operating in a dwelling mode at a low grazing angle. The data consist of 50,000 samples, of which the first 40,000 are used for network training, while the remaining 10,000 samples are reserved for testing. An embedding dimension of $d_E = 6$ and embedding delay of $\tau = 10$ were calculated for sea clutter.

Open-Loop Evaluation The results for the modeling of sea clutter by the EKF-RMLP method are shown in Figure 4.26. A 6-8R-7R-1 network trained with $p_r = 0.5$ was used. The SER over the unexposed 10,000 test samples was found to be 36.65 dB, which is certainly impressive, since in the EKF-RMLP based method presented here, the synaptic weights remain fixed after the training was completed.

Closed-Loop Evaluation The iterated continuation from where the training data end, corresponding to an index of $N_0 = 0$, is shown in Figure

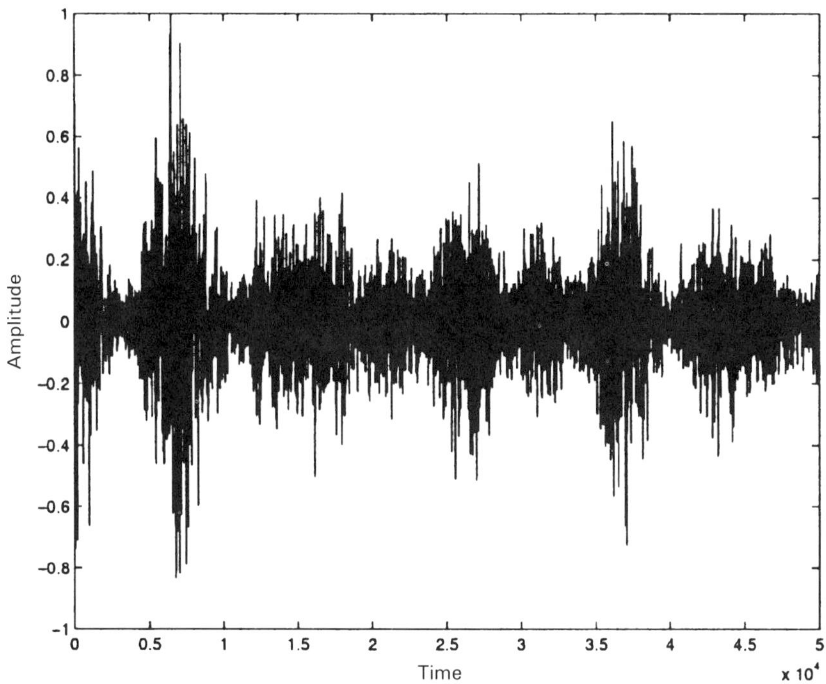

Figure 4.25 In-phase component of typical sea-clutter data.

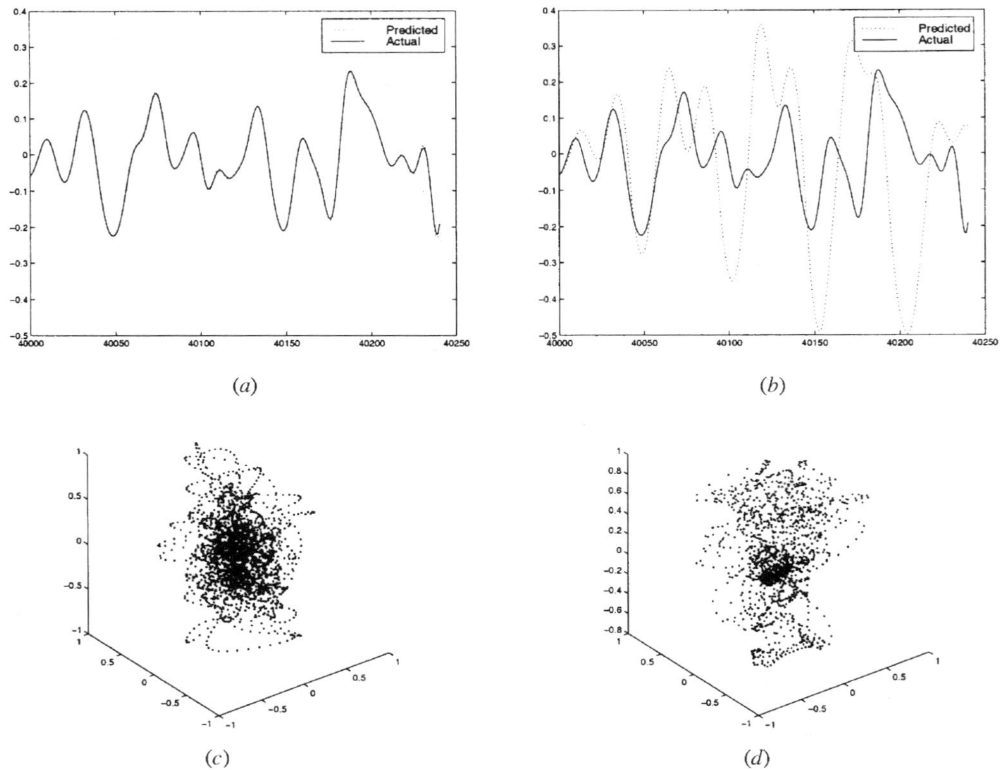

Figure 4.26 Results for the dynamic reconstruction of sea clutter. (*a*) One-step prediction. (*b*) Iterated prediction. (*c*) Attractor of original signal. (*d*) Attractor of iteratively reconstructed signal.

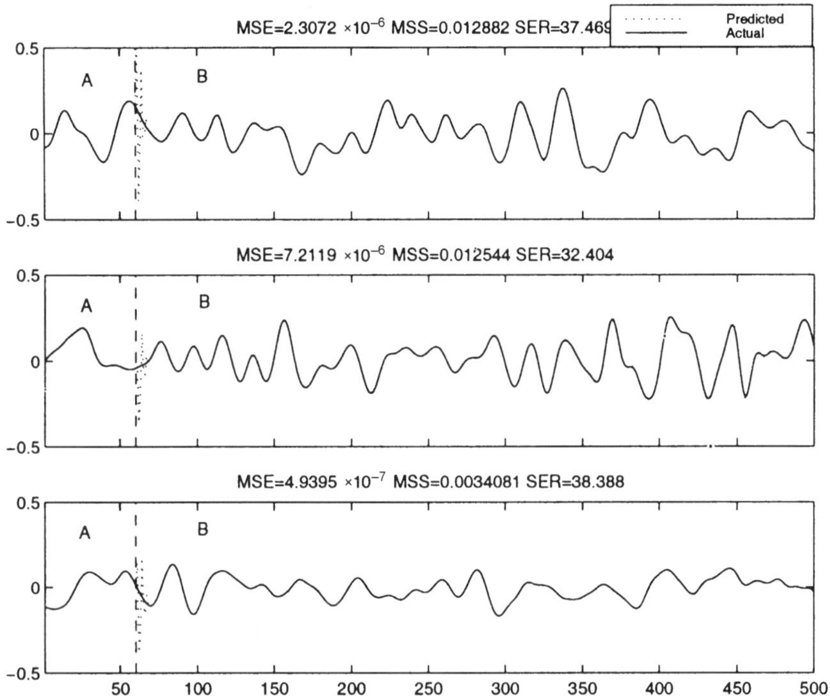

Figure 4.27 One-step prediction of sea clutter from different starting points. Note that A = initialization and B = one-step mode of operation.

4.26*b*. Observe that the prediction is close to the actual data for roughly less than 10 steps (which is significantly less than the average theoretical horizon of predictability of 17 calculated from the Lyapunov spectrum), after which it diverges into a distinct trajectory of its own. The network, when iterated in this fashion for several thousand steps, produces an output whose attractor is plotted in Figure 4.26*d*. It is difficult to compare the reconstructed attractor with the original attractor in Figure 4.26*c* because we are looking at a five-dimensional attractor in three dimensions.

Since it is impossible to plot attractors in five dimensions, we may determine the success of dynamic reconstruction by a numerical comparison of the dynamic invariants of the reconstructed signal with those of the actual signal. As shown in Table 4.8, the Lyapunov exponents of the reconstructed signal are in close agreement with those obtained from the actual data. In addition, the correlation dimension and Kolmogorov entropy of both signals are also close, except for the slightly lower than expected value of D_{ML} for the reconstructed signal. Furthermore, D_{KY} and

KE_{LE} are close to D_{ML} and KE_{ML}, respectively. Figure 4.27 plots one-step predictions of sea clutter for three different starting points.

Stability and Robustness Furthermore, the more difficult case of autonomous prediction from different starting points chosen from the test data is shown in Figure 4.28. This shows that stable reconstruction is also possible from indices of $N_0 = 5160$, 6160, and 7160 on the test data. The dynamic invariants calculated from signals reconstructed at these indices are in close agreement with the actual Table 4.8. Again, there seems to be a slight disagreement in the values of D_{ML}.

This shows that the EKF-RMLP scheme was able to reconstruct sea-clutter dynamics on several occasions in a reasonable manner, depending on the data used. However, it is important to point out that *robust reconstruction was not possible* from all initialization points – for this and other types of networks used. The different types of reconstruction

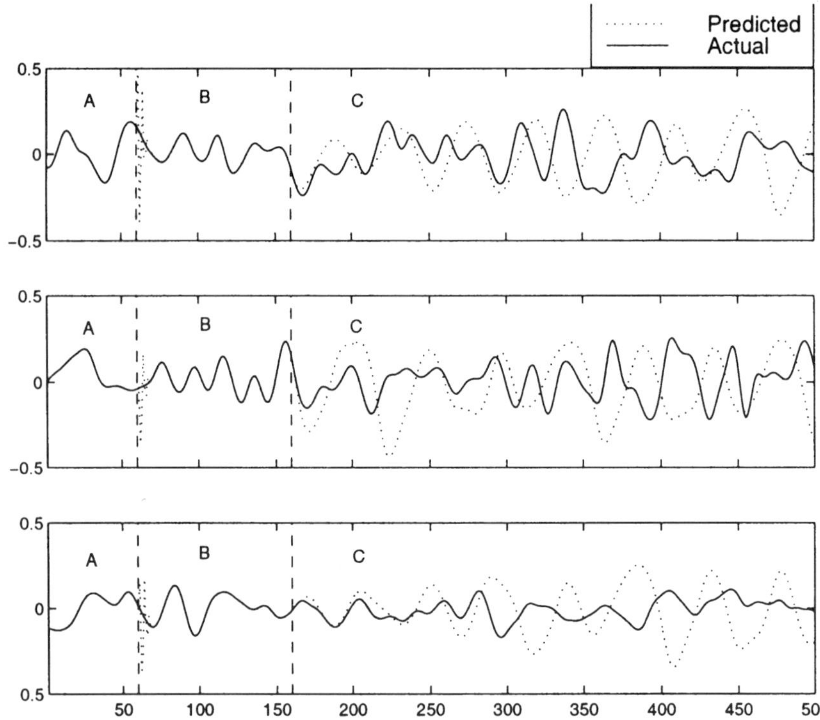

Figure 4.28 Iterative prediction of sea clutter from different starting points, corresponding to indices of $N_0 = 5160$, 6160, and 7160, respectively, with $p_l = 100$. Note that A = initialization, B = priming phase, and C = autonomous phase.

failures encountered are illustrated in Figure 4.29. In the first type of reconstruction failure, the output becomes constant after a few steps. In other words, the output behaves as a fixed-point attractor. In the second type of failure, the output becomes periodic after a few time steps. In the third type of failure, the output breaks down after a few steps of good prediction. How do we explain the reconstruction failures of Figure 4.29? Dynamic reconstruction, in accordance with Taken's embedding theorem tolerates a moderate level of measurement noise, but the process equation has to be noise-free. The presence of dynamical noise in the process equation is therefore in violation of Taken's theorem — hence the difficulty in devising a robust dynamic reconstruction system for sea clutter.

4.6 DISCUSSION

Takens' theorem provides the theoretical basis for an experimental approach for the autoregressive modeling of a nonlinear dynamical system. Specifically, the state evolution of the system with a d-dimensional state space can be reconstructed by observing $2d + 1$ time lags of a single output of the system for an arbitrary delay. The theorem requires that the observable time series be noiseless and of infinite length. Unfortunately, neither of these two conditions can be satisfied in practice. To mitigate these difficulties, the recommended procedure is to do the following:

- Use an "optimum" time delay for the delay coordinate method, for which the adjacent samples are as statistically independent as practically possible.

For the experimental study presented here, we minimized the mutual information between the observable time series and its delayed version as the basis for estimating the optimum time delay.

In this chapter, we have presented experimental results demonstrating the ability of the EKF-RMLP scheme to reconstruct the dynamics of chaotic processes in accordance with Takens' embedding theorem. The results presented for (1) the numerically generated logistic map, Ikeda map, and Lorenz attractor, and (2) the experimentally generated laser pulsations are indeed remarkable in that the reconstructed data closely match all the important chaotic characteristics of the original time series. Moreover, the dynamic reconstruction performed here is not only stable

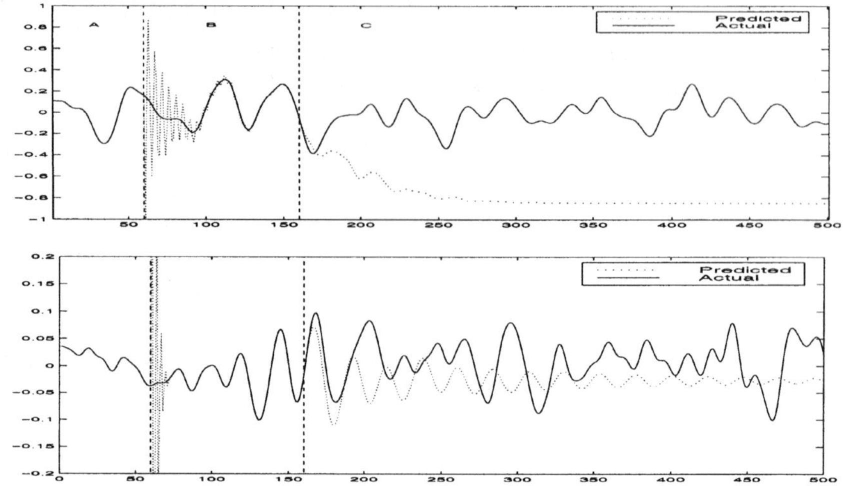
A
B
C
Predicted
Actual

(*a*)

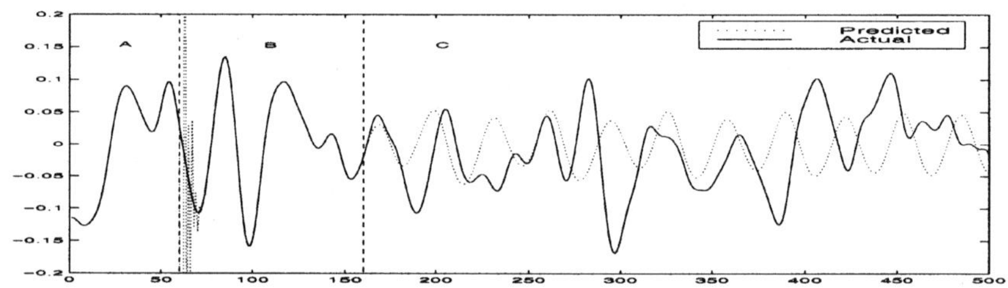
A
B
C
Predicted
Actual

(*b*)

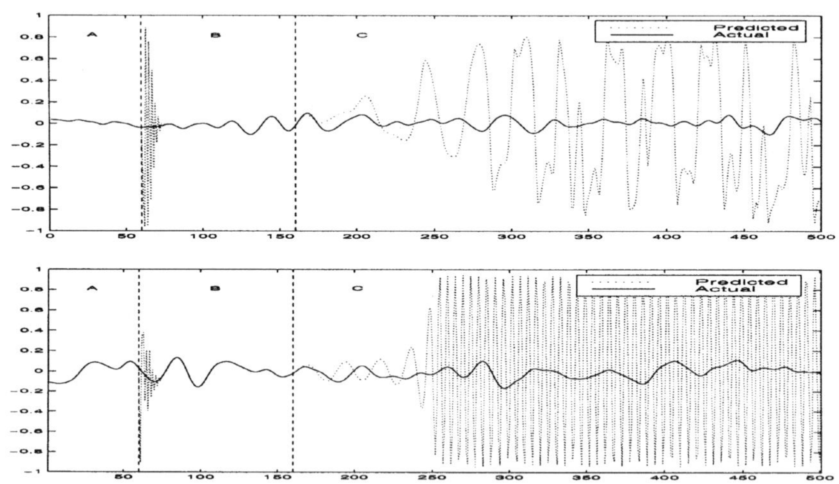
A
B
C
Predicted
Actual

(*c*)

Figure 4.29 Illustrative failure modes in iterative prediction of sea clutter by various networks. (*a*) Output converges to a constant (fixed point attractor); $N_0 = 4160$ and 160, respectively. (*b*) Output becomes periodic; $N_0 = 7160$. (*c*) Output breaks down after a few time steps; $N_0 = 160$ and 7160, respectively. Note that A = initialization, B = priming phase, and C = autonomous phase.

but also robust with respect to the choice of initial conditions and the presence of additive noise.

The results of this study also show that dynamic reconstruction of sea clutter is possible using the EKF-RMLP-based method depending on the data set used, but unfortunately as the results obtained with the EKF-MLP method were prone to numerous failures and therefore not as robust as one would expect from a good model. The results of this study clearly demonstrate that the dynamics of sea clutter is difficult to reconstruct. We attribute the difficulty to the unavoidable presence of dynamical noise in the process equation.

REFERENCES

[1] H.D.I. Abarbanel, *Analysis of Observed Chaotic Data*. New York: Springer-Verlag, 1995.

[2] J.C. Schouten, F. Takens, and C.M. Van den Bleek, "Estimation of the Dimension of a Noisy Attractor," *Physical Review A*, **33**, 1134–1140 (1994).

[3] N.H. Packard, J.P. Crutchfield, J.D. Farmer, and R.S. Shaw, "Geometry from a time series," *Physical Review Letters*, **45**, 712–716 (1980).

[4] F. Takens, "Detecting Strange Actuators in Turbulence," in D.A. Rand and L.S. Young, Eds. *Dynamical Systems and Turbulence, Warwick, 1980*, Lecture Notes in Mathematics, Vol. 898.

[5] R. Mañé, "On the dimension of compact invariant sets of certain nonlinear maps," in D.A. Rand and L.S. Young, Eds. *Dynamical Systems and Turbulence, Warwick 1980, Lecture Notes in Mathematics* Vol. 898. 1981, p. 230. Berlin: Springer-Verlag.

[6] A.M. Fraser, "Information and entropy in strange attractors," *IEEE Transactions on Information Theory* **35**, 245–262 (1989).

[7] A.S. Weigend, "Paradigm change in prediction," *Philosophical Transactions of the Royal Society of London, Ser. A*, 405–420 (1994).

[8] U. Hubner, W. Klische, and C.O. Weiss, "Generalized dimensions of laser attractors," *Physical Review A*, **45**, 2128–2130 (1992).

[9] S. Haykin and S. Puthusserypady, "Chaotic dynamics of sea clutter" *Chaos*, **7**, 777–802 (1997).

[10] S. Haykin and Xiao bo Li, "Detection of signals in chaos," *Proceedings of the IEEE*, **83** (1995).

[11] H. Leung and S. Haykin, "Is there a Radar Clutter Attractor?" *Appl. Phys. Letters*, **56**, 393–395 (1990).

[12] H. Leung and T. Lo, "Chaotic Radar Signal-Processing Over the Sea," *IEEE J. Oceanic Engineering*, **18**, 287–295 (1993).

[13] A.J. Palmer, R.A. Kropfli, and C.W. Fairall, "Signature of Deterministic Chaos in Radar Sea Clutter and Ocean Surface Waves," *Chaos*, **6**, 613–616 (1995).

[14] S. Haykin, R. Bakker, and B. Currie, "Uncovering Nonlinear Dynamics: The Case Study of Sea Clutter," *Proc. IEEE*, under review.

[15] C.P. Unsworth, M.R. Cowper, B. Mulgrew, and S. McLaughlin, "False Detection of Chaotic Behaviour in Stochastic Computed K-distribution Model of Radar Sea Clutter," *IEEE Workshop on Statistical Signal and Array Processing*, pp. 296–300 (2000).

[16] G. Sugihara, "Nonlinear Forecasting for the Classification of Natural Time Series," *Phil. Trans. R. Soc. London*, A348, pp. 477–495 (1994).

5

DUAL EXTENDED KALMAN FILTER METHODS

Eric A. Wan and Alex T. Nelson

Department of Electrical and Computer Engineering, Oregon Graduate Institute of Science and Technology, Beaverton, Oregon, U.S.A.

5.1 INTRODUCTION

The Extended Kalman Filter (EKF) provides an efficient method for generating approximate maximum-likelihood estimates of the state of a discrete-time nonlinear dynamical system (see Chapter 1). The filter involves a recursive procedure to optimally combine noisy observations with predictions from the known dynamic model. A second use of the EKF involves estimating the parameters of a model (e.g., neural network) given clean training data of input and output data (see Chapter 2). In this case, the EKF represents a modified-Newton type of algorithm for on-line system identification. In this chapter, we consider the *dual estimation* problem, in which both the states of the dynamical system and its parameters are estimated simultaneously, given only noisy observations.

Kalman Filtering and Neural Networks, Edited by Simon Haykin
ISBN 0-471-36998-5

To be more specific, we consider the problem of learning both the hidden states $\mathbf{x}_k$ and parameters $\mathbf{w}$ of a discrete-time nonlinear dynamical system,

$$\begin{aligned} \mathbf{x}_{k+1} &= \mathbf{F}(\mathbf{x}_k, \mathbf{u}_k, \mathbf{w}) + \mathbf{v}_k, \\ \mathbf{y}_k &= \mathbf{H}(\mathbf{x}_k, \mathbf{w}) + \mathbf{n}_k, \end{aligned} \tag{5.1}$$

where both the system states $\mathbf{x}_k$ and the set of model parameters $\mathbf{w}$ for the dynamical system must be simultaneously estimated from only the observed noisy signal $\mathbf{y}_k$. The *process* noise $\mathbf{v}_k$ drives the dynamical system, *observation* noise is given by $\mathbf{n}_k$, and $\mathbf{u}_k$ corresponds to observed exogenous inputs. The model structure, $\mathbf{F}(\cdot)$ and $\mathbf{H}(\cdot)$, may represent multilayer neural networks, in which case $\mathbf{w}$ are the weights.

The problem of dual estimation can be motivated either from the need for a model to estimate the signal or (in other applications) from the need for good signal estimates to estimate the model. In general, applications can be divided into the tasks of modeling, estimation, and prediction. In *estimation*, all noisy data up to the current time is used to approximate the current value of the clean state. *Prediction* is concerned with using all available data to approximate a *future* value of the clean state. *Modeling* (sometimes referred to as *identification*) is the process of approximating the underlying dynamics that generated the states, again given only the noisy observations. Specific applications may include noise reduction (e.g., speech or image enhancement), or prediction of financial and economic time series. Alternatively, the model may correspond to the explicit equations derived from first principles of a robotic or vehicle system. In this case, $\mathbf{w}$ corresponds to a set of unknown parameters. Applications include adaptive control, where parameters are used in the design process and the estimated states are used for feedback.

Heuristically, dual estimation methods work by alternating between using the model to estimate the signal, and using the signal to estimate the model. This process may be either *iterative* or *sequential*. *Iterative* schemes work by repeatedly estimating the signal using the current model and all available data, and then estimating the model using the estimates and all the data (see Fig. 5.1*a*). Iterative schemes are necessarily restricted to off-line applications, where a batch of data has been previously collected for processing. In contrast, *sequential* approaches use each individual measurement as soon as it becomes available to update both the signal and model estimates. This characteristic makes these algorithms useful in either on-line or off-line applications (see Fig. 5.1*b*).

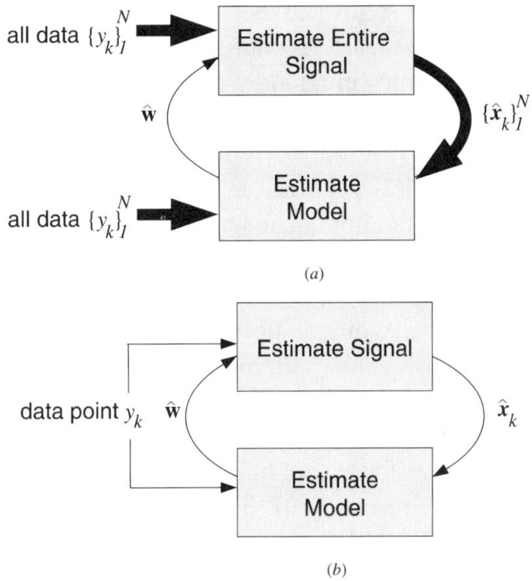

Figure 5.1 Two approaches to the dual estimation problem. (*a*) Iterative approaches use large blocks of data repeatedly. (*b*) Sequential approaches are designed to pass over the data one point at a time.

The vast majority of work on dual estimation has been for linear models. In fact, one of the first applications of the EKF combines both the state vector $\mathbf{x}_k$ and unknown parameters $\mathbf{w}$ in a joint bilinear state-space representation. An EKF is then applied to the resulting nonlinear estimation problem [1, 2]; we refer to this approach as the *joint extended Kalman filter*. Additional improvements and analysis of this approach are provided in [3, 4]. An alternative approach, proposed in [5], uses two separate Kalman filters: one for signal estimation, and another for model estimation. The signal filter uses the current estimate of $\mathbf{w}$, and the weight filter uses the signal estimates $\hat{\mathbf{x}}_k$ to minimize a prediction error cost. In [6], this *dual Kalman* approach is placed in a general family of recursive prediction error algorithms. Apart from these sequential approaches, some iterative methods developed for linear models include maximum-likelihood approaches [7–9] and expectation-maximization (EM) algorithms [10–13]. These algorithms are suitable only for off-line applications, although sequential EM methods have been suggested.

Fewer papers have appeared in the literature that are explicitly concerned with dual estimation for nonlinear models. One algorithm (proposed in [14]) alternates between applying a robust form of the

EKF to estimate the time-series and using these estimates to train a neural network via gradient descent. A joint EKF is used in [15] to model partially unknown dynamics in a model reference adaptive control framework. Furthermore, iterative EM approaches to the dual estimation problem have been investigated for radial basis function networks [16] and other nonlinear models [17]; see also Chapter 6. Errors-in-variables (EIV) models appear in the nonlinear statistical regression literature [18], and are used for regressing on variables related by a nonlinear function, but measured with some error. However, errors-in-variables is an iterative approach involving batch computation; it tends not to be practical for dynamical systems because the computational requirements increase in proportion to N^2, where N is the length of the data. A heuristic method known as *Clearning* minimizes a simplified approximation to the EIV cost function. While it allows for sequential estimation, the simplification can lead to severely biased results [19]. The dual EKF [19] is a nonlinear extension of the linear dual Kalman approach of [5], and recursive prediction error algorithm of [6]. Application of the algorithm to speech enhancement appears in [20], while extensions to other cost functions have been developed in [21] and [22]. The crucial, but often overlooked issue of sequential variance estimation is also addressed in [22].

Overview The goal of this chapter is to present a unified probabilistic and algorithmic framework for nonlinear dual estimation methods. In the next section, we start with the basic dual EKF prediction error method. This approach is the most intuitive, and involves simply running two EKF filters in parallel. The section also provides a quick review of the EKF for both state and weight estimation, and introduces some of the complications in coupling the two. An example in noisy time-series prediction is also given. In Section 5.3, we develop a general probabilistic framework for dual estimation. This allows us to relate the various methods that have been presented in the literature, and also provides a general algorithmic approach leading to a number of different dual EKF algorithms. Results on additional example data sets are presented in Section 5.5.

5.2 DUAL EKF–PREDICTION ERROR

In this section, we present the basic dual EKF prediction error algorithm. For completeness, we start with a quick review of the EKF for state estimation, followed by a review of EKF weight estimation (see Chapters

1 and 2 for more details). We then discuss coupling the state and weight filters to form the dual EKF algorithm.

5.2.1 EKF–State Estimation

For a *linear* state-space system with *known* model and Gaussian noise, the Kalman filter [23] generates optimal estimates and predictions of the state $\mathbf{x}_k$. Essentially, the filter recursively updates the (posterior) mean $\hat{\mathbf{x}}_k$ and covariance $\mathbf{P}_{\mathbf{x}_k}$ of the state by combining the predicted mean $\hat{\mathbf{x}}_k^-$ and covariance $\mathbf{P}_{\mathbf{x}_k}^-$ with the current noisy measurement $\mathbf{y}_k$. These estimates are optimal in both the MMSE and MAP senses. Maximum-likelihood signal estimates are obtained by letting the initial covariance $\mathbf{P}_{\mathbf{x}_0}$ approach infinity, thus causing the filter to ignore the value of the initial state $\hat{\mathbf{x}}_0$.

For nonlinear systems, the extended Kalman filter provides *approximate* maximum-likelihood estimates. The mean and covariance of the state are again recursively updated; however, a first-order linearization of the dynamics is necessary in order to analytically propagate the Gaussian random-variable representation. Effectively, the nonlinear dynamics are approximated by a time-varying linear system, and the linear Kalman filters equations are applied. The full set of equations are given in Table 5.1. While there are more accurate methods for dealing with the nonlinear dynamics (e.g., particle filters [24, 25], second-order EKF, etc.), the standard EKF remains the most popular approach owing to its simplicity. Chapter 7 investigates the use of the unscented Kalman filter as a potentially superior alternative to the EKF [26–29].

Another interpretation of Kalman filtering is that of an optimization algorithm that recursively determines the state $\mathbf{x}_k$ in order to minimize a cost function. It can be shown that the cost function consists of a weighted prediction error and estimation error components given by

$$\begin{aligned} J(\mathbf{x}_1^k) = \sum_{t=1}^{k} \{ & [\mathbf{y}_t - \mathbf{H}(\mathbf{x}_t, \mathbf{w})]^T (\mathbf{R}^\mathbf{n})^{-1} [\mathbf{y}_t - \mathbf{H}(\mathbf{x}_t, \mathbf{w})] \\ & + (\mathbf{x}_t - \mathbf{x}_t^-)^T (\mathbf{R}^\mathbf{v})^{-1} (\mathbf{x}_t - \mathbf{x}_t^-) \} \end{aligned} \tag{5.10}$$

where $\mathbf{x}_t^- = \mathbf{F}(\mathbf{x}_{t-1}, \mathbf{w})$ is the predicted state, and $\mathbf{R}^\mathbf{n}$ and $\mathbf{R}^\mathbf{v}$ are the additive noise and innovations noise covariances, respectively. This interpretation will be useful when dealing with alternate forms of the dual EKF in Section 5.3.3.

Table 5.1 Extended Kalman filter (EKF) equations

Initialize with:

$$\hat{\mathbf{x}}_0 = E[\mathbf{x}_0], \tag{5.2}$$

$$\mathbf{P}_{\mathbf{x}_0} = E[(\mathbf{x}_0 - \hat{\mathbf{x}}_0)(\mathbf{x}_0 - \hat{\mathbf{x}}_0)^T]. \tag{5.3}$$

For $k \in \{1, \ldots, \infty\}$, the time-update equations of the extended Kalman filter are

$$\hat{\mathbf{x}}_k^- = \mathbf{F}(\hat{\mathbf{x}}_{k-1}, \mathbf{u}_k, \mathbf{w}), \tag{5.4}$$

$$\mathbf{P}_{\mathbf{x}_k}^- = \mathbf{A}_{k-1}\mathbf{P}_{\mathbf{x}_{k-1}}\mathbf{A}_{k-1}^T + \mathbf{R}^{\mathrm{v}}, \tag{5.5}$$

and the measurement-update equations are

$$\mathbf{K}_k^{\mathbf{x}} = \mathbf{P}_{\mathbf{x}_k}^-\mathbf{C}_k^T(\mathbf{C}_k\mathbf{P}_{\mathbf{x}_k}^-\mathbf{C}_k^T + \mathbf{R}^{\mathrm{n}})^{-1}, \tag{5.6}$$

$$\hat{\mathbf{x}}_k = \hat{\mathbf{x}}_k^- + \mathbf{K}_k^{\mathbf{x}}[\mathbf{y}_k - \mathbf{H}(\hat{\mathbf{x}}_k^-, \mathbf{w})], \tag{5.7}$$

$$\mathbf{P}_{\mathbf{x}_k} = (\mathbf{I} - \mathbf{K}_k^{\mathbf{x}}\mathbf{C}_k)\mathbf{P}_{\mathbf{x}_k}^-, \tag{5.8}$$

where

$$\mathbf{A}_k \triangleq \left.\frac{\partial \mathbf{F}(\mathbf{x}, \mathbf{u}_k, \mathbf{w})}{\partial \mathbf{x}}\right|_{\hat{\mathbf{x}}_k}, \qquad \mathbf{C}_k \triangleq \left.\frac{\partial \mathbf{H}(\mathbf{x}, \mathbf{w})}{\partial \mathbf{x}}\right|_{\hat{\mathbf{x}}_k}, \tag{5.9}$$

and where $\mathbf{R}^{\mathrm{v}}$ and $\mathbf{R}^{\mathrm{n}}$ are the covariances of $\mathbf{v}_k$ and $\mathbf{n}_k$, respectively.

5.2.2 EKF–Weight Estimation

As proposed initially in [30], and further developed in [31] and [32], the EKF can also be used for estimating the parameters of nonlinear models (i.e., training neural networks) from *clean* data. Consider the general problem of learning a mapping using a parameterized nonlinear function $\mathbf{G}(\mathbf{x}_k, \mathbf{w})$. Typically, a training set is provided with sample pairs consisting of known input and desired output, $\{\mathbf{x}_k, \mathbf{d}_k\}$. The error in the model is defined as $\mathbf{e}_k = \mathbf{d}_k - \mathbf{G}(\mathbf{x}_k, \mathbf{w})$, and the goal of learning involves solving for the parameters $\mathbf{w}$ in order to minimize the expected squared error. The EKF may be used to estimate the parameters by writing a new state-space representation

$$\mathbf{w}_{k+1} = \mathbf{w}_k + \mathbf{r}_k, \tag{5.11}$$

$$\mathbf{d}_k = \mathbf{G}(\mathbf{x}_k, \mathbf{w}_k) + \mathbf{e}_k, \tag{5.12}$$

where the parameters $\mathbf{w}_k$ correspond to a stationary process with identity state transition matrix, driven by process noise $\mathbf{r}_k$. The output $\mathbf{d}_k$

Table 5.2 The extended Kalman weight filter equations

Initialize with:

$$\hat{\mathbf{w}}_0 = E[\mathbf{w}] \tag{5.13}$$

$$\mathbf{P}_{\mathbf{w}_0} = E[(\mathbf{w} - \hat{\mathbf{w}}_0)(\mathbf{w} - \hat{\mathbf{w}}_0)^T] \tag{5.14}$$

For $k \in \{1, \ldots, \infty\}$, the time update equations of the Kalman filter are:

$$\hat{\mathbf{w}}_k^- = \hat{\mathbf{w}}_{k-1} \tag{5.15}$$

$$\mathbf{P}_{\mathbf{w}_k}^- = \mathbf{P}_{\mathbf{w}_{k-1}} + \mathbf{R}_{k-1}^{\mathbf{r}} \tag{5.16}$$

and the measurement update equations:

$$\mathbf{K}_k^{\mathbf{w}} = \mathbf{P}_{\mathbf{w}_k}^-(\mathbf{C}_k^{\mathbf{w}})^T(\mathbf{C}_k^{\mathbf{w}}\mathbf{P}_{\mathbf{w}_k}^-(\mathbf{C}_k^{\mathbf{w}})^T + \mathbf{R}^{\mathbf{e}})^{-1} \tag{5.17}$$

$$\hat{\mathbf{w}}_k = \hat{\mathbf{w}}_k^- + \mathbf{K}_k^{\mathbf{w}}(\mathbf{d}_k - \mathbf{G}(\hat{\mathbf{w}}_k^-, \mathbf{x}_{k-1})) \tag{5.18}$$

$$\mathbf{P}_{\mathbf{w}_k} = (\mathbf{I} - \mathbf{K}_k^{\mathbf{w}}\mathbf{C}_k^{\mathbf{w}})\mathbf{P}_{\mathbf{w}_k}^-. \tag{5.19}$$

where

$$\mathbf{C}_k^{\mathbf{w}} \triangleq \left.\frac{\partial \mathbf{G}(\mathbf{x}_{k-1}, \mathbf{w})^T}{\partial \mathbf{w}}\right|_{\mathbf{w}=\hat{\mathbf{w}}_k^-} \tag{5.20}$$

---

corresponds to a nonlinear observation on $\mathbf{w}_k$. The EKF can then be applied directly, with the equations given in Table 5.2. In the linear case, the relationship between the Kalman filter (KF) and the popular recursive least-squares (RLS) is given [33] and [34]. In the nonlinear case, the EKF training corresponds to a modified-Newton optimization method [22].

As an optimization approach, the EKF minimizes the *prediction error* cost:

$$J(\mathbf{w}) = \sum_{t=1}^{k} [\mathbf{d}_t - \mathbf{G}(\mathbf{x}_t, \mathbf{w})]^T(\mathbf{R}^{\mathbf{e}})^{-1}[\mathbf{d}_t - \mathbf{G}(\mathbf{x}_t, \mathbf{w})]. \tag{5.21}$$

If the "noise" covariance $\mathbf{R}^{\mathbf{e}}$ is a constant diagonal matrix, then, in fact, it cancels out of the algorithm (this can be shown explicitly), and hence can be set arbitrarily (e.g., $\mathbf{R}^{\mathbf{e}} = 0.5\mathbf{I}$). Alternatively, $\mathbf{R}^{\mathbf{e}}$ can be set to specify a weighted MSE cost. The innovations covariance $E[\mathbf{r}_k\mathbf{r}_k^T] = \mathbf{R}_k^{\mathbf{r}}$, on the other hand, affects the convergence rate and tracking performance. Roughly speaking, the larger the covariance, the more quickly older data are discarded. There are several options on how to choose $\mathbf{R}_k^{\mathbf{r}}$:

- Set $\mathbf{R}_k^{\mathbf{r}}$ to an arbitrary diagonal value, and anneal this towards zeroes as training continues.

- Set $\mathbf{R}_k^{\mathbf{r}} = (\lambda^{-1} - 1)\mathbf{P}_{\mathbf{w}_k}$, where $\lambda \in (0, 1]$ is often referred to as the "forgetting factor." This provides for an approximate exponentially decaying weighting on past data and is described more fully in [22].
- Set $\mathbf{R}_k^{\mathbf{r}} = (1 - \alpha)\mathbf{R}_{k-1}^{\mathbf{r}} + \alpha\mathbf{K}_k^{\mathbf{w}}[\mathbf{d}_k - \mathbf{G}(\mathbf{x}_k, \hat{\mathbf{w}})][\mathbf{d}_k - \mathbf{G}(\mathbf{x}_k, \hat{\mathbf{w}})]^T(\mathbf{K}_k^{\mathbf{w}})^T$, which is a Robbins–Monro stochastic approximation scheme for estimating the innovations [6]. The method assumes that the covariance of the Kalman update model is consistent with the actual update model.

Typically, $\mathbf{R}_k^{\mathbf{r}}$ is also constrained to be a diagonal matrix, which implies an independence assumption on the parameters.

Study of the various trade-offs between these different approaches is still an area of open research. For the experiments performed in this chapter, the forgetting factor approach is used.

Returning to the dynamic system of Eq. (5.1), the EKF weight filter can be used to estimate the model parameters for either $\mathbf{F}$ or $\mathbf{H}$. To learn the state dynamics, we simply make the substitutions $\mathbf{G} \rightarrow \mathbf{F}$ and $\mathbf{d}_k \rightarrow \mathbf{x}_{k+1}$. To learn the measurement function, we make the substitutions $\mathbf{G} \rightarrow \mathbf{H}$ and $\mathbf{d}_k \rightarrow \mathbf{y}_k$. Note that for both cases, it is assumed that the noise-free state $\mathbf{x}_k$ is available for training.

5.2.3 Dual Estimation

When the clean state is *not* available, a dual estimation approach is required. In this section, we introduce the basic dual EKF algorithm, which combines the Kalman state and weight filters. Recall that the task is to estimate *both* the state and model from only noisy observations. Essentially, two EKFs are run concurrently. At every time step, an EKF state filter estimates the state using the current model estimate $\hat{\mathbf{w}}_k$, while the EKF weight filter estimates the weights using the current state estimate $\hat{\mathbf{x}}_k$. The system is shown schematically in Figure 5.2. In order to simplify the presentation of the equations, we consider the slightly less general state-space model:

$$\mathbf{x}_{k+1} = \mathbf{F}(\mathbf{x}_k, \mathbf{u}_k, \mathbf{w}) + \mathbf{v}_k, \tag{5.22}$$

$$y_k = \mathbf{C}\mathbf{x}_k + n_k, \qquad \mathbf{C} = [1 \quad 0 \quad \ldots \quad 0], \tag{5.23}$$

in which we take the scalar observation y_k to be one of the states. Thus, we only need to consider estimating the parameters associated with a single

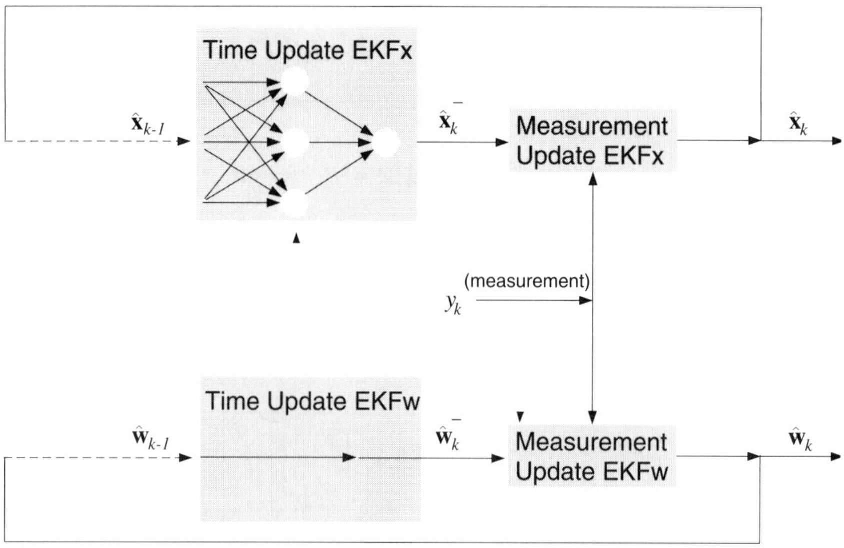

Figure 5.2 The dual extended Kalman filter. The algorithm consists of two EKFs that run concurrently. The top EKF generates state estimates, and requires $\hat{\mathbf{w}}_{k-1}$ for the time update. The bottom EKF generates weight estimates, and requires $\hat{\mathbf{x}}_{k-1}$ for the measurement update.

nonlinear function $\mathbf{F}$. The dual EKF equations for this system are presented in Table 5.3. Note that for clarity, we have specified the equations for the additive white-noise case. The case of colored measurement noise n_k is treated in Appendix B.

Recurrent Derivative Computation While the dual EKF equations appear to be a simple concatenation of the previous state and weight EKF equations, there is actually a necessary modification of the linearization $\mathbf{C}_k^{\mathbf{w}} = \mathbf{C}\partial\hat{\mathbf{x}}_k^- / \partial\hat{\mathbf{w}}_k^-$ associated with the weight filter. This is due to the fact that the signal filter, whose parameters are being estimated by the weight filter, has a recurrent architecture, i.e., $\hat{\mathbf{x}}_k$ is a function of $\hat{\mathbf{x}}_{k-1}$, and both are functions of $\mathbf{w}$.[1] Thus, the linearization must be computed using recurrent derivatives with a routine similar to real-time recurrent learning

[1] Note that a linearization is also required for the state EKF, but this derivative, $\partial\mathbf{F}(\hat{\mathbf{x}}_{k-1}, \hat{\mathbf{w}}_k^-)/\partial\hat{\mathbf{x}}_{k-1}$, can be computed with a simple technique (such as backpropagation) because $\hat{\mathbf{w}}_k^-$ is not itself a function of $\hat{\mathbf{x}}_{k-1}$.

Table 5.3 The dual extended Kalman filter equations. The definitions of $\mathfrak{e}_k$ and $\mathbf{C}_k^{\mathbf{w}}$ depend on the particular form of the weight filter being used. See the text for details

Initialize with:

$$\hat{\mathbf{w}}_0 = E[\mathbf{w}], \qquad \mathbf{P}_{\mathbf{w}_0} = E[(\mathbf{w} - \hat{\mathbf{w}}_0)(\mathbf{w} - \hat{\mathbf{w}}_0)^T],$$

$$\hat{\mathbf{x}}_0 = E[\mathbf{x}_0], \qquad \mathbf{P}_{\mathbf{x}_0} = E[(\mathbf{x}_0 - \hat{\mathbf{x}}_0)(\mathbf{x}_0 - \hat{\mathbf{x}}_0)^T].$$

For $k \in \{1, \ldots, \infty\}$, the time-update equations for the weight filter are

$$\hat{\mathbf{w}}_k^- = \hat{\mathbf{w}}_{k-1}, \tag{5.24}$$

$$\mathbf{P}_{\mathbf{w}_k}^- = \mathbf{P}_{\mathbf{w}_{k-1}} + \mathbf{R}_{k-1}^{\mathbf{r}} = \lambda^{-1}\mathbf{P}_{\mathbf{w}_{k-1}}, \tag{5.25}$$

and those for the state filter are

$$\hat{\mathbf{x}}_k^- = \mathbf{F}(\hat{\mathbf{x}}_{k-1}\mathbf{u}_k, \hat{\mathbf{w}}_k^-), \tag{5.26}$$

$$\mathbf{P}_{\mathbf{x}_k}^- = \mathbf{A}_{k-1}\mathbf{P}_{\mathbf{x}_{k-1}}\mathbf{A}_{k-1}^T + \mathbf{R}^{\mathbf{v}}. \tag{5.27}$$

The measurement-update equations for the state filter are

$$\mathbf{K}_k^{\mathbf{x}} = \mathbf{P}_{\mathbf{x}_k}^-\mathbf{C}^T(\mathbf{C}\mathbf{P}_{\mathbf{x}_k}^-\mathbf{C}^T + \mathbf{R}^{\mathbf{n}})^{-1}, \tag{5.28}$$

$$\hat{\mathbf{x}}_k = \hat{\mathbf{x}}_k^- + \mathbf{K}_k^{\mathbf{x}}(y_k - \mathbf{C}\hat{\mathbf{x}}_k^-), \tag{5.29}$$

$$\mathbf{P}_{\mathbf{x}_k} = (\mathbf{I} - \mathbf{K}_k^{\mathbf{x}}\mathbf{C})\mathbf{P}_{\mathbf{x}_k}^-, \tag{5.30}$$

and those for the weight filter are

$$\mathbf{K}_k^{\mathbf{w}} = \mathbf{P}_{\mathbf{w}_k}^-(\mathbf{C}_k^{\mathbf{w}})^T[\mathbf{C}_k^{\mathbf{w}}\mathbf{P}_{\mathbf{w}_k}^-(\mathbf{C}_k^{\mathbf{w}})^T + \mathbf{R}^{\mathbf{e}}]^{-1}, \tag{5.31}$$

$$\hat{\mathbf{w}}_k = \hat{\mathbf{w}}_k^- + \mathbf{K}_k^{\mathbf{w}} \cdot \mathfrak{e}$$

where

$$\mathbf{A}_{k-1} \triangleq \left.\frac{\partial \mathbf{F}(\mathbf{x}, \hat{\mathbf{w}}_k^-)}{\partial \mathbf{x}}\right|_{\hat{\mathbf{x}}_{k-1}}, \qquad \mathfrak{e}_k = (y_k - \mathbf{C}\hat{\mathbf{x}}_k^-), \qquad \mathbf{C}_k^{\mathbf{w}} \triangleq -\frac{\partial \mathfrak{e}_k}{\partial \mathbf{w}} = \mathbf{C}\left.\frac{\partial \hat{\mathbf{x}}_k^-}{\partial \mathbf{w}}\right|_{\mathbf{w}=\hat{\mathbf{w}}_k^-}. \tag{5.34}$$

(RTRL) [35]. Taking the derivative of the signal filter equations results in the following system of recursive equations:

$$\frac{\partial \hat{\mathbf{x}}_{k+1}^-}{\partial \hat{\mathbf{w}}} = \frac{\partial \mathbf{F}(\hat{\mathbf{x}}, \hat{\mathbf{w}})}{\partial \hat{\mathbf{x}}_k}\frac{\partial \hat{\mathbf{x}}_k}{\partial \hat{\mathbf{w}}} + \frac{\partial \mathbf{F}(\hat{\mathbf{x}}, \hat{\mathbf{w}})}{\partial \hat{\mathbf{w}}_k}, \tag{5.35}$$

$$\frac{\partial \hat{\mathbf{x}}_k}{\partial \hat{\mathbf{w}}} = (\mathbf{I} - \mathbf{K}_k^{\mathbf{x}}\mathbf{C})\frac{\partial \hat{\mathbf{x}}_k^-}{\partial \hat{\mathbf{w}}} + \frac{\partial \mathbf{K}_k^{\mathbf{x}}}{\partial \hat{\mathbf{w}}}(y_k - \mathbf{C}\hat{\mathbf{x}}_k^-), \tag{5.36}$$

where $\partial\mathbf{F}(\hat{\mathbf{x}}, \hat{\mathbf{w}})/\partial\hat{\mathbf{x}}_k$ and $\partial\mathbf{F}(\hat{\mathbf{x}}, \hat{\mathbf{w}})/\partial\hat{\mathbf{w}}_k$ are evaluated at $\hat{\mathbf{w}}_k$ and contain static linearizations of the nonlinear function.

The last term in Eq. (5.36) may be dropped if we assume that the Kalman gain $\mathbf{K}_k^{\mathbf{x}}$ is independent of $\mathbf{w}$. Although this greatly simplifies the algorithm, the exact computation of $\partial\mathbf{K}_k^{\mathbf{x}}/\partial\hat{\mathbf{w}}$ may be computed, as shown in Appendix A. Whether the computational expense in calculating the recursive derivatives (especially that of calculating $\partial\mathbf{K}_k^{\mathbf{x}}/\partial\hat{\mathbf{w}}$) is worth the improvement in performance is clearly a design issue. Experimentally, the recursive derivatives appear to be more critical when the signal is highly nonlinear, or is corrupted by a high level of noise.

Example As an example application, consider the noisy time-series $\{x_k\}_1^N$ generated by a nonlinear autoregression:

$$\begin{aligned} x_k &= f(x_{k-1}, \ldots x_{k-M}, \mathbf{w}) + v_k, \\ y_k &= x_k + n_k \qquad \forall k \in \{1, \ldots, N\}. \end{aligned} \tag{5.37}$$

The observations of the series y_k contain measurement noise n_k in addition to the signal. The dual EKF requires reformulating this model into a state-space representation. One such representation is given by

$$\mathbf{x}_k = \mathbf{F}(\mathbf{x}_{k-1}, \mathbf{w}) + \mathbf{B}v_k, \tag{5.38}$$

$$\begin{bmatrix} x_k \\ x_{k-1} \\ \vdots \\ x_{k-M+1} \end{bmatrix} = \begin{bmatrix} f(x_{k-1}, \ldots, x_{k-M}, \mathbf{w}) \\ \begin{bmatrix} 1 & 0 & 0 & 0 \\ 0 & \ddots & 0 & \vdots \\ 0 & 0 & 1 & 0 \end{bmatrix} \cdot \begin{bmatrix} x_{k-1} \\ \vdots \\ x_{k-M} \end{bmatrix} \end{bmatrix} + \begin{bmatrix} 1 \\ 0 \\ \vdots \\ 0 \end{bmatrix} v_k,$$

$$\begin{aligned} y_k &= \mathbf{C}\mathbf{x}_k + n_k, \\ &= [1 \quad 0 \quad \ldots \quad 0]\mathbf{x}_k + n_k, \end{aligned} \tag{5.39}$$

where the state $\mathbf{x}_k$ is chosen to be lagged values of the time series, and the state transition function $\mathbf{F}(\cdot)$ has its first element given by $f(\cdot)$, with the remaining elements corresponding to shifted values of the previous state.

The results of a controlled time-series experiment are shown in Figure 5.3. The clean signal, shown by the thin curve in Figure 5.3*a*, is generated by a neural network (10-5-1) with chaotic dynamics, driven by white Gaussian-process noise ($\sigma_v^2 = 0.36$). Colored noise generated by a linear autoregressive model is added at 3 dB signal-to-noise ratio (SNR) to produce the noisy data indicated by + symbols. Figure 5.3*b* shows the

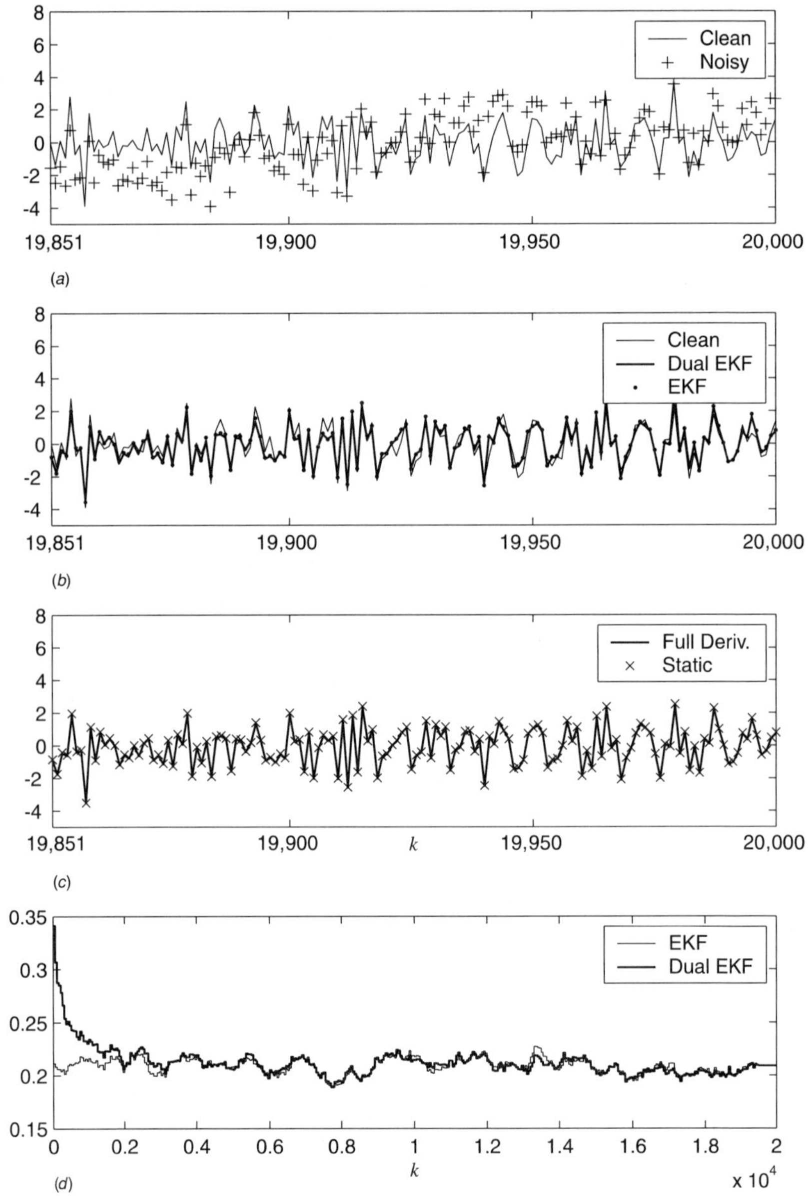

Figure 5.3 The dual EKF estimate (heavy curve) of a signal generated by a neural network (thin curve) and corrupted by adding colored noise at 3 dB (+). For clarity, the last 150 points of a 20,000-point series are shown. Only the noisy data are available: both the signal and weights are estimated by the dual EKF. (*a*) Clean neural network signal and noisy measurements. (*b*) Dual EKF estimates versus EKF estimates. (*c*) Estimates with full and static derivatives. (*d*) MSE profiles of EKF versus dual EKF.

time series estimated by the dual EKF. The algorithm estimates both the clean time series and the neural network weights. The algorithm is run sequentially over 20,000 points of data; for clarity, only the last 150 points are shown. For comparison, the estimates using an EKF with the *known* neural network model are also shown. The MSE for the dual EKF, computed over the final 1000 points of the series, is 0.2171, whereas the EKF produces an MSE of 0.2153, indicating that the dual algorithm has successfully learned both the model and the states estimates.[2]

Figure 5.3*c* shows the estimate when the static approximation to recursive derivatives is used. In this example, this static derivative actually provides a slight advantage, with an MSE of 0.2122. The difference, however, is not statistically significant. Finally, Figure 5.3*d* assesses the convergence behavior of the algorithm. The mean-squared error (MSE) is computed over 500 point segments of the time series at 50 point intervals to produce the *MSE profile* (dashed line). For comparison, the solid line is the MSE profile of the EKF signal estimation algorithm, which uses the true neural network model. The dual EKF appears to converge to the optimal solution after only about 2000 points.

5.3 A PROBABILISTIC PERSPECTIVE

In this section, we present a unified framework for dual estimation. We start by developing a probabilistic perspective, which leads to a number of possible cost functions that can be used in the estimation process. Various approaches in the literature, which may differ in their actual optimization procedure, can then be related based on the underlying cost function. We then show how a Kalman-based optimization procedure can be used to provide a common algorithmic framework for minimizing each of the cost functions.

MAP Estimation Dual estimation can be cast as a *maximum a posteriori* (MAP) solution. The statistical information contained in the sequence of data $\{y_k\}_1^N$ about the signal and parameters is embodied by the joint conditional probability density of the sequence of states $\{\mathbf{x}_k\}_1^N$ and weights

[2] A surprising result is that the dual EKF sometimes actually outperforms the EKF, even though the EKF appears to have an unfair advantage of knowing the true model. Our explanation is that the EKF, even with the known model, is still an *approximate* estimation algorithm. While the dual EKF also learns an approximate model, this model can actually be better matched to the state estimation approximation.

$\mathbf{w}$, given the noisy data $\{y_k\}_1^N$. For notational convenience, define the column vectors $\mathbf{x}_1^N$ and $\mathbf{y}_1^N$, with elements from $\{\mathbf{x}_k\}_1^N$ and $\{y_k\}_1^N$, respectively. The joint conditional density function is written as

$$\rho_{\mathbf{x}_1^N \mathbf{w}|\mathbf{y}_1^N}(\mathbf{X} = \mathbf{x}_1^N, \ \mathbf{W} = \mathbf{w}|\mathbf{Y} = \mathbf{y}_1^N), \tag{5.40}$$

where $\mathbf{X}$, $\mathbf{Y}$, and $\mathbf{W}$ are the vectors of random variables associated with $\mathbf{x}_1^N$, $\mathbf{y}_1^N$, and $\mathbf{w}$, respectively. This joint density is abbreviated as $\rho_{\mathbf{x}_1^N \mathbf{w}|\mathbf{y}_1^N}$. The MAP estimation approach consists of determining instances of the states and weights that maximize this conditional density. For Gaussian distributions, the MAP estimate also corresponds to the minimum mean-squared error (MMSE) estimator. More generally, as long as the density is unimodal and symmetric around the mean, the MAP estimate provides the *Bayes estimate* for a broad class of loss functions [36].

Taking MAP as the starting point allows dual estimation approaches to be divided into two basic classes. The first, referred to here as *joint estimation methods*, attempt to maximize $\rho_{\mathbf{x}_1^N \mathbf{w}|\mathbf{y}_1^N}$ directly. We can write this optimization problem explicitly as

$$(\hat{\mathbf{x}}_1^N, \hat{\mathbf{w}}) = \arg\max_{\mathbf{x}_1^N, \mathbf{w}} \rho_{\mathbf{x}_1^N \mathbf{w}|\mathbf{y}_1^N}. \tag{5.41}$$

The second class of methods, which will be referred to as *marginal estimation methods*, operate by expanding the joint density as

$$\rho_{\mathbf{x}_1^N \mathbf{w}|\mathbf{y}_1^N} = \rho_{\mathbf{x}_1^N|\mathbf{w}\mathbf{y}_1^N} \rho_{\mathbf{w}|\mathbf{y}_1^N} \tag{5.42}$$

and maximizing the two terms separately, that is,

$$\hat{\mathbf{x}}_1^N = \arg\max_{\mathbf{x}_1^N} \rho_{\mathbf{x}_1^N|\mathbf{w}\mathbf{y}_1^N}, \qquad \hat{\mathbf{w}} = \arg\max_{\mathbf{w}} \rho_{\mathbf{w}|\mathbf{y}_1^N}. \tag{5.43}$$

The cost functions associated with joint and marginal approaches will be discussed in the following sections.

5.3.1 Joint Estimation Methods

Using Bayes' rule, the joint conditional density can be expressed as

$$\rho_{\mathbf{x}_1^N \mathbf{w}|\mathbf{y}_1^N} = \frac{\rho_{\mathbf{y}_1^N|\mathbf{x}_1^N \mathbf{w}} \rho_{\mathbf{x}_1^N \mathbf{w}}}{\rho_{\mathbf{y}_1^N}} = \frac{\rho_{\mathbf{y}_1^N|\mathbf{x}_1^N \mathbf{w}} \rho_{\mathbf{x}_1^N|\mathbf{w}} \rho_{\mathbf{w}}}{\rho_{\mathbf{y}_1^N}}. \tag{5.44}$$

Although $\{y_k\}_1^N$ is *statistically* dependent on $\{\mathbf{x}_k\}_1^N$ and $\mathbf{w}$, the prior $\rho_{\mathbf{y}_1^N}$ is nonetheless *functionally* independent of $\{\mathbf{x}_k\}_1^N$ and $\mathbf{w}$. Therefore, $\rho_{\mathbf{x}_1^N \mathbf{w}|\mathbf{y}_1^N}$ can be maximized by maximizing the terms in the numerator alone. Furthermore, if no prior information is available on the weights, $\rho_{\mathbf{w}}$ can be dropped, leaving the maximization of

$$\rho_{\mathbf{y}_1^N \mathbf{x}_1^N|\mathbf{w}} = \rho_{\mathbf{y}_1^N|\mathbf{x}_1^N \mathbf{w}} \rho_{\mathbf{x}_1^N|\mathbf{w}} \tag{5.45}$$

with respect to $\{\mathbf{x}_k\}_1^N$ and $\mathbf{w}$.

To derive the corresponding cost function, we assume $\mathbf{v}_k$ and n_k are both zero-mean white Gaussian noise processes. It can then be shown (see [22]), that

$$\rho_{\mathbf{y}_1^N \mathbf{x}_1^N|\mathbf{w}} = \frac{1}{\sqrt{(2\pi)^N (\sigma_n^2)^N}} \exp\left[-\sum_{k=1}^N \frac{(y_k - \mathbf{C}\mathbf{x}_k)^2}{2\sigma_n^2}\right]$$
$$\times \frac{1}{\sqrt{(2\pi)^N |\mathbf{R}^{\mathbf{v}}|^N}} \exp\left[-\sum_{k=1}^N \frac{1}{2}(\mathbf{x}_k - \mathbf{x}_k^-)^T (\mathbf{R}^{\mathbf{v}})^{-1} (\mathbf{x}_k - \mathbf{x}_k^-)\right], \tag{5.46}$$

where

$$\mathbf{x}_k^- \triangleq E[\mathbf{x}_k | \{\mathbf{x}_t\}_1^{k-1}, \mathbf{w}] = \mathbf{F}(\mathbf{x}_{k-1}, \mathbf{w}). \tag{5.47}$$

Here we have used the structure given in Eq. (5.37) to compute the prediction $\mathbf{x}_k^-$ using the model $\mathbf{F}(\cdot, \mathbf{w})$. Taking the logarithm, the corresponding cost function is given by

$$J = \sum_{k=1}^{N} \left[\log(2\pi\sigma_n^2) + \frac{(y_k - \mathbf{C}\mathbf{x}_k)^2}{\sigma_n^2} \right. \tag{5.48}$$

$$\left. + \log(2\pi|\mathbf{R}^{\mathbf{v}}|) + (\mathbf{x}_k - \mathbf{x}_k^-)^T (\mathbf{R}^{\mathbf{v}})^{-1} (\mathbf{x}_k - \mathbf{x}_k^-) \right]. \tag{5.49}$$

This cost function can be minimized with respect to any of the unknown quantities (including the variances, which we will consider in Section 5.4). For the time being, consider only the optimization of $\{\mathbf{x}_k\}_1^N$ and $\mathbf{w}$. Because the log terms in the above cost are independent of the signal and weights, they can be dropped, providing a more specialized cost function:

$$J^j(\mathbf{x}_1^N, \mathbf{w}) = \sum_{k=1}^{N} \left[\frac{(y_k - \mathbf{C}\mathbf{x}_k)^2}{\sigma_n^2} + (\mathbf{x}_k - \mathbf{x}_k^-)^T (\mathbf{R}^{\mathbf{v}})^{-1} (\mathbf{x}_k - \mathbf{x}_k^-) \right]. \tag{5.50}$$

The first term is a soft constraint keeping $\{\mathbf{x}_k\}_1^N$ close to the observations $\{y_k\}_1^N$. The smaller the measurement noise variance σ_n^2, the stronger this constraint will be. The second term keeps the state estimates and model estimates mutually consistent with the model structure. This constraint will be strong when the state is highly deterministic (i.e., $\mathbf{R}^{\mathbf{v}}$ is small).

$J^j(\mathbf{x}_1^N, \mathbf{w})$ should be minimized with respect to both $\{\mathbf{x}_k\}_1^N$ and $\mathbf{w}$ to find the estimates that maximize the joint density $\rho_{\mathbf{y}_1^N \mathbf{x}_1^N | \mathbf{w}}$. This is a difficult optimization problem because of the high degree of coupling between the unknown quantities $\{\mathbf{x}_k\}_1^N$ and $\mathbf{w}$. In general, we can classify approaches as being either *direct* or *decoupled*. In direct approaches, both the signal and the state are determined jointly as a multivariate optimization problem. Decoupled approaches optimize one variable at a time while the other variable is fixed, and then alternating. Direct algorithms include the joint EKF algorithm (see Section 5.1), which attempts to minimize the cost sequentially by combining the signal and weights into a single (joint) state vector. The decoupled approaches are elaborated below.

Decoupled Estimation To minimize $J^j(\mathbf{x}_1^N, \mathbf{w})$ with respect to the signal, the cost function is evaluated using the current estimate $\hat{\mathbf{w}}$ of the weights to generate the predictions. The simplest approach is to substitute the predictions $\hat{\mathbf{x}}_k^- \triangleq \mathbf{F}(\mathbf{x}_{k-1}, \hat{\mathbf{w}})$ directly into Eq. (5.50), obtaining

$$J^j(\mathbf{x}_1^N, \hat{\mathbf{w}}) = \sum_{k=1}^{N} \left[\frac{(y_k - \mathbf{C}\mathbf{x}_k)^2}{\sigma_n^2} + (\mathbf{x}_k - \hat{\mathbf{x}}_k^-)^T (\mathbf{R}^{\mathbf{v}})^{-1} (\mathbf{x}_k - \hat{\mathbf{x}}_k^-) \right]. \tag{5.51}$$

This cost function is then minimized with respect to $\{\mathbf{x}_k\}_1^N$. To minimize the joint cost function with respect to the weights, $J^j(\mathbf{x}_1^N, \mathbf{w})$ is evaluated using the current signal estimate $\{\hat{\mathbf{x}}_k\}_1^N$ and the associated (redefined) predictions $\hat{\mathbf{x}}_k^- \triangleq \mathbf{F}(\hat{\mathbf{x}}_{k-1}, \mathbf{w})$. Again, this results in a straightforward substitution in Eq. (5.50):

$$J^j(\hat{\mathbf{x}}_1^N, \mathbf{w}) = \sum_{k=1}^{N} \left[\frac{(y_k - \mathbf{C}\hat{\mathbf{x}}_k)^2}{\sigma_n^2} + (\hat{\mathbf{x}}_k - \hat{\mathbf{x}}_k^-)^T (\mathbf{R}^{\mathbf{v}})^{-1} (\hat{\mathbf{x}}_k - \hat{\mathbf{x}}_k^-) \right]. \tag{5.52}$$

An alternative simplified cost function can be used if it is assumed that only $\hat{\mathbf{x}}_k^-$ is a function of the weights:

$$J_i^j(\hat{\mathbf{x}}_1^N, \mathbf{w}) = \sum_{k=1}^{N} (\hat{\mathbf{x}}_k - \hat{\mathbf{x}}_k^-)^T (\mathbf{R}^{\mathbf{v}})^{-1} (\hat{\mathbf{x}}_k - \hat{\mathbf{x}}_k^-). \tag{5.53}$$

This is essentially a type of prediction error cost, where the model is trained to predict the *estimated* state. Effectively, the method maximizes $\rho_{\mathbf{x}_1^N|\mathbf{w}}$, evaluated at $\mathbf{x}_1^N = \hat{\mathbf{x}}_1^N$. A potential problem with this approach is that it is not directly constrained by the actual data $\{y_k\}_1^N$. An inaccurate (yet self-consistent) pair of estimates $(\hat{\mathbf{x}}_1^N, \hat{\mathbf{w}})$ could conceivably be obtained as a solution. Nonetheless, this is essentially the approach used in [14] for robust prediction of time series containing outliers.

In the decoupled approach to joint estimation, by separately minimizing each cost with respect to its argument, the values are found that maximize (at least locally) the joint conditional density function. Algorithms that fall into this class include a sequential two-observation form of the dual EKF algorithm [21], and the errors-in-variables (EIV) method applied to batch-style minimization [18, 19]. An alternative approach, referred to as *error coupling*, makes the extra step of taking the errors in the estimates into account. However, this error-coupled approach (investigated in [22]) does not appear to perform reliably, and is not described further in this chapter.

5.3.2 Marginal Estimation Methods

Recall that in marginal estimation, the joint density function is expanded as

$$\rho_{\mathbf{x}_1^N \mathbf{w}|\mathbf{y}_1^N} = \rho_{\mathbf{x}_1^N|\mathbf{w}\mathbf{y}_1^N} \rho_{\mathbf{w}|\mathbf{y}_1^N}, \tag{5.54}$$

and $\hat{\mathbf{x}}_1^N$ is found by maximizing the first factor on the right-hand side, while $\hat{\mathbf{w}}$ is found by maximizing the second factor. Note that only the first factor ($\rho_{\mathbf{x}_1^N|\mathbf{w}\mathbf{y}_1^N}$) is dependent on the state. Hence, maximizing this factor for the state will yield the same solution as when maximizing the joint density (assuming the optimal weights have been found). However, because both factors also depend on $\mathbf{w}$, maximizing the second ($\rho_{\mathbf{w}|\mathbf{y}_1^N}$) alone with respect to $\mathbf{w}$ is *not* the same as maximizing the joint density $\rho_{\mathbf{x}_1^N \mathbf{w}|\mathbf{y}_1^N}$ with respect to $\mathbf{w}$. Nonetheless, the resulting estimates $\hat{\mathbf{w}}$ are consistent and unbiased, if conditions of sufficient excitation are met [37].

The marginal estimation approach is exemplified by the maximum-likelihood approaches [8, 9] and EM approaches [11, 12]. Motivation for these methods usually comes from considering only the marginal density $\rho_{\mathbf{w}|\mathbf{y}_1^N}$ to be the relevant quantity to maximize, rather than the joint density $\rho_{\mathbf{x}_1^N \mathbf{w}|\mathbf{y}_1^N}$. However, in order to maximize the marginal density, it is necessary to generate signal estimates that are invariably produced by maximizing the first term $\rho_{\mathbf{x}_1^N|\mathbf{w}\mathbf{y}_1^N}$.

Maximum-Likelihood Cost To derive a cost function for weight estimation, we further expand the marginal density as

$$\rho_{\mathbf{w}|\mathbf{y}_1^N} = \frac{\rho_{\mathbf{y}_1^N|\mathbf{w}} \rho_{\mathbf{w}}}{\rho_{\mathbf{y}_1^N}}. \tag{5.55}$$

If there is no prior information on $\mathbf{w}$, maximizing this posterior density is equivalent to maximizing the likelihood function $\rho_{\mathbf{y}_1^N|\mathbf{w}}$. Assuming Gaussian statistics, the chain rule for conditional probabilities can be used to express this likelihood function as:

$$\rho_{\mathbf{y}_1^N|\mathbf{w}} = \prod_{k=1}^{N} \frac{1}{\sqrt{2\pi\sigma_{\varepsilon_k}^2}} \exp\left[-\frac{(y_k - \overline{y_{k|k-1}})^2}{2\sigma_{\varepsilon_k}^2} \right], \tag{5.56}$$

where

$$\overline{y_{k|k-1}} \triangleq E[y_k | \{y_t\}_1^{k-1}, \mathbf{w}] \tag{5.57}$$

is the conditional mean (and optimal prediction), and $\sigma_{\varepsilon_k}^2$ is the prediction error variance. Taking the logarithm yields the following *maximum-likelihood* cost function:

$$J^{ml}(\mathbf{w}) = \sum_{k=1}^{N} \left[\log(2\pi\sigma_{\varepsilon_k}^2) + \frac{(y_k - \overline{y_{k|k-1}})^2}{\sigma_{\varepsilon_k}^2} \right]. \tag{5.58}$$

Note that the log-likelihood function takes the same form whether the measurement noise is colored or white. In evaluating this cost function, the term $\overline{y_{k|k-1}} = \mathbf{C}\hat{\mathbf{x}}_k^-$ must be computed. Thus, the signal estimate must be determined as a step to weight estimation. For linear models, this can be done exactly using an ordinary Kalman filter. For nonlinear models, however, the expectation is approximated by an extended Kalman filter, which equivalently attempts to minimize the joint cost $J^j(\mathbf{x}_1^k, \hat{\mathbf{w}})$ defined in Section 5.3.1 by Eq. (5.51).

An iterative maximum-likelihood approach for linear models is described in [7] and [8]; this chapter presents a sequential maximum-likelihood approach for nonlinear models, developed in [21].

Prediction Error Cost Often the variance $\sigma_{\varepsilon_k}^2$ in the maximum-likelihood cost is assumed (incorrectly) to be independent of the weights $\mathbf{w}$ and the time index k. Under this assumption, the log likelihood can be maximized by minimizing the squared prediction error cost function:

$$J^{pe}(\mathbf{w}) = \sum_{k=1}^{N} (y_k - \overline{y_{k|k-1}})^2. \tag{5.59}$$

The basic dual EKF algorithm described in the previous section minimizes this simplified cost function with respect to the weights $\mathbf{w}$, and is an example of a recursive prediction error algorithm [6, 19]. While questionable from a theoretical perspective, these algorithms have been shown in the literature to be quite useful. In addition, they benefit from reduced computational cost, because the derivative of the variance $\sigma_{\varepsilon_k}^2$ with respect to $\mathbf{w}$ is not computed.

EM Algorithm Another approach to maximizing $\rho_{\mathbf{w}|\mathbf{y}_1^N}$ is offered by the expectation-maximization (EM) algorithm [10, 12, 38]. The EM algorithm can be derived by first expanding the log-likelihood as

$$\log \rho_{\mathbf{y}_1^N|\mathbf{w}} = \log \rho_{\mathbf{y}_1^N \mathbf{x}_1^N|\mathbf{w}} - \log \rho_{\mathbf{x}_1^N|\mathbf{w}\mathbf{y}_1^N}. \tag{5.60}$$

Taking the conditional expectation of both sides using the conditional density $\rho_{\mathbf{x}_1^N|\mathbf{w}\mathbf{y}_1^N}$ gives

$$\log \rho_{\mathbf{y}_1^N|\mathbf{w}} = E_{\mathbf{X}|\mathbf{Y}\mathbf{W}}[\log \rho_{\mathbf{y}_1^N \mathbf{x}_1^N|\mathbf{w}}|\mathbf{y}_1^N, \hat{\mathbf{w}}] - E_{\mathbf{X}|\mathbf{Y}\mathbf{W}}[\log \rho_{\mathbf{x}_1^N|\mathbf{w}\mathbf{y}_1^N}|\mathbf{y}_1^N, \hat{\mathbf{w}}], \tag{5.61}$$

where the expectation over $\mathbf{X}$ of the left-hand side has no effect, because $\mathbf{X}$ does not appear in $\log \rho_{\mathbf{y}_1^N|\mathbf{w}}$. Note that the expectation is conditional on a previous estimate of the weights, $\hat{\mathbf{w}}$. The second term on the right is concave by Jensen's inequality [39],[3] so choosing $\mathbf{w}$ to maximize the first term on the right-hand side alone will always increase the log-likelihood on the left-hand side. Thus, the EM algorithm repeatedly maximizes $E_{\mathbf{X}|\mathbf{Y}\mathbf{W}}[\log \rho_{\mathbf{y}_1^N \mathbf{x}_1^N|\mathbf{w}}|\mathbf{y}_1^N, \hat{\mathbf{w}}]$ with respect to $\mathbf{w}$, each time setting $\hat{\mathbf{w}}$ to the new maximizing value. The procedure results in maximizing the original marginal density $\rho_{\mathbf{y}_1^N|\mathbf{w}}$.

For the white-noise case, it can be shown (see [12, 22]) that the EM cost function is

$$J^{em} = E_{\mathbf{X}|\mathbf{Y}\mathbf{W}}\left[\sum_{k=1}^{N}\left\{\log(2\pi\sigma_n^2) + \frac{(y_k - \mathbf{C}\mathbf{x}_k)^2}{\sigma_n^2} + \log(2\pi|\mathbf{R}^{\mathbf{v}}|) + (\mathbf{x}_k - \mathbf{x}_k^-)^T(\mathbf{R}^{\mathbf{v}})^{-1}(\mathbf{x}_k - \mathbf{x}_k^-)\right\}z \,\middle|\, \mathbf{y}_1^N, \hat{\mathbf{w}}\right], \tag{5.62}$$

where $\mathbf{x}_k^- \triangleq \mathbf{F}(\mathbf{x}_{k-1}, \mathbf{w})$, as before. The evaluation of this expectation is computable on a term-by-term basis (see [12] for the linear case). However, for the sake of simplicity, we present here the resulting

[3] Jensen's inequality states that $E[g(x)] \leq g(E[x])$ for a concave function $g(\cdot)$.

expression for the special case of time-series estimation, represented in Eq. (5.37). As shown in [22], the expectation evaluates to

$$J^{em} = N\log(4\pi^2\sigma_v^2\sigma_n^2) + \sum_{k=1}^{N}\left[\frac{(y_k - \hat{x}_{k|N})^2 + p_{k|N}}{\sigma_n^2} + \frac{(\hat{x}_{k|N} - \hat{x}_{k|N}^-)^2 + p_{k|N} - 2p_{k|N}^\dagger + p_{k|N}^-}{\sigma_v^2}\right], \tag{5.63}$$

where $\hat{x}_{k|N}$ and $p_{k|N}$ are defined as the conditional mean and variance of x_k given $\hat{\mathbf{w}}$ and *all* the data, $\{y_k\}_1^N$. The terms $\hat{x}_{k|N}^-$ and $p_{k|N}^-$ are the conditional mean and variance of $x_k^- = f(\mathbf{x}_{k-1}, \mathbf{w})$, given all the data. The additional term $p_{k|N}^\dagger$ represents the cross-variance of x_k and x_k^-, conditioned on all the data. Again we see that determining state estimates is a necessary step to determining the weights. In this case, the estimates $\hat{x}_{k|N}$ are found by minimizing the joint cost $J^j(\mathbf{x}_1^N, \hat{\mathbf{w}})$, which can be approximated using an extended Kalman smoother. A sequential version of EM can be implemented by replacing $\hat{x}_{k|N}$ with the usual causal estimates $\hat{x}_k$, found using the EKF.

Summary of Cost Functions The various cost functions given in this section are summarized in Table 5.4. No explicit signal estimation cost is given for the marginal estimation methods, because signal estimation is only an *implicit* step of the marginal approach, and uses the joint cost $J^j(\mathbf{x}_1^N, \hat{\mathbf{w}})$. These cost functions, combined with specific optimization methods, lead to the variety of algorithms that appear in the literature.

Table 5.4 Summary of dual estimation cost functions

	Symbol	Name of cost	Density	Eq.
Joint	$J^j(\mathbf{x}_1^N, \mathbf{w})$	Joint	$\rho_{\mathbf{x}_1^N \mathbf{w}\|\mathbf{y}_1^N}$	(5.50)
	$J^j(\mathbf{x}_1^N, \hat{\mathbf{w}})$	Joint signal	$\rho_{\mathbf{x}_1^N \mathbf{w}\|\mathbf{y}_1^N}$	(5.51)
	$J^j(\hat{\mathbf{x}}_1^N, \mathbf{w})$	Joint weight	$\rho_{\mathbf{x}_1^N \mathbf{w}\|\mathbf{y}_1^N}$	(5.52)
	$J_i^j(\hat{\mathbf{x}}_1^N, \mathbf{w})$	Joint weight (independent)	$\rho_{\mathbf{x}_1^N\|\mathbf{w}}$	(5.53)
Marginal	$J^{pe}(\mathbf{w})$	Prediction error	$\sim\rho_{\mathbf{w}\|\mathbf{y}_1^N}$	(5.59)
	$J^{ml}(\mathbf{w})$	Maximum likelihood	$\rho_{\mathbf{w}\|\mathbf{y}_1^N}$	(5.58)
	$J^{em}(\mathbf{w})$	EM	*n.a.*	(5.62)

In the next section, we shall show how each of these cost functions can be minimized using a general dual EKF-based approach.

5.3.3 Dual EKF Algorithms

In this section, we show how the dual EKF algorithm can be modified to minimize any of the cost functions discussed earlier. Recall that the basic dual EKF as presented in Section 5.2.3 minimized the prediction error cost of Eq. (5.59). As was shown in the last section, all approaches use the same joint cost function for the state-estimation component. Thus, the state EKF remains unchanged. Only the weight EKF must be modified. We shall show that this involves simply redefining the error term $\mathbf{e}_k$.

To develop the method, consider again the general state-space formulation for weight estimation (Eq. (5.11)):

$$\mathbf{w}_{k+1} = \mathbf{w}_k + \mathbf{r}_k, \tag{5.64}$$

$$\mathbf{d}_k = \mathbf{G}(\mathbf{x}_k, \mathbf{w}_k) + \mathbf{e}_k. \tag{5.65}$$

We may reformulate this state-space representation as

$$\mathbf{w}_k = \mathbf{w}_{k-1} + \mathbf{r}_k, \tag{5.66}$$

$$0 = -\mathbf{e}_k, + \mathbf{e}_k, \tag{5.67}$$

where $\mathbf{e}_k = \mathbf{d}_k - \mathbf{G}(\mathbf{x}_k, \mathbf{w}_k)$ and the target "observation" is fixed at zero. This *observed error* formulation yields the exact same set of Kalman equations as before, and hence minimizes the same prediction error cost, $J(\mathbf{w}) = \sum_{t=1}^{k} [\mathbf{d}_t - \mathbf{G}(\mathbf{x}_t, \mathbf{w})]^T (\mathbf{R}^e)^{-1} [\mathbf{d}_t - \mathbf{G}(\mathbf{x}_t, \mathbf{w})] = \sum_{t=1}^{k} J_t$. However, if we consider the modified-Newton algorithm interpretation, it can be shown [22] that the EKF weight filter is also equivalent to the recursion

$$\hat{\mathbf{w}}_k = \hat{\mathbf{w}}_k^- + \mathbf{P}_{\mathbf{w}_k} (\mathbf{C}_k^{\mathbf{w}})^T (\mathbf{R}^e)^{-1} (0 + \mathbf{e}_k), \tag{5.68}$$

where

$$\mathbf{C}_k^{\mathbf{w}} \triangleq \left. \frac{\partial(-\mathbf{e}_k)}{\partial \mathbf{w}} \right|_{\mathbf{w}=\mathbf{w}_k}^{T} \tag{5.69}$$

and

$$\mathbf{P}_{\mathbf{w}_k}^{-1} = (\lambda^{-1}\mathbf{P}_{\mathbf{w}_{k-1}})^{-1} + (\mathbf{C}_k^{\mathbf{w}})^T(\mathbf{R}^{\mathrm{e}})^{-1}\mathbf{C}_k^{\mathbf{w}}. \tag{5.72}$$

The weight update in Eq. (5.68) is of the form

$$\hat{\mathbf{w}}_k = \hat{\mathbf{w}}_k^- - \mathbf{S}_k \nabla_{\mathbf{w}} J(\hat{\mathbf{w}}_k^-)^T, \tag{5.73}$$

where $\nabla_{\mathbf{w}} J$ is the gradient of the cost J with respect to $\mathbf{w}$, and $\mathbf{S}_k$ is a symmetric matrix that approximates the inverse Hessian of the cost. Both the gradient and Hessian are evaluated at the previous value of the weight estimate. Thus, we see that by using the observed error formulation, it is possible to redefine the error term $\mathfrak{e}_k$, which in turn allows us to minimize an arbitrary cost function that can be expressed as a sum of instantaneous terms $J_k = \mathfrak{e}_k^T \mathfrak{e}_k$. This basic idea was presented by Puskorius and Feldkamp [40] for minimizing an entropic cost function; see also Chapter 2. Note that $J_k = \mathfrak{e}_k^T \mathfrak{e}_k$ does not uniquely specify $\mathfrak{e}_k$, which can be vector-valued. The error must be chosen such that the gradient and inverse Hessian approximations (Eqs. (5.70) and (5.72)) are consistent with the desired batch cost.

In the following sections, we give the exact specification of the error term (and corresponding gradient $\mathbf{C}_k^{\mathbf{w}}$) necessary to modify the dual EKF algorithm to minimize the different cost functions. The original set of dual EKF equations given in Table 5.3 remains the same, with only $\mathfrak{e}_k$ being redefined. Note that for each case, the full evaluation of $\mathbf{C}_k^{\mathbf{w}}$ requires taking recursive gradients. The procedure for this is analogous to that taken in Section 5.2.3. Furthermore, we restrict ourselves to the autoregressive time-series model with state-space representation given in Eqs. (5.38) and (5.39).

Joint Estimation Forms The corresponding weight cost function (see also Eq. (5.52)) and error terms are given in Table 5.5. Note that this represents a special *two-observation* form of the weight filter, where

$$\hat{x}_t^- = f(\hat{\mathbf{x}}_{t-1}, \mathbf{w}), \qquad e_k \triangleq (y_k - \hat{x}_k), \qquad \tilde{\hat{x}}_k \triangleq (\hat{x}_k - \hat{x})_k^-,$$

Note that this dual EKF algorithm represents a sequential form of the *decoupled* approach to joint optimization; that is, the two EKFs minimize the overall joint cost function by alternately optimizing one argument at a

Table 5.5 Joint cost function observed error terms for the dual EKF weight filter

$$J^j(\hat{\mathbf{x}}_1^k, \mathbf{w}) = \sum_{t=1}^{k} \left[\frac{(y_t - \hat{x}_t)^2}{\sigma_n^2} + \frac{(\hat{x}_t - \hat{x}_t^-)^2}{\sigma_v^2} \right], \tag{5.74}$$

$$\mathbf{e}_k \triangleq \begin{bmatrix} \sigma_n^{-1} e_k \\ \sigma_v^{-1} \tilde{x}_k \end{bmatrix}, \quad \text{with} \quad \mathbf{C}_k^{\mathbf{w}} = -\begin{bmatrix} \sigma_n^{-1} \nabla_{\mathbf{w}}^T e_k \\ \sigma_v^{-1} \nabla_{\mathbf{w}}^T \tilde{x}_k] \end{bmatrix}.$$

time while the other argument is fixed. A *direct* approach found using the *joint* EKF is described later in Section 5.3.4.

Marginal Estimation Forms–Maximum-Likelihood Cost The corresponding weight cost function (see Eq. (5.58)) and error terms are given in Table 5.6, where

$$\varepsilon_k = y_k - \hat{x}_k^-, \qquad l_{\varepsilon,k} = \frac{\sigma_{\varepsilon,k}^2}{3\varepsilon_k^2 - 2\sigma_{\varepsilon,k}^2}.$$

Note that the prediction error variance is given by

$$\sigma_{\varepsilon_k}^2 = E[(y_k - \overline{y_{k|k-1}})^2 | \{y_t\}_1^{k-1}, \mathbf{w}] \tag{5.75a}$$

$$= E[(n_k + x_k - \hat{x}_k^-)^2 | \{y_t\}_1^{k-1}, \mathbf{w}] \tag{5.75b}$$

$$= \sigma_n^2 + \mathbf{C}\mathbf{P}_k^- \mathbf{C}^T, \tag{5.75c}$$

where $\mathbf{P}_k^-$ is computed by the Kalman signal filter (see [22] for a discussion of the selection and interpretation of $l_{\varepsilon,k}$).

Table 5.6 Maximum-likelihood cost function observed error terms for dual EKF weight filter

$$J^{ml}(\mathbf{w}) = \sum_{k=1}^{N} \left[\log(2\pi\sigma_{\varepsilon_k}^2) + \frac{(y_k - \hat{x}_k^-)^2}{\sigma_{\varepsilon_k}^2} \right],$$

$$\mathbf{e}_k \triangleq \begin{bmatrix} (l_{\varepsilon,k})^{1/2} \\ \sigma_{\varepsilon_k}^{-1} \varepsilon_k \end{bmatrix}, \quad \mathbf{C}_k^{\mathbf{w}} = \begin{bmatrix} -\frac{1}{2} \frac{(l_{\varepsilon,k})^{-1/2}}{\sigma_{\varepsilon_k}^2} \nabla_{\mathbf{w}}^T (\sigma_{\varepsilon_k}^2) \\ -\frac{1}{\sigma_{\varepsilon_k}} \nabla_{\mathbf{w}}^T \varepsilon_k + \frac{\varepsilon_k}{2(\sigma_{\varepsilon_k}^2)^{3/2}} \nabla_{\mathbf{w}}^T (\sigma_{\varepsilon_k}^2) \end{bmatrix}.$$

Table 5.7 Prediction error cost function observed error terms for the dual EKF weight filter

$$J^{pe}(\mathbf{w}) = \sum_{k=1}^{N} \varepsilon_k^2 = (y_k - \hat{x}_k^-)^2, \tag{5.76}$$

$$\mathfrak{e}_k \stackrel{\Delta}{=} \varepsilon_k = (y_k - \hat{x}_k^-), \qquad \mathbf{C}_k^{\mathbf{w}} = -\nabla_{\mathbf{w}} \varepsilon_k = \mathbf{C} \frac{\partial \hat{\mathbf{x}}_k^-}{\partial \mathbf{w}} \bigg|_{\mathbf{w}=\hat{\mathbf{w}}_k^-}.$$

Marginal Estimation Forms–Prediction Error Cost If $\sigma_{\varepsilon_k}^2$ is assumed to be independent of $\mathbf{w}$, then we are left with the formulas corresponding to the original basic dual EKF algorithm (for the time-series case); see Table 5.7.

Marginal Estimation Forms–EM Cost The dual EKF can be modified to implement a sequential EM algorithm. Note that the M-step, which relates to the weight filter, corresponds to a *generalized* M-step, in which the cost function is *decreased* (but not necessarily minimized) at each iteration. The formulation is given in Table 5.8, where $\tilde{\hat{x}}_{k|k} = \hat{x}_k - \hat{x}_{k|k}^-$. Note that $J_k^{em}(\mathbf{w})$ was specified by dropping terms in Eq. (5.63) that are independent of the weights (see [22]). While $\hat{x}_k$ are found by the usual state EKF, the variance terms $p_{k|k}^{\dagger}$, and $p_{k|k}^-$, as well as $\hat{x}_{k|k}^-$ (a noncausal prediction), are not typically computed in the normal implementation of the state EKF. To compute these, the state vector is augmented by one additional lagged value of the signal:

$$\mathbf{x}_k^+ = \begin{bmatrix} \mathbf{x}_k \\ x_{k-M} \end{bmatrix} = \begin{bmatrix} x_k \\ \mathbf{x}_{k-1} \end{bmatrix}, \tag{5.78}$$

Table 5.8 EM cost function observed error terms for the dual EKF weight filter

$$J_k^{em}(\mathbf{w}) = \frac{(\hat{x}_k - \hat{x}_{k|k}^-)^2 - 2p_{k|k}^{\dagger} + p_{k|k}^-}{\sigma_v^2}, \tag{5.77}$$

$$\mathfrak{e}_k = \begin{bmatrix} \sigma_v^{-1} \tilde{\hat{x}}_{k|k} \\ \sqrt{-2}\sigma_v^{-1}(p_{k|k}^{\dagger})^{1/2} \\ \sigma_v^{-1}(p_{k|k}^-)^{1/2} \end{bmatrix}, \qquad \mathbf{C}_k^{\mathbf{w}} = \begin{bmatrix} -\frac{1}{\sigma_v} \nabla_{\mathbf{w}}^T \tilde{\hat{x}}_{k|k} \\ -\frac{\sqrt{-2}(p_{k|k}^{\dagger})^{-1/2}}{2\sigma_v} \nabla_{\mathbf{w}}^T p_{k|k}^{\dagger} \\ -\frac{(p_{k|k}^-)^{-1/2}}{2\sigma_v} \nabla_{\mathbf{w}}^T p_{k|k}^- \end{bmatrix}.$$

The state Kalman filter is then modified by adding a final zero element to the vectors **B** and **C** (see Eqs. (5.38) and (5.39)), and the linearized state transition matrix is redefined as

$$\mathbf{A}_k \triangleq \begin{bmatrix} & \nabla_{\mathbf{x}}^T f \\ \mathbf{I} & 0 \end{bmatrix}.$$

Now the estimate $\hat{\mathbf{x}}_k^+$ will contain $\hat{\mathbf{x}}_{k-1|k}$ in its last M elements. As shown in [22], the variance terms are then approximated by

$$p_{k|k}^- = \mathbf{C}\mathbf{A}_{k-1|k}(\mathbf{P}_{k-1|k})\mathbf{A}_{k-1|k}^T\mathbf{C}^T, \qquad p_{k|k}^\dagger = \mathbf{C}(\mathbf{P}_k^\sharp)\mathbf{A}_{k-1|k}^T\mathbf{C}^T, \tag{5.79}$$

where the covariance $\mathbf{P}_{k-1|k}$ is provided as the lower right block of the augmented covariance $\mathbf{P}_k^+$, and $\mathbf{P}_k^\sharp$ is the upper right block of $\mathbf{P}_k^+$. The usual error covariance $\mathbf{P}_k$ is provided in the upper left block of $\mathbf{P}_k^+$. Furthermore, $\mathbf{A}_{k-1|k}$ is found by linearizing $f(\cdot)$ at $\hat{\mathbf{x}}_{k-1|k}$. The noncausal prediction $\hat{x}_{k|k}^- = E[f(\mathbf{x}_{k-1}, \mathbf{w})|\{y_t\}_1^k, \hat{\mathbf{w}}] \approx f(\hat{\mathbf{x}}_{k-1|k}, \mathbf{w})$.

Finally, the necessary gradient terms in the dual EKF algorithm can be evaluated as follows:

$$\nabla_{\mathbf{w}}^T \tilde{\hat{x}}_{k|k} = -\nabla_{\mathbf{w}} \hat{x}_{k|k}^- = -\nabla_{\mathbf{w}} f(\hat{\mathbf{x}}_{k-1|k}, \mathbf{w}), \tag{5.80}$$

which is evaluated at $\hat{\mathbf{w}}_k$. The ith element of the gradient vector $\nabla_{\mathbf{w}} p_{k|k}^-$ is constructed from the expression

$$\frac{\partial p_{k|k}^-}{\partial w^{(i)}} = \mathbf{C}\left[\frac{\partial \mathbf{A}_{k-1|k}}{\partial w^{(i)}}(\mathbf{P}_{k-1|k})\mathbf{A}_{k-1|k}^T + \mathbf{A}_{k-1|k}(\mathbf{P}_{k-1|k})\frac{\partial \mathbf{A}_{k-1|k}^T}{\partial w^{(i)}}\right]\mathbf{C}^T, \tag{5.81}$$

and the elements of the gradient $\nabla_{\mathbf{w}} p_{k|k}^\dagger$ are given by

$$\frac{\partial p_{k|k}^\dagger}{\partial w^{(i)}} = \mathbf{C}\left[\frac{\partial \mathbf{A}_{k-1|k}}{\partial w^{(i)}}(\mathbf{P}_k^\sharp)^T\right]\mathbf{C}^T. \tag{5.82}$$

Note that all components in the cost function are recursive functions of $\hat{\mathbf{w}}$, but not of $\mathbf{w}$. Hence, no recurrent derivative computations are required for the EM algorithm. Furthermore, it can be shown that the actual error variances $p_{k|k}^-$ and $p_{k|k}^\dagger$ (not their gradients) cancel out in the formula for $\mathbf{C}_k^{\mathbf{w}}$, and thus should be replaced with large constant values to obtain a good Hessian approximation.

5.3.4 Joint EKF

In the previous section, the dual EKF was modified to minimize the joint cost function. This implementation represented a *decoupled* type of approach, in which *separate* state-space representations were used to estimate x_k and $\mathbf{w}$. An alternative *direct* approach is given by the joint EKF, which generates *simultaneous* MAP estimates of $\mathbf{x}_k$ and $\mathbf{w}$. This is accomplished by defining a new joint state-space representation with concatenated state:

$$\mathbf{z}_k = \begin{bmatrix} \mathbf{x}_k \\ \mathbf{w}_k \end{bmatrix}. \tag{5.83}$$

It is clear that maximizing the density $\rho_{\mathbf{z}_k|\mathbf{y}_1^k}$ is equivalent to maximizing $\rho_{\mathbf{x}_k \mathbf{w}|\mathbf{y}_1^k}$. Hence, the MAP-optimal estimate of $\mathbf{z}_k$ will contain the values of x_k and $\mathbf{w}_k$ that minimize the batch cost $J(\mathbf{x}_1^k, \mathbf{w})$. Running the single EKF with this state vector provides a sequential estimation algorithm. The joint EKF first appeared in the literature for the estimation of linear systems (in which there is a bilinear relation between the states and weights) [1, 2]. The general equations for nonlinear systems are given in Table 5.9.

Note that because the gradient of $f(\mathbf{z})$ with respect to $\mathbf{w}$ is taken with the other elements (namely, $\hat{\mathbf{x}}_k$) fixed, it will *not* involve recursive derivatives of $\hat{\mathbf{x}}_k$ with respect to $\mathbf{w}$. This fact is cited in [3] and [6] as a potential source of convergence problems for the joint EKF. Additional results and citations in [5] corroborate the difficulties of the approach, although the cause of divergence is linked there to the linearization of the coupled system, rather than the lack of recurrent derivatives (note that no divergence problems were encountered in preparing the experimental results in this chapter). Although the use of recurrent derivatives is suggested in [3] and [6], there is no theoretical justification for this. In summary, the joint EKF provides an alternative to the dual EKF for sequential minimization of the joint cost function. Note that the joint EKF cannot be readily adapted to minimize other cost functions discussed in this chapter.

5.4 DUAL EKF VARIANCE ESTIMATION

The implementation of the EKF requires the noise variance, σ_v^2 and σ_n^2 as parameters in the algorithm. Often these can be determined from physical knowledge of the problem (e.g., sensor accuracy or ambient noise

Table 5.9 The joint extended Kalman filter equations (time-series case)

Initialize with

$$\hat{\mathbf{z}}_0 = E[\mathbf{z}_0], \tag{5.84}$$

$$\mathbf{P}_0 = E[(\mathbf{z}_0 - \hat{\mathbf{z}}_0)(\mathbf{z}_0 - \hat{\mathbf{z}}_0)^T]. \tag{5.85}$$

For $k \in \{1, \ldots, \infty\}$, the time-update equations of the Kalman filter are

$$\hat{\mathbf{z}}_k^- = \bar{\mathbf{F}}(\hat{\mathbf{z}}_{k-1}), \tag{5.86}$$

$$\mathbf{P}_k^- = \bar{\mathbf{A}}_{k-1}\mathbf{P}_{k-1}\bar{\mathbf{A}}_{k-1}^T + \bar{\mathbf{V}}_k, \tag{5.87}$$

and the measurement update equations are

$$\bar{\mathbf{K}}_k = \mathbf{P}_k^- \bar{\mathbf{C}}^T (\bar{\mathbf{C}}\mathbf{P}_k^- \bar{\mathbf{C}}^T + \sigma_n^2)^{-1}, \tag{5.88}$$

$$\hat{\mathbf{z}}_k = \hat{\mathbf{z}}_k^- + \bar{\mathbf{K}}_k(y_k - \bar{\mathbf{C}}\hat{\mathbf{z}}_k^-), \tag{5.89}$$

$$\mathbf{P}_k = (\mathbf{I} - \bar{\mathbf{K}}_k\bar{\mathbf{C}})\mathbf{P}_k^-, \tag{5.90}$$

where

$$\bar{\mathbf{V}}_k \triangleq \operatorname{Cov}\begin{bmatrix} \mathbf{B}v_k \\ \mathbf{u}_k \end{bmatrix} = \begin{bmatrix} \mathbf{B}\sigma_v^2\mathbf{B}^T & 0 \\ 0 & \mathbf{R}^{\mathrm{r}} \end{bmatrix}, \tag{5.91}$$

$$\bar{\mathbf{F}}(\mathbf{z}_{k-1}) \triangleq \begin{bmatrix} \mathbf{F}(\mathbf{x}_{k-1}, \mathbf{w}_{k-1}) \\ \mathbf{I}\cdot\mathbf{w}_{k-1} \end{bmatrix}, \qquad \bar{\mathbf{C}} \triangleq [\mathbf{C} \quad 0 \quad \ldots \quad 0], \tag{5.92}$$

$$\bar{\mathbf{A}}_k \triangleq \left.\frac{\partial \bar{\mathbf{F}}(\mathbf{z})}{\partial \mathbf{z}}\right|_{\mathbf{z}=\hat{\mathbf{z}}_k} = \begin{bmatrix} \begin{bmatrix} \dfrac{\partial f(\hat{\mathbf{x}}_k, \mathbf{w})^T}{\partial \mathbf{x}} \\ \mathbf{I} \quad 0 \end{bmatrix} & \begin{bmatrix} \dfrac{\partial f(\hat{\mathbf{x}}_k, \mathbf{w})^T}{\partial \mathbf{w}} \\ 0 \quad 0 \end{bmatrix} \\ 0 & \mathbf{I} \end{bmatrix}. \tag{5.93}$$

measurements). However, if the variances are unknown, their values can be estimated within the dual EKF framework using cost functions similar to those derived in Section 5.3. A full treatment of the variance estimation filter is presented in [22]; here we focus on the maximum-likelihood cost function

$$J^{ml}(\sigma^2) = \sum_{k=1}^{N}\left[\log(2\pi\sigma_{\varepsilon_k}^2) + \frac{(y_k - \overline{y_{k|k-1}})^2}{\sigma_{\varepsilon_k}^2}\right]. \tag{5.94}$$

Note that this cost function is identical to the weight-estimation cost function, except that the argument is changed to emphasize the estimation of the unknown variances.

As with the weight filter, a modified-Newton algorithm can be found for each variance by using an observed error form of the Kalman filter and modeling the variances as

$$\sigma_{k+1}^2 = \sigma_k^2 + r_k, \tag{5.95}$$

$$0 = \mathfrak{e}_k + e_k, \tag{5.96}$$

which gives a one-dimensional state-space representation. However, a peculiar difficulty in the estimation of variances is that these quantities must be positive-valued. Because this constraint is not built explicitly into the cost functions, it is conceivable that negative values can be obtained.

One solution to this problem (inspired by [41]) is to estimate $l \triangleq \log(\sigma^2)$ instead. Negative values of l map to small positive values of σ^2, and $l = -\infty$ maps to $\sigma^2 = 0$. The logarithm is a monotonic function, so a one-to-one mapping exists between the optimal value of l and the optimal value of σ^2. An additional benefit of the logarithmic function is that it expands the dynamic range near $\sigma^2 = 0$, where the solution is more likely to reside; this can improve the numerical properties of the optimization.

Of course, this new formulation requires computing the gradients and Hessians of the cost J with respect to l. Fortunately, the change is fairly straightforward; these expressions are simple functions of the derivatives with respect to σ^2. The variance estimation filter is given in Table 5.10. Note that the dimension of the state space is 1 in the case of variance estimation, while the observation $\mathfrak{e}_k$ is generally multidimensional. For this reason, the covariance form of the KF is more efficient than the forms shown earlier for signal or weight estimation, which employ the matrix inversion lemma and use a Kalman gain term. This form of the variance filter is used in the experiments in the next section.

In Table 5.10, the mean $\overline{y_{k|k-1}}_k$ and variance $\sigma_{\varepsilon_k}^2$ are computed in the same way as before (see Eq. (5.75)), except that the unknown variance σ^2 is now an additional conditioning argument in the expectations. Hence, the derivatives are

$$\frac{\partial \overline{y_{k|k-1}}_k}{\partial \sigma^2} = -\frac{\partial \hat{x}_k^-}{\partial \sigma^2}, \tag{5.103}$$

$$\frac{\partial \sigma_{\varepsilon_k}^2}{\partial \sigma^2} = \frac{\partial \sigma_n^2}{\partial \sigma^2} + \frac{\partial \mathbf{P}_k-(1,1)}{\partial \sigma^2}, \tag{5.104}$$

where $\mathbf{P}_k^-(1,1)$ is the upper left element of the matrix, and $\partial\sigma_n^2/\partial\sigma^2$ is either 1 or 0, depending on whether σ^2 is σ_n^2 or σ_v^2. The other derivatives

Table 5.10 Variance estimation filter of the dual EKF

Initialize with

$$\hat{\sigma}_0^2 = E[\sigma^2], \qquad p_{\sigma_0} = E[(\sigma^2 - \hat{\sigma}_0^2)(\sigma^2 - \hat{\sigma}_0^2)^T].$$

For $k \in \{1, \ldots, \infty\}$, the time-update equations for the variance filter are

$$\hat{\sigma}_k^{2-} = \hat{\sigma}_{k-1}^2, \tag{5.97}$$

$$p_{\sigma_k}^- = p_{\sigma_{k-1}} + \sigma_u^2, \qquad \sigma_u^2 = \left(\frac{1}{\lambda} - 1\right) p_{\sigma_{k-1}}, \tag{5.98}$$

and the measurement equations are

$$p_{\sigma_k} = \left[(p_{\sigma_k}^-)^{-1} + (\mathbf{C}_k^\sigma)^T \sigma_r^{-2} \mathbf{C}_k^\sigma (\hat{\sigma}_k^{2-})^2 + (\mathbf{C}_k^\sigma)^T \sigma_r^{-2} \mathbf{e}_k \hat{\sigma}_k^{2-} \right]^{-1}, \tag{5.99}$$

$$\hat{l}_k^- = \log(\hat{\sigma}_k^{2-}), \tag{5.100}$$

$$\hat{l}_k = \hat{l}_k^- + p_{\sigma_k} (\mathbf{C}_k^\sigma)^T \sigma_r^{-2} \mathbf{e}_k \hat{\sigma}_k^{2-}, \tag{5.101}$$

$$\hat{\sigma}_k^2 = e^{\hat{l}_k}. \tag{5.102}$$

The observed error term and gradient are

$$\mathbf{e}_k \triangleq \begin{bmatrix} (l_{\varepsilon,k})^{1/2} \\ \sigma_{\varepsilon_k}^{-1} \varepsilon_k \end{bmatrix}, \qquad \mathbf{C}_k^\sigma = \begin{bmatrix} -\dfrac{1}{2} \dfrac{(l_{\varepsilon,k})^{-1/2}}{\sigma_{\varepsilon_k}^2} \dfrac{\partial \sigma_{\varepsilon_k}^2}{\partial \sigma^2} \\ -\dfrac{1}{\sigma_{\varepsilon_k}} \dfrac{\partial \varepsilon_k}{\partial \sigma^2} + \dfrac{\varepsilon_k}{2(\sigma_{\varepsilon_k}^2)^{(3/2)}} \dfrac{\partial \sigma_{\varepsilon_k}^2}{\partial \sigma^2} \end{bmatrix},$$

where σ^2 represents either σ_n^2 or σ_v^2, and where $\dfrac{\partial \varepsilon_k}{\partial \sigma^2} = -\dfrac{\partial \overline{y_{k|k-1}}_k}{\partial \sigma^2}$.

---

are computed by taking the derivative of the Kalman filter equations with respect to either variance term (represented by σ^2). This results in the following system of recursive equations:

$$\frac{\partial \hat{\mathbf{x}}_{k+1}^-}{\partial \sigma^2} = \frac{\partial \mathbf{F}(\hat{\mathbf{x}}, \hat{\mathbf{w}})}{\partial \hat{\mathbf{x}}_k} \frac{\partial \hat{\mathbf{x}}_k}{\partial \sigma^2}, \tag{5.105}$$

$$\frac{\partial \hat{\mathbf{x}}_k}{\partial \sigma^2} = (\mathbf{I} - \mathbf{K}_k^{\mathbf{x}} \mathbf{C}) \frac{\partial \hat{\mathbf{x}}_k^-}{\partial \sigma^2} + \frac{\partial \mathbf{K}_k^{\mathbf{x}}}{\partial \sigma^2} (y_k - \mathbf{C}\hat{\mathbf{x}}_k^-), \tag{5.106}$$

where $\partial \mathbf{F}(\hat{\mathbf{x}}, \hat{\mathbf{w}})/\partial \hat{\mathbf{x}}_k$ is evaluated at $\hat{\mathbf{w}}_k$, and represents a static linearization of the neural network. Note that $[\partial \mathbf{F}(\hat{\mathbf{x}}, \hat{\mathbf{w}})/\partial \hat{\mathbf{w}}_k][\partial \hat{\mathbf{w}}/\partial \sigma^2]$ does not appear in Eq. (5.105), under the assumption that $\partial \hat{\mathbf{w}}/\partial \sigma^2 = 0$. The last term in

Eq. (5.106) may be dropped if we assume that the Kalman gain $\mathbf{K}^{\mathbf{x}}$ is independent of σ^2. However, for accurate computation of the recursive derivatives, $\partial \mathbf{K}_k^{\mathbf{x}} / \partial \sigma^2$ must be calculated; this is shown along with the derivative $\partial \mathbf{P}_k^-(1, 1)/\partial \sigma^2$ in Appendix A.

5.5 APPLICATIONS

In this section, we present the results of using the dual EKF methods on a number of different applications.

5.5.1 Noisy Time-Series Estimation and Prediction

For the first example, we return to the noisy time-series example of Section 5.2. Figure 5.4 compares the performance of the various dual EKF and joint EKF methods presented in this chapter. Box plots are used to show the mean, median, and quartile values based on 10 different runs (note that the higher mean for the maximum-likelihood cost is a consequence of a single outlier). The figure also compares performances for both known variances and estimated variances. The results in the example are fairly consistent with our findings on a number of controlled

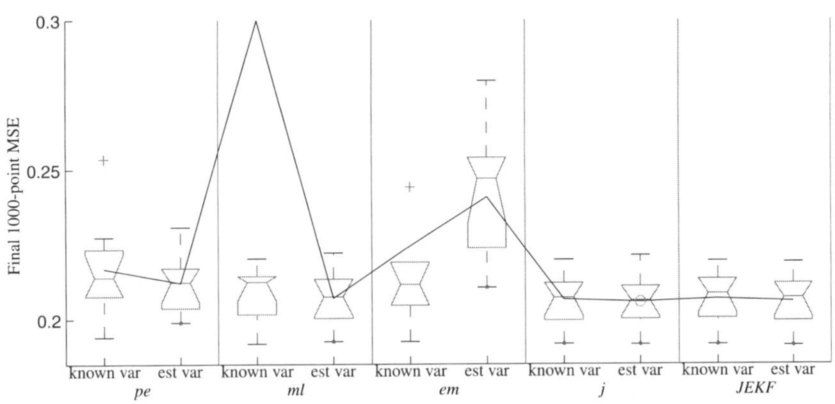

Figure 5.4 Performance of the dual EKF and joint EKF on a chaotic neural network time series described by Eqs. (5.37). MSE values are computed for the final 1000 points of the series. For each cost function, the horizontal lines indicate the median, upper and lower quartile values, and range of the overall NMSE. The mean values are linked by thin lines, and the algorithm with the lowest mean MSE is indicated by a superimposed circle.

experiments using different time series and parameter settings [22]. For both white and colored noises, the maximum-likelihood weight-estimation cost $J^{ml}(\mathbf{w})$ generally produces the best results, but often exhibits instability. This fact is analyzed in [22], and reduces the desirability of using this cost function. The joint cost function $J^{j}(\mathbf{w})$ has better stability properties and produces excellent results for colored measurement noise. For white noise, the prediction error cost performs nearly as well as $J^{ml}(\mathbf{w})$, but without stability problems. Hence, the dual EKF cost functions $J^{pe}(\mathbf{w})$ and $J^{j}(\mathbf{w})$ are generally the best choices for white and colored measurement noise, respectively. The joint EKF and dual EKF perform similarly in many cases, although the joint EKF is considerably less robust to inaccuracies in the assumed model structure and noise variances.

Chaotic Hénon Map As a second time-series example, we consider modeling the long-term behavior of the chaotic Hénon map:

$$a_{k+1} = 1 - 1.4a_k^2 + b_k, \tag{5.107}$$

$$b_{k+1} = 0.3b_k. \tag{5.108}$$

To obtain a one-dimensional time series for the following experiment, the signal is defined as $x_k = a_k$. The phase plot of x_{k+1} versus x_k (Fig. 5.5*a*) shows the chaotic attractor. A neural network (5-7-1) can easily be trained as a single-step predictor on this clean signal. The network is then iterated–feeding back the predictions of the network as future inputs–to produce the attractor shown in Figure 5.5*b*. However, if the signal is corrupted by white noise at 10 dB SNR (Fig. 5.5*c*), and a neural network with the same architecture is trained on these noisy data, the dynamics are not adequately captured. The iterated predictions exhibit limit-cycle behavior with far less complexity.

In contrast, using the dual EKF to train the neural network on the noisy data captures significantly more of the chaotic dynamics, as shown in Figure 5.5*d*. Here, $J^{pe}(\mathbf{w})$ is used for weight estimation, and the maximum-likelihood cost is used for estimating σ_v^2. The measurement-noise variance is assumed to be known. Parameter covariances are initialized at 0.1, and the initial signal covariance is $\mathbf{P}_{\mathbf{x}_0} = \mathbf{I}$. Forgetting factors are $\lambda_\mathbf{w} = 0.9999$ and $\lambda_{\sigma_v^2} = 0.9993$. Although the attractor is not reproduced with total fidelity, its general structure has been extracted from the noisy data.

This example also illustrates an interesting interpretation of dual EKF prediction training. During the training process, estimations from the output of the predictor are fed back as inputs, which are optimally

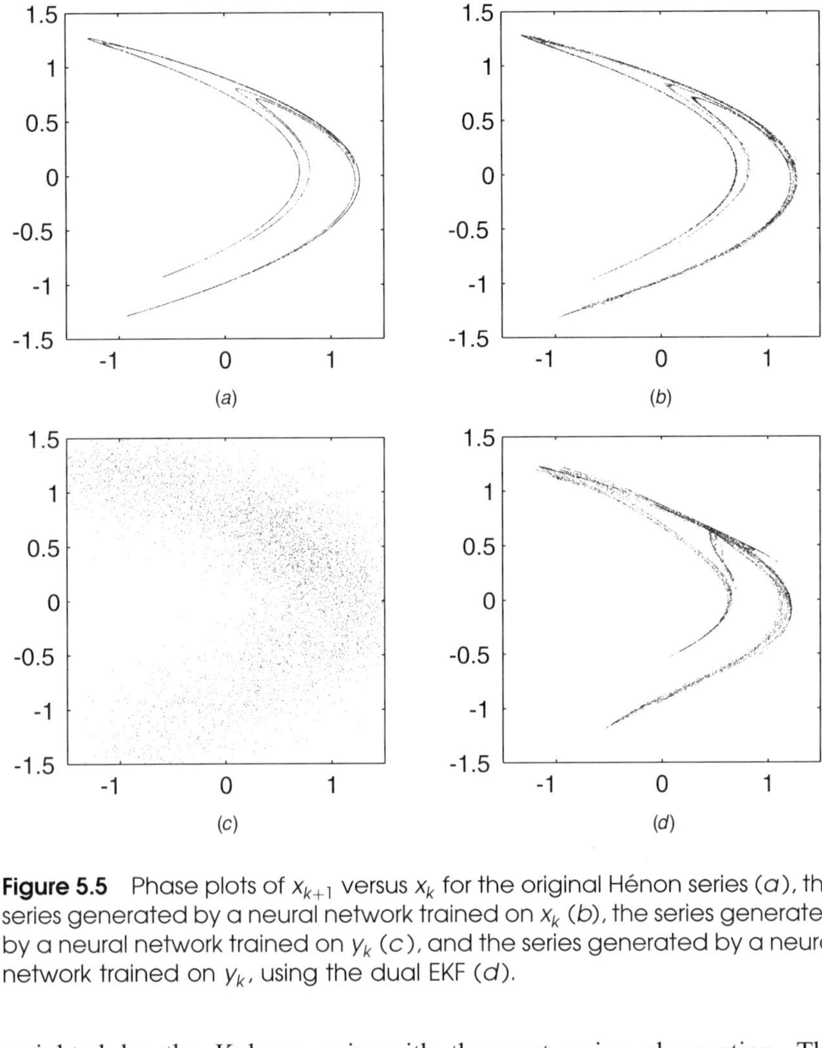

Figure 5.5 Phase plots of x_{k+1} versus x_k for the original Hénon series (*a*), the series generated by a neural network trained on x_k (*b*), the series generated by a neural network trained on y_k (*c*), and the series generated by a neural network trained on y_k, using the dual EKF (*d*).

weighted by the Kalman gain with the next noisy observation. The effective recurrent learning that takes place is analogous to *compromise* methods [42], which use the same principle of combining new observations with output predictions during training, in order to improve robustness of iterated performance.

5.5.2 Economic Forecasting–Index of Industrial Production

Economic and financial data are inherently noisy. As an example, we consider the *index of industrial production* (IP). As with most macroeconomic series, the IP is a composite index of many different economic indicators, each of which is generally measured by a survey of some kind. The monthly IP data are shown in Figure 5.6*a* for January 1950 to January 1990. To remove the trend, the differences between the $\log_2$ values for

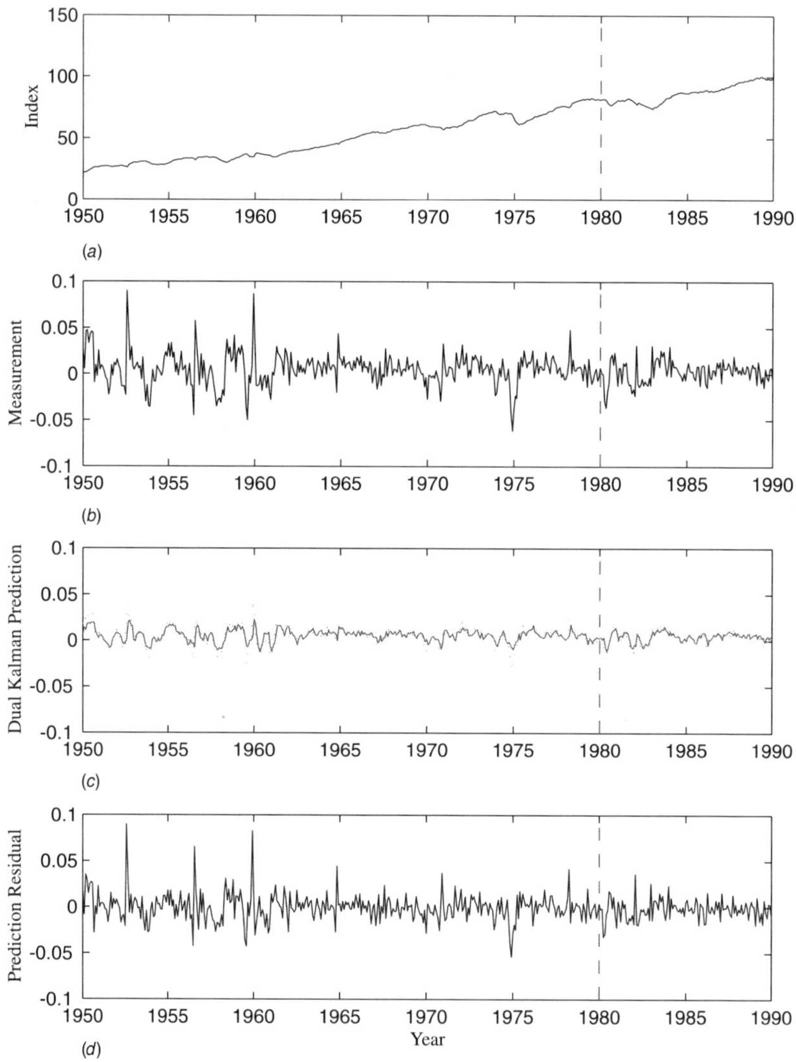

Figure 5.6 (*a*) Index of industrial production (IP) in the United States from January 1940 through March 2000. Data available from Federal Reserve (43). (*b*) Monthly rate of return of the IP in the United States, 1950–1990. (*c*) The dual KF prediction for a typical run (middle), along with the signal estimates (dotted line). (*d*) The prediction residual.

adjacent months are computed; this is called the IP monthly rate of return, and is shown in Figure 5.6*b*.

An important baseline approach is to predict the IP from its past values, using a standard linear autoregressive model. Results with an AR-14 model are reported by Moody et al. [44]. For comparison, both a linear AR-14 model and neural network (14 input, 4 hidden unit) model are tested using the dual EKF methods. Consistent with experiments reported in [44], data from January 1950 to December 1979 are used for a training set, and the remainder of the data is reserved for testing. The dual KF (or dual EKF) is iterated over the training set for several epochs, and the resultant model–consisting of $\hat{\mathbf{w}}$, $\hat{\sigma}_v^2$, and $\hat{\sigma}_n^2$–is used with a standard KF (or EKF) to produce causal predictions on the test set.

All experiments are repeated 10 times with different initial weights $\hat{\mathbf{w}}_0$ to produce the boxplots in Figure 5.7*a*. The advantage of dual estimation on linear models in comparison to the benchmark AR-14 model trained with least-squares (LS) is clear. For the nonlinear model, overtraining is a serious concern, because the algorithm is being run repeatedly over a very short training set (only 360 points). This effect is shown in Figure 5.7. Based on experience with other types of data (which may not have been optimal in this case) all runs were halted after only 5 training epochs. Nevertheless, the dual EKF with $J^j(\mathbf{w})$ cost produces significantly better results.

Although better results are reported on this problem in [44] using models with external inputs from other series, the dual EKF single-time-series results are quite competitive. Future work to incorporate additional inputs would, of course, be a straightforward extension within the dual EKF framework.

5.5.3 Speech Enhancement

Speech enhancement is concerned with the processing of noisy speech in order to improve the *quality* or *intelligibility* of the signal. Applications range from front-ends for automatic speech recognition systems, to telecommunications in aviation, military, teleconferencing, and cellular environments. While there exist a broad array of traditional enhancement techniques (e.g., spectral subtraction, signal-subspace embedding, time-domain iterative approaches, etc. [45]), such methods frequently result in audible distortion of the signal, and are somewhat unsatisfactory in real-world noisy environments. Different neural-network-based approaches are reviewed in [45].

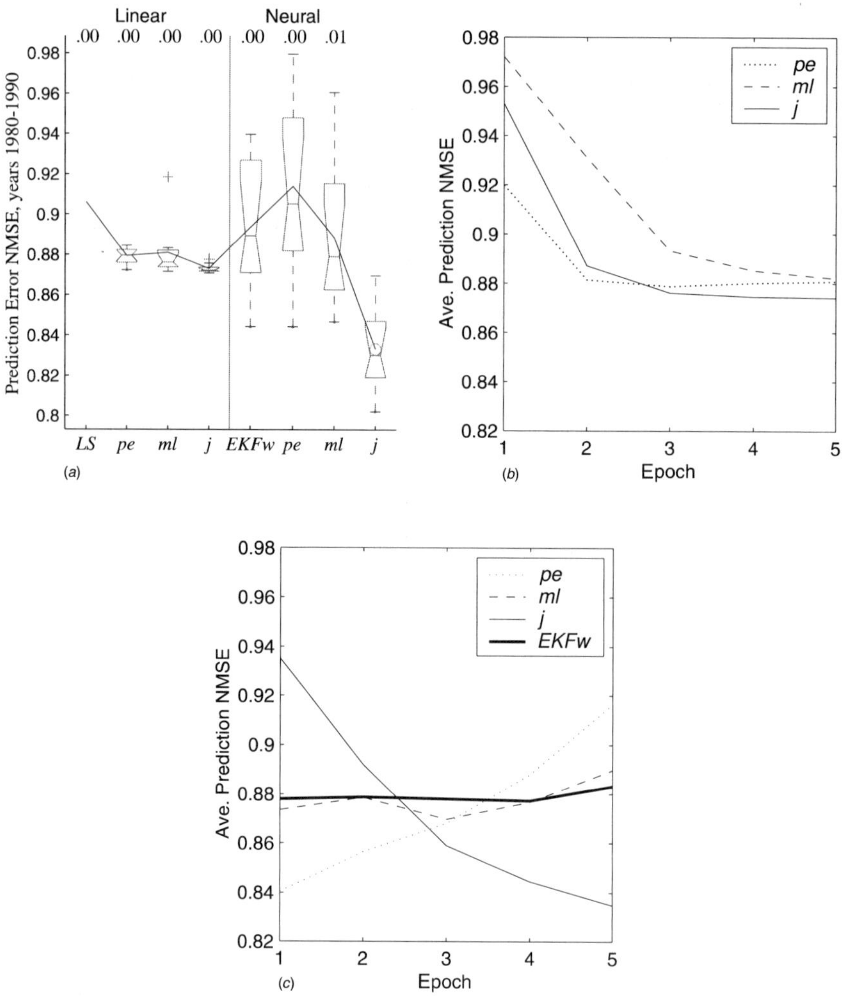

Figure 5.7 (*a*) Boxplots of the prediction NMSE on the test set (1980-1990). (*b*) and (*c*) show the average convergence behavior of linear and neural network model structures, respectively.

The dual estimation algorithms presented in this chapter have the advantage of generating estimates using only the noisy signal itself. To address its nonstationarity, the noisy speech is segmented into short overlapping windows. The dual estimation algorithms are then iterated over each window to generate the signal estimate. Parameters learned from one window are used to initialize the algorithm for the next window. Effectively, a *sequence* of time-series models is trained on the specific noisy speech signal of interest, resulting in a nonstationary model that can be used to remove noise from the given signal.

A number of controlled experiments using 8 kHz sampled speech have been performed in order to compare the different algorithms (joint EKF versus dual EKF with various costs). It was generally concluded that the best results were obtained with the dual EKF with $J^{ml}(\mathbf{w})$ cost, using a 10-4-1 neural network (versus a linear model), and window length set at 512 samples (overlap of 64 points). Preferred nominal settings were found to be: $\mathbf{P}_{\mathbf{x}_0} = \mathbf{I}$, $\mathbf{P}_{\mathbf{w}_0} = 0.01\mathbf{I}$, $p_{\sigma_0} = 0.1$, $\lambda_{\mathbf{w}} = 0.9997$, and $\lambda_{\sigma_v^2} = 0.9993$. The process-noise variance σ_v^2 is estimated with the dual EKF using $J^{ml}(\sigma_v^2)$, and is given a lower limit (e.g., 10^{-8}) to avoid potential divergence of the filters during silence periods. While, in practice, the additive-noise variance σ_n^2 could be estimated as well, we used the common procedure of estimating this from the start of the recording (512 points) where no speech is assumed present. In addition, linear AR-12 (or 10) filters were used to model colored additive noise. Using the settings found from these controlled experiments, several enhancement applications are reviewed below.

SpEAR Database The dual-EKF algorithm was applied to a portion of CSLU's Speech Enhancement Assessment Resource (SpEAR [47]). As opposed to artificially adding noise, the database is constructed by *acoustically* combining prerecorded speech (e.g., TIMIT) and noise (e.g., SPIB database [48]). Synchronous playback and recording in a room is used to provide exact time-aligned references to the clean speech such that objective measures can still be calculated. Table 5.11 presents sample results in terms of average *segmental* SNR.[4]

Car Phone Speech In this example, the dual EKF is used to process an actual recording of a woman talking on her cellular telephone while driving on the highway. The signal contains a significant level of road and engine noise, in addition to the distortion introduced by the telephone channel. The results appear in Figure 5.8, along with the noisy signal.

[4] Segmental SNR is considered to be a more perceptually relevant measure than standard SNR, and is computed as the average of the SNRs computed within 240-point windows, or *frames* of speech: SSNR $= (\#\text{ frames})^{-1} \sum_i \max(\text{SNR}_i, -10\,\text{dB})$. Here, SNR_i is the SNR of the ith frame (weighted by a Hanning window), which is thresholded from below at -10 dB. The thresholding reduces the contribution of portions of the series where no speech is present (i.e., where the SNR is strongly negative) [49], and is expected to improve the measure's perceptual relevance.

Table 5.11 Dual EKF enhancement results using a portion of the SpEAR database[a]

	Male voice (segmental SNR)			Female voice (segmental SNR)		
Noise	Before	After	Static	Before	After	Static
F-16	−2.27	2.65	1.69	0.16	4.51	3.46
Factory	−1.63	2.58	2.48	1.07	4.19	4.24
Volvo	1.60	5.60	6.42	4.10	6.78	8.10
Pink	−2.59	1.44	1.06	−0.23	4.39	3.54
White	−1.35	2.87	2.68	1.05	4.96	5.05
Bursting	1.60	5.05	4.24	7.82	9.36	9.61

[a] Different noise sources are used for the same male and female speaker. All results are in dB, and represent the segmental SNR averaged over the length of the waveform. Results labeled "static" were obtained using the static approximation to the derivatives. For reference, in this range of values, an improvement of 3 dB in segmental SNR relates to approximately an improvement of 5 dB in normal SNR.

Spectrograms of both the noisy speech and estimated speech are included to aid in the comparison. The noise reduction is most successful in nonspeech portions of the signal, but is also apparent in the visibility of formants of the estimated signal, which are obscured in the noisy signal. The perceptual quality of the result is quite good, with an absence of the "musical noise" artifacts often present in spectral subtraction results.

Seminar Recording The next example comes from an actual recording made of a lecture at the Oregon Graduate Institute. In this instance, the audio recording equipment was configured improperly, resulting in a very loud buzzing noise throughout the entire recording. The noise has a fundamental frequency of 60 Hz (indicating that improper grounding was the likely culprit), but many other harmonics and frequencies are present as well owing to some additional nonlinear clipping. As suggested by Figure 5.9, the SNR is extremely low, making for an unusually difficult audio enhancement problem.

Digit Recognition As the final example, we consider the application of speech enhancement for use as a front-end to automatic speech recognition (ASR) systems. The effectiveness of the dual EKF in this

[5] The authors wish to thank Edward Kaiser for his invaluable assistance in this experiment.

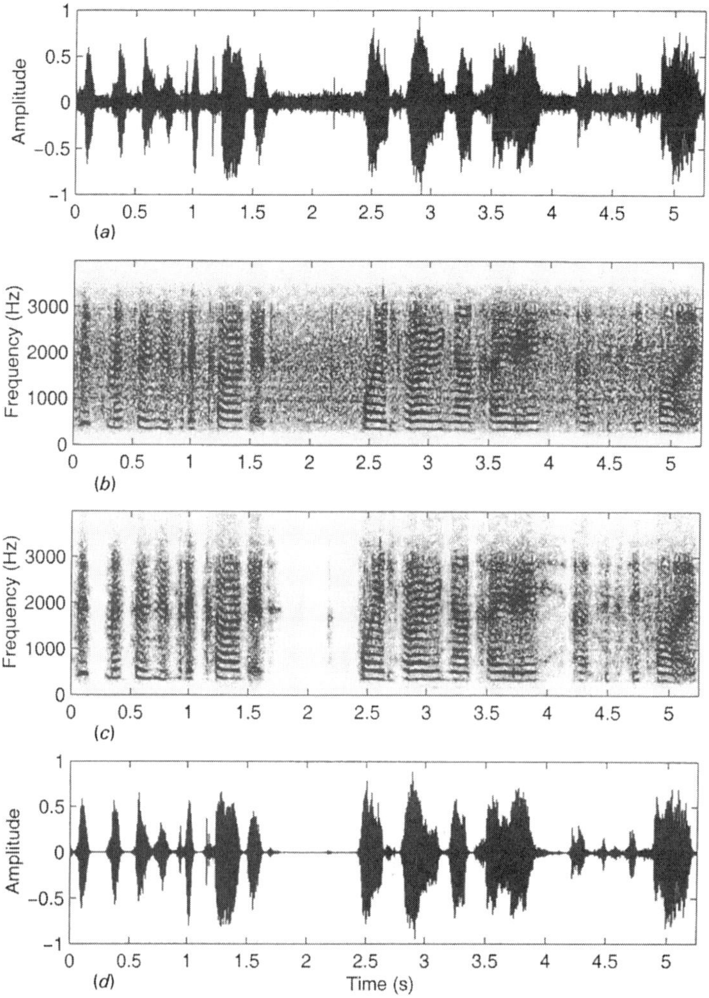

Figure 5.8 Enhancement of car phone speech. The noisy waveform appears in (*a*), with its spectrogram in (*b*). The spectrogram and waveform of the dual EKF result are shown in (*c*) and (*d*), respectively. To make the spectrograms easier to view, the spectral tilt is removed, and their histograms are equalized according to the range of intensities of the enhanced speech spectrogram.

application is demonstrated using a speech corpus and ASR system[5] developed at the Oregon Graduate Institute's Center for Spoken Language Understanding (CSLU). The speech corpus consists of zip-codes, addresses, and other digits read over the telephone by various people; the

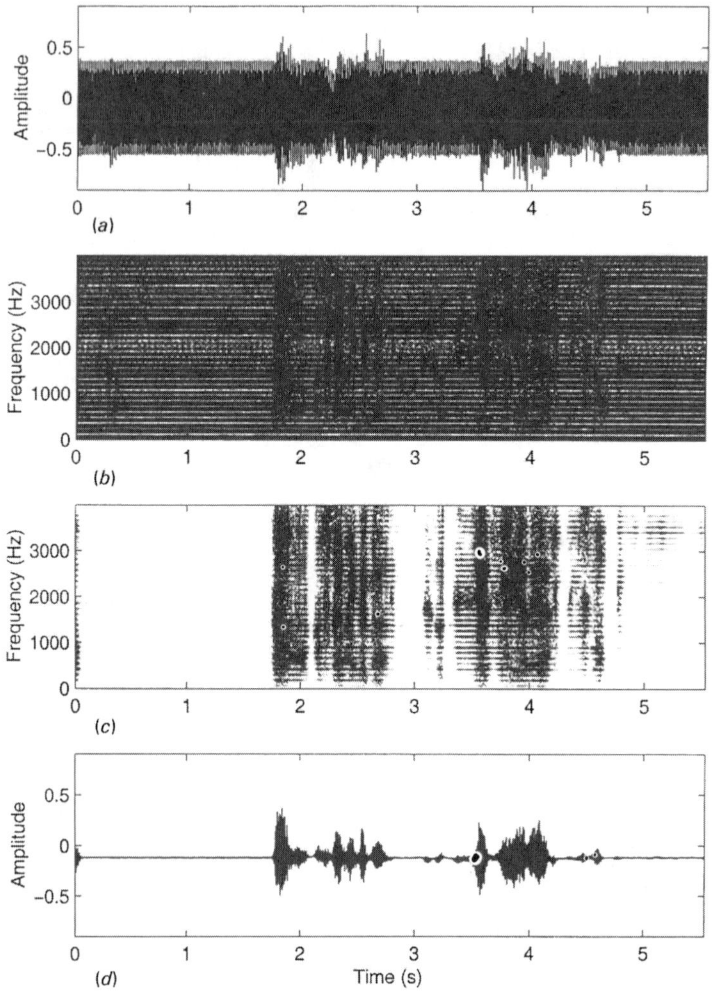

Figure 5.9 Enhancement of high-noise seminar recording. The noisy waveform appears in (*a*), with its spectrogram in (*b*). The spectrogram and waveform of the dual EKF result are shown in (*c*) and (*d*), respectively.

ASR system is a speaker-independent digit recognizer, trained exclusively to recognize numbers from zero to nine when read over the phone.

A subset of 599 sentences was used in this experiment. As seen in Table 5.12, the recognition rates on the clean telephone speech are quite good. However, adding white Gaussian noise to the speech at 6 dB significantly reduces the performance. In addition to the dual EKF, a standard spectral subtraction routine and an enhancement algorithm built into the speech codec TIA/EIA/IS-718 for digital cellular phones (published by the

Table 5.12 Automatic speech recognition rates for clean recordings of telephone speech (spoken digits), as compared with the same speech corrupted by white noise, and subsequently processed by spectral subtraction (SSUB), a cellular phone enhancement standard (IS-718), and the dual EKF

	Correct words	Correct sentences	
Clean	96.37%	85.81%	(514/599)
Noisy	59.21%	21.37%	(128/599)
SSUB	77.45%	38.06%	(228/599)
IS-718	67.32%	29.22%	(175/599)
Dual EKF	82.19%	52.92%	(317/599)

Telecommunications Industry Association) was used for comparison. As shown by Table 5.12, the dual EKF outperforms both the IS-718 and spectral subtraction recognition rates by a significant amount.

5.6 CONCLUSIONS

This chapter has detailed a unified approach to dual estimation based on a maximum *a posteriori* perspective. By maximizing the joint conditional density $\rho_{\mathbf{x}_1^N,\mathbf{w}|\mathbf{y}_1^N}$, the most probable values of the signal and parameters are sought, given the noisy observations. This probabilistic perspective elucidates the relationships between various dual estimation methods proposed in the literature, and allows their categorization in terms of methods that maximize the *joint* conditional density function directly, and those that maximize a related *marginal* conditional density function.

Cost functions associated with the joint and marginal densities have been derived under a Gaussian assumption. This approach offers a number of insights about previously developed methods. For example, the prediction error cost is viewed as an approximation to the maximum-likelihood cost; moreover, both are classified as marginal estimation cost functions. Thus, the recursive prediction error method of [5] and [6] is quite different from the joint EKF approach of [1] and [2], which minimizes a joint estimation cost.[6] Furthermore, the joint EKF and errors-in-variables algorithms are shown to offer two different ways of minimizing the same joint cost function; one is a sequential method and the other is iterative.

The dual EKF algorithm has been presented, which uses two extended Kalman filters run concurrently–one for state estimation and one for

[6] This fact is overlooked in [6], which emphasizes the similarity of these two algorithms.

weight estimation. By modification of the weight filter into an *observed error* form, it is possible to minimize each of the cost functions that are developed.[7] This provided a common algorithmic platform for the implementation of a broad variety of methods. In general, the dual EKF algorithm represents a *sequential* approach, which is applicable to both linear and nonlinear models, and which can be used in the presence of white or colored measurement noise. In addition, the algorithm has been extended to provide estimation of noise variance parameters within the same theoretical framework; this contribution is crucial in applications for which this information is not otherwise available.

Finally, a number of examples have been presented to illustrate the performance of the dual EKF methods. The ability of the dual EKF to capture the underlying dynamics of a noisy time series has been illustrated using the chaotic Hénon map. The application of the algorithm to the IP series demonstrates its potential in a real-world prediction context. On speech enhancement problems, the lack of musical noise in the enhanced speech underscores the advantages of a time-domain approach; the usefulness of the dual EKF as a front-end to a speech recognizer has also been demonstrated. In general, the state-space formulation of the algorithm makes it applicable to a much wider variety of contexts than has been explored here. The intent of this chapter was to show the utility of the dual EKF as a fundamental method for solving a range of problems in signal processing and modeling.

ACKNOWLEDGMENTS

This work was sponsored in part by the NSF under Grants ECS-0083106 and IRI-9712346.

APPENDIX A: RECURRENT DERIVATIVE OF THE KALMAN GAIN

(1) With Respect to the Weights

In the state-estimation filter, the derivative of the Kalman gain with respect to the weights $\mathbf{w}$ is computed as follows. Denoting the derivative of $\mathbf{K}_k^{\mathbf{x}}$

[7] Note that Kalman algorithms are *approximate* MAP optimization procedures for nonlinear systems. Hence, future work considers alternative optimization procedures (e.g., unscented Kalman filters [29]), which can still be cast within the same theoretically motivated dual estimation framework.

with respect to the ith element of $\hat{\mathbf{w}}$ by $\partial \mathbf{K}_k^{\mathbf{x}}/\partial \hat{w}(i)$ (the ith column of $\partial \mathbf{K}_k^{\mathbf{x}}/\partial \hat{\mathbf{w}}$) gives

$$\frac{\partial \mathbf{K}_k^{\mathbf{x}}}{\partial \hat{w}(i)} = \frac{\mathbf{I} - \mathbf{K}_k^{\mathbf{x}}\mathbf{C}}{\mathbf{C}\mathbf{P}_{\mathbf{x}_k}^{-}\mathbf{C}^T + \sigma_n^2} \frac{\partial \mathbf{P}_{\mathbf{x}_k}^{-}}{\partial \hat{w}(i)} \mathbf{C}^T, \tag{5.109}$$

where the derivatives of the error covariances are

$$\frac{\partial \mathbf{P}_{\mathbf{x}_k}^{-}}{\partial \hat{w}(i)} = \frac{\partial \mathbf{A}_{k-1}}{\partial \hat{w}(i)} \mathbf{P}_{\mathbf{x}_{k-1}} \mathbf{A}_{k-1}^T + \mathbf{A}_{k-1} \frac{\partial \mathbf{P}_{\mathbf{x}_{k-1}}}{\partial \hat{w}(i)} \mathbf{A}_{k-1}^T + \mathbf{A}_{k-1}\mathbf{P}_{\mathbf{x}_{k-1}} \frac{\partial \mathbf{A}_{k-1}^T}{\partial \hat{w}(i)}, \tag{5.110}$$

$$\frac{\partial \mathbf{P}_{\mathbf{x}_{k-1}}}{\partial \hat{w}(i)} = -\frac{\partial \mathbf{K}_{k-1}^{\mathbf{x}}}{\partial \hat{w}(i)} \mathbf{C}\mathbf{P}_{k-1}^{-} + (\mathbf{I} - \mathbf{K}_{k-1}^{\mathbf{x}}\mathbf{C}) \frac{\partial \mathbf{P}_{\mathbf{x}_{k-1}}^{-}}{\partial \hat{w}(i)}. \tag{5.111}$$

Note that $\mathbf{A}_{k-1}$ depends not only on the weights $\hat{\mathbf{w}}$, but also on the point of linearization, $\hat{\mathbf{x}}_{k-1}$. Therefore,

$$\frac{\partial \mathbf{A}_{k-1}}{\partial \hat{w}(i)} = \frac{\partial^2 \mathbf{F}}{\partial \hat{\mathbf{x}}_{k-1} \partial \hat{w}(i)} + \frac{\partial^2 \mathbf{F}}{(\partial \hat{\mathbf{x}}_{k-1})^2} \frac{\partial \hat{\mathbf{x}}_{k-1}}{\partial \hat{w}(i)}, \tag{5.112}$$

where the first term is the static derivative of $\mathbf{A}_{k-1} = \partial \mathbf{F}/\partial \mathbf{x}_{k-1}$ with $\hat{\mathbf{x}}_{k-1}$ fixed, and the second term includes the recurrent derivative of $\hat{\mathbf{x}}_{k-1}$. The term $\partial^2 \mathbf{F}/(\partial \hat{\mathbf{x}}_{k-1})^2$ actually represents a three-dimensional tensor (rather than a matrix), and care must be taken with this calculation. However, when $\mathbf{A}_{k-1}$ takes on a sparse structure, as with time-series applications, its derivative with respect to $\mathbf{x}$ contains mostly zeroes, and is in fact entirely zero for linear models.

(2) With Respect to the Variances

In the variance-estimation filter, the derivatives $\partial \mathbf{K}_k^{\mathbf{x}}/\partial \sigma^2$ may be calculated as follows:

$$\frac{\partial \mathbf{K}_k^{\mathbf{x}}}{\partial \sigma^2} = \frac{\mathbf{I} - \mathbf{K}_k^{\mathbf{x}}\mathbf{C}}{\mathbf{C}\mathbf{P}_k^{-}\mathbf{C}^T + \sigma_n^2} \frac{\partial \mathbf{P}_{\mathbf{x}_k}^{-}}{\partial \sigma^2} \mathbf{C}^T, \tag{5.113}$$

where

$$\frac{\partial \mathbf{P}_{\mathbf{x}_k}^-}{\partial \sigma^2} = \frac{\partial \mathbf{A}_{k-1}}{\partial \sigma^2}\mathbf{P}_{k-1}\mathbf{A}_{k-1}^T + \mathbf{A}_{k-1}\frac{\partial \mathbf{P}_{\mathbf{x}_{k-1}}}{\partial \sigma^2}\mathbf{A}_{k-1}^T + \mathbf{A}_{k-1}\mathbf{P}_{k-1}\frac{\partial \mathbf{A}_{k-1}}{\partial \sigma^2}, \quad (5.114)$$

$$\frac{\partial \mathbf{P}_{\mathbf{x}_{k-1}}}{\partial \sigma^2} = -\frac{\partial \mathbf{K}_{k-1}^{\mathbf{x}}}{\partial \sigma^2}\mathbf{C}\mathbf{P}_{k-1}^- + (\mathbf{I} - \mathbf{K}_{k-1}^{\mathbf{x}}\mathbf{C})\frac{\partial \mathbf{P}_{\mathbf{x}_{k-1}}^-}{\partial \sigma^2}. \quad (5.115)$$

Because $\mathbf{A}_{k-1}$ depends on the linearization point, $\hat{\mathbf{x}}_{k-1}$, its derivative is

$$\frac{\partial \mathbf{A}_{k-1}}{\partial \sigma^2} = \frac{\partial \mathbf{A}_{k-1}}{\partial \hat{\mathbf{x}}_{k-1}}\frac{\partial \hat{\mathbf{x}}_{k-1}}{\partial \sigma^2}, \quad (5.116)$$

where again the derivative $\partial\hat{\mathbf{w}}/\partial\sigma^2$ is assumed to be zero.

APPENDIX B: DUAL EKF WITH COLORED MEASUREMENT NOISE

In this appendix, we give dual EKF equations for additive *colored* noise. Colored noise is modeled as a linear AR process:

$$n_k = \sum_{i=1}^{M_n} a_n^{(i)} n_{k-i} + v_{n,k}, \quad (5.128)$$

where the parameters $a_n^{(i)}$ are assumed to be known, and v_{nk} is a white Gaussian process with (possibly unknown) variance $\sigma_{v_n}^2$. The noise n_k can now be thought of as a second signal added to the first, but with the distinction that it has been generated by a known system. Note that the constraint $y_k = x_k + n_k$, requires that the estimates $\hat{x}_k$ and $\hat{n}_k$ must also sum to y_k. To enforce this constraint, both the signal and noise are incorporated into a combined state-space representation:

$$\boldsymbol{\xi}_k = \mathbf{F}_c(\boldsymbol{\xi}_{k-1}, \mathbf{w}, \mathbf{a}_n) + \mathbf{B}_c\mathbf{v}_{c,k}, \quad (5.129)$$

$$\begin{bmatrix} \mathbf{x}_k \\ \mathbf{n}_k \end{bmatrix} = \begin{bmatrix} \mathbf{F}(\mathbf{x}_{k-1}, \mathbf{w}) \\ \mathbf{A}_n \cdot \mathbf{n}_{k-1} \end{bmatrix} + \begin{bmatrix} \mathbf{B} & 0 \\ 0 & \mathbf{B}_n \end{bmatrix}\begin{bmatrix} v_k \\ v_{n,k} \end{bmatrix}, \quad (5.130)$$

$$y_k = \mathbf{C}_c\boldsymbol{\xi}_k,$$

$$y_k = [\mathbf{C} \quad \mathbf{C}_n]\begin{bmatrix} \mathbf{x}_k \\ \mathbf{n}_k \end{bmatrix},$$

where

$$\mathbf{A}_n \triangleq \begin{bmatrix} a_n^{(1)} & a_n^{(2)} & \dots & a_n^{(M_n)} \\ 1 & 0 & 0 & 0 \\ 0 & \ddots & 0 & \vdots \\ 0 & 0 & 1 & 0 \end{bmatrix}, \qquad \mathbf{C}_n = \mathbf{B}_n^T = [1 \quad 0 \quad \dots \quad 0].$$

The *effective* measurement noise is zero, and the process noise $\mathbf{v}_{c,k}$ is white, as required, with covariance

$$\mathbf{V}_c = \begin{bmatrix} \sigma_v^2 & 0 \\ 0 & \sigma_{v_n}^2 \end{bmatrix}.$$

Because n_k can be viewed as a second signal, it should be estimated on an equal footing with x_k. Consider, therefore, maximizing $\rho_{\mathbf{x}_1^N \mathbf{n}_1^N \mathbf{w}|\mathbf{y}_1^N}$ (where $\mathbf{n}_1^N$ is a vector comprising elements in $\{n_k\}_1^N$) instead of $\rho_{\mathbf{x}_1^N \mathbf{w}|\mathbf{y}_1^N}$. We can write this term as

$$\rho_{\mathbf{x}_1^N \mathbf{n}_1^N \mathbf{w}|\mathbf{y}_1^N} = \frac{\rho_{\mathbf{y}_1^N \mathbf{x}_1^N \mathbf{n}_1^N|\mathbf{w}} \rho_{\mathbf{w}}}{\rho_{\mathbf{y}_1^N}}, \tag{5.131}$$

and (in the absence of prior information about $\mathbf{w}$) maximize $\rho_{\mathbf{y}_1^N \mathbf{x}_1^N \mathbf{n}_1^N|\mathbf{w}}$ alone. As before, the cost functions that result from this approach can be categorized as joint or marginal costs, and their derivations are similar to those for the white noise case. The associated dual EKF algorithm for colored noise is given in Table 5.13. Minimization of different cost function is again achieved by simply redefining the error term. These modifications are presented without derivation below.

Joint Estimation Forms

The corresponding weight cost function and error terms for a *decoupled* approach is given in Table 5.14.

Table 5.13 The dual extended Kalman filter equations for colored measurement noise. The definitions of $\mathfrak{e}_k$ and $\mathbf{C}_k^{\mathbf{w}}$ will depend on the cost function used for weight estimation

---

Initialize with

$$\hat{\mathbf{w}}_0 = E[\mathbf{w}], \qquad \mathbf{P}_{\mathbf{w}_0} = E[(\mathbf{w} - \hat{\mathbf{w}}_0)(\mathbf{w} - \hat{\mathbf{w}}_0)^T],$$

$$\hat{\boldsymbol{\xi}}_0 = E[\boldsymbol{\xi}_0], \qquad \mathbf{P}_{\boldsymbol{\xi}_0} = E[(\boldsymbol{\xi}_0 - \hat{\boldsymbol{\xi}}_0)(\boldsymbol{\xi}_0 - \hat{\boldsymbol{\xi}}_0)^T].$$

For $k \in \{1, \ldots, \infty\}$, the time-update equations for the weight filter are

$$\hat{\mathbf{w}}_k^- = \hat{\mathbf{w}}_{k-1}, \tag{5.117}$$

$$\mathbf{P}_{\mathbf{w}_k}^- = \mathbf{P}_{\mathbf{w}_{k-1}} + \mathbf{R}_{k-1}^{\mathbf{r}}, \tag{5.118}$$

and those for the signal filter are

$$\hat{\boldsymbol{\xi}}_k^- = \mathbf{F}(\hat{\boldsymbol{\xi}}_{k-1}, \hat{\mathbf{w}}_k^-), \tag{5.119}$$

$$\mathbf{P}_{\boldsymbol{\xi}_k}^- = \mathbf{A}_{k-1}\mathbf{P}_{\boldsymbol{\xi}_{k-1}}\mathbf{A}_{k-1}^T + \mathbf{B}_c\mathbf{R}_c^{\mathbf{v}}\mathbf{B}_c^T. \tag{5.120}$$

The measurement-update equations for the signal filter are

$$\mathbf{K}_k^{\boldsymbol{\xi}} = \mathbf{P}_{\boldsymbol{\xi}_k}^- \mathbf{C}_c^T(\mathbf{C}_c\mathbf{P}_{\boldsymbol{\xi}_k}^-\mathbf{C}_c^T)^{-1}, \tag{5.121}$$

$$\hat{\boldsymbol{\xi}}_k = \hat{\boldsymbol{\xi}}_k^- + \mathbf{K}_k^{\boldsymbol{\xi}}(y_k - \mathbf{C}_c\hat{\boldsymbol{\xi}}_k^-), \tag{5.122}$$

$$\mathbf{P}_{\boldsymbol{\xi}_k} = (\mathbf{I} - \mathbf{K}_k^{\boldsymbol{\xi}}\mathbf{C})\mathbf{P}_{\boldsymbol{\xi}_k}^-, \tag{5.123}$$

and those for the weight filter are

$$\mathbf{K}_k^{\mathbf{w}} = \mathbf{P}_{\mathbf{w}_k}^-(\mathbf{C}_k^{\mathbf{w}})^T[\mathbf{C}_k^{\mathbf{w}}\mathbf{P}_{\mathbf{w}_k}^-(\mathbf{C}_k^{\mathbf{w}})^T + \mathbf{R}^{\mathbf{e}}]^{-1}, \tag{5.124}$$

$$\hat{\mathbf{w}}_k = \hat{\mathbf{w}}_k^- + \mathbf{K}_k^{\mathbf{w}} \cdot \mathfrak{e}_k, \tag{5.125}$$

$$\mathbf{P}_{\mathbf{w}_k} = (\mathbf{I} - \mathbf{K}_k^{\mathbf{w}}\mathbf{C}_k^{\mathbf{w}})\mathbf{P}_{\mathbf{w}_k}^-, \tag{5.126}$$

where

$$\mathbf{A}_{k-1} \triangleq \left.\frac{\partial \mathbf{F}(\boldsymbol{\xi}, \hat{\mathbf{w}}_k^-, \mathbf{a}_n)}{\partial \boldsymbol{\xi}}\right|_{\hat{\boldsymbol{\xi}}_{k-1}}, \qquad \mathfrak{e}_k = y_k - \mathbf{C}_c\hat{\boldsymbol{\xi}}_k^-, \qquad \mathbf{C}_k^{\mathbf{w}} \triangleq -\frac{\partial \mathfrak{e}_k}{\partial \mathbf{w}} = \mathbf{C}_c \left.\frac{\partial \hat{\boldsymbol{\xi}}_k^-}{\partial \mathbf{w}}\right|_{\mathbf{w}=\hat{\mathbf{w}}_k^-}. \tag{5.127}$$

Marginal Estimation–Maximum-Likelihood Cost

The corresponding weight cost function, and error terms are given in Table 5.15, where

$$\varepsilon_k \triangleq y_k - (\hat{x}_k^- + \hat{n}_k^-), \qquad l_{\varepsilon,k} = \frac{\sigma_{\varepsilon,k}^2}{3\varepsilon_k^2 - 2\sigma_{\varepsilon,k}^2}$$

Table 5.14 Colored-noise joint cost function: observed error terms for the dual EKF weight filter

$$J(\hat{\mathbf{x}}_1^k, \hat{\mathbf{n}}_1^k, \mathbf{w}) + \sum_{t=1}^{k}\left[\frac{(\hat{x}_t - \hat{x}_t^-)^2}{\sigma_v^2} + \frac{(\hat{n}_t - \hat{n}_t^-)^2}{\sigma_{v_n}^2}\right], \quad \text{or} \quad J_k = \sum_{t=1}^{k}\left(\frac{\tilde{\tilde{x}}_k^2}{\sigma_v^2} + \frac{\tilde{\tilde{n}}_k^2}{\sigma_{v_n}^2}\right),$$

$$\mathbf{e}_k \triangleq \begin{bmatrix} \sigma_v^{-1}\tilde{\tilde{x}}_k \\ \sigma_{v_n}^{-1}\tilde{\tilde{n}}_k \end{bmatrix}, \qquad \mathbf{C}_k^{\mathbf{w}} = -\begin{bmatrix} \sigma_v^{-1}\nabla_{\mathbf{w}}^T\tilde{\tilde{x}}_k \\ \sigma_{v_n}^{-1}\nabla_{\mathbf{w}}^T\tilde{\tilde{n}}_k \end{bmatrix}. \tag{5.132}$$

Table 5.15 Maximum-likelihood cost function: observed error terms for the dual EKF weight filter

$$J_c^{ml}(\mathbf{w}) = \sum_{k=1}^{N}\left[\log(2\pi\sigma_{\varepsilon_k}^2) + \frac{(y_k - \hat{x}_k^- - \hat{n}_k^-)^2}{\sigma_{\varepsilon_k}^2}\right],$$

$$\mathbf{e}_k \triangleq \begin{bmatrix} (l_{\varepsilon,k}) \\ \sigma_{\varepsilon_k}^{-1}\varepsilon_k \end{bmatrix}^{1/2}, \qquad \mathbf{C}_k^{\mathbf{w}} = \begin{bmatrix} -\frac{1}{2}\frac{(l_{\varepsilon,k})^{-1/2}}{\sigma_{\varepsilon_k}^2}\nabla_{\mathbf{w}}^T(\sigma_{\varepsilon_k}^2) \\ -\frac{1}{\sigma_{\varepsilon_k}}\nabla_{\mathbf{w}}^T\varepsilon_k + \frac{\varepsilon_k}{2(\sigma_{\varepsilon_k}^2)^{3/2}}\nabla_{\mathbf{w}}^T(\sigma_{\varepsilon_k}^2) \end{bmatrix}.$$

Marginal Estimation Forms–Prediction Error Cost

If $\sigma_{\varepsilon_k}^2$ is assumed to be independent of **w**, then we have the prediction error cost shown in Table 5.16. Note that an alternative prediction error form may be derived by including $\nabla_{\mathbf{w}}\hat{n}_k^-$ in the calculation of $\mathbf{C}_k^{\mathbf{w}}$. However, the performance appears superior if this term is neglected.

Table 5.16 Colored-noise prediction-error cost function: observed error terms for the dual EKF weight filter

$$J_c^{pe}(\mathbf{w}) = \sum_{k=1}^{N}\varepsilon_k^2 = \sum_{k=1}^{N}(y_k - \hat{x}_k^- - \hat{n}_k^-)^2, \tag{5.133}$$

$$\mathbf{e}_k \triangleq y_k - \hat{n}_k - \hat{x}_k^-, \qquad \mathbf{C}_k^{\mathbf{w}} = -\nabla_{\mathbf{w}}\hat{x}_k^-.$$

Marginal Estimation–EM Cost

The cost and observed error terms for weight estimation with colored noise are identical to those for the white-noise case, shown in Table 5.8. In this case, the on-line statistics are found by augmenting the combined state vector with one additional lagged value for both the signal and noise. Specifically,

$$\boldsymbol{\xi}_k^+ = \begin{bmatrix} \mathbf{x}_k \\ x_{k-M} \\ \mathbf{n}_k \\ n_{k-M_n} \end{bmatrix} = \begin{bmatrix} x_k \\ \mathbf{x}_{k-1} \\ n_k \\ \mathbf{n}_{k-1} \end{bmatrix}, \tag{5.134}$$

so that the estimate $\hat{\boldsymbol{\xi}}_k^+$ produced by a Kalman filter will contain $\hat{\mathbf{x}}_{k-1|k}$ in elements $2, \ldots, 1 + M$, and $\hat{\mathbf{n}}_{k-1|k}$ in its last M_n elements. Furthermore, the error variances $p_{\bar{k}|k}$ and $p_{k|k}^{\dagger}$ can be obtained from the covariance $\mathbf{P}_{c,k}^+$ of $\boldsymbol{\xi}_k^+$ produced by the KF.

REFERENCES

[1] R.E. Kopp and R.J. Orford, "Linear regression applied to system identification for adaptive control systems," *AIAA Journal*, **1**, 2300–2006 (1963).

[2] H. Cox, "On the estimation of state variables and parameters for noisy dynamic systems," *IEEE Transactions on Automatic Control*, **9**, 5–12 (1964).

[3] L. Ljung, "Asymptotic behavior of the extended Kalman filter as a parameter estimator for linear systems," *IEEE Transactions on Automatic Control*, **24**, 36–50 (1979).

[4] M. Niedźwiecki and K. Cisowski, "Adaptive scheme of elimination of broadband noise and impulsive disturbances from AR and ARMA signals," *IEEE Transactions on Signal Processing*, **44**, 528–537 (1996).

[5] L.W. Nelson and E. Stear, "The simultaneous on-line estimation of parameters and states in linear systems," *IEEE Transactions on Automatic Control*, Vol. AC-12, 438–442 (1967).

[6] L. Ljung and T. Söderström, *Theory and Practice of Recursive Identification*. Cambridge, MA: MIT Press, 1983.

[7] H. Akaike, "Maximum likelihood identification of Gaussian autoregressive moving average models," *Biometrika*, **60**, 255–265 (1973).

[8] N.K. Gupta and R.K. Mehra, "Computational aspects of maximum likelihood estimation and reduction in sensitivity function calculations," *IEEE Transaction on Automatic Control*, **19**, 774–783 (1974).

[9] J.S. Lim and A.V. Oppenheim, "All-pole modeling of degraded speech," *IEEE Transactions on Acoustics, Speech and Signal Processing*, **26**, 197–210 (1978).

[10] A. Dempster, N.M. Laird, and D.B. Rubin, "Maximum-likelihood from incomplete data via the EM algorithm," *Journal of the Royal Statistical Society, Ser. B*, **39**, 1–38 (1977).

[11] B.R. Musicus and J.S. Lim, "Maximum likelihood parameter estimation of noisy data," in *Proceedings of the International Conference on Acoustics, Speech, and Signal Processing, IEEE, April 1979*, pp. 224–227.

[12] R.H. Shumway and D.S. Stoffer, "An approach to time series smoothing and forecasting using the EM algorithm," *Journal of Time Series Analysis*, **3**, 253–264 (1982).

[13] E. Weinstein, A.V. Oppenheim, M. Feder, and J.R. Buck, "Iterative and sequential algorithms for multisensor signal enhancement," *IEEE Transactions on Signal Processing*, **42**, 846–859 (1994).

[14] J.T. Connor, R.D. Martin, and L.E. Atlas, "Recurrent neural networks and robust time series prediction," *IEEE Transactions on Neural Networks*, **5**(2), 240–254 (1994).

[15] S.C. Stubberud and M. Owen, "Artificial neural network feedback loop with on-line training, in *Proceedings of International Symposium on Intelligent Control, IEEE, September* 1996, pp. 514–519.

[16] Z. Ghahramani and S.T. Roweis, "Learning nonlinear dynamical systems using an EM algorithm," in *Advances in Neural Information Processing Systems 11: Proceedings of the 1998 Conference*. Cambridge, MA: MIT Press, 1999.

[17] T. Briegel and V. Tresp, "Fisher scoring and a mixture of modes approach for approximate inference and learning in nonlinear state space models," in *Advances in Neural Information Processing Systems 11: Proceedings of the 1998 Conference*. Cambridge, MA: MIT Press, 1999.

[18] G. Seber and C. Wild, *Nonlinear Regression*. New York: Wiley, 1989, pp. 491–527.

[19] E.A. Wan and A.T. Nelson, "Dual Kalman filtering methods for nonlinear prediction, estimation, and smoothing," in *Advances in Neural Information Processing Systems 9*. Cambridge, MA: MIT Press, 1997.

[20] E.A. Wan and A.T. Nelson, "Neural dual extended Kalman filtering: Applications in speech enhancement and monaural blind signal separation," in *Proceedings of IEEE Workshop on Neural Networks for Signal Processing, 1997*.

[21] A.T. Nelson and E.A. Wan, "A two-observation Kalman framework for maximum-likelihood modeling of noisy time series," in *Proceedings of International Joint Conference on Neural Networks, IEEE/INNS, May 1998.*

[22] A.T. Nelson, "Nonlinear Estimation and Modeling of Noisy Time-Series by Dual Kalman Filter Methods," PhD thesis, Oregon Graduate Institute of Science and Technology, 2000.

[23] R.E. Kalman, "A new approach to linear filtering and prediction problems," *Transactions of the ASME, Ser. D, Journal of Basic Engineering*, **82**, 35–45 (1960).

[24] J.F.G. de Freitas, M. Niranjan, A.H. Gee, and A. Doucet, "Sequential Monte Carlo methods for optimisation of neural network models," Technical Report TR-328, Cambridge University Engineering Department, November 1998.

[25] R. van der Merwe, N. de Freitas, A. Doucet, and E. Wan, "The unscented particle filter," Technical Report CUED/F-INFENG/TR 380, Cambridge University Engineering Department, August 2000.

[26] S.J. Julier, J.K. Uhlmann, and H. Durrant-Whyte, "A new approach for filtering nonlinear systems," in *Proceedings of the American Control Conference, 1995*, pp. 1628–1632.

[27] S.J. Julier and J.K. Uhlmann, "A general method for approximating nonlinear transformations of probability distributions," Technical Report, RRG, Department of Engineering Science, University of Oxford, November 1996. http://www.robots.ox.ac.uk/~siju/work/publications/letter_size/Unscented.zip.

[28] E.A. Wan and R. van der Merwe, "The unscented Kalman filter for nonlinear estimation," in *Proceedings of Symposium 2000 on Adaptive Systems for Signal Processing, Communication and Control (AS-SPCC), Lake Louise, Alberta, Canada, IEEE, October 2000.*

[29] E.A. Wan, R. van der Merwe, and A.T. Nelson, "Dual estimation and the unscented transformation," in *Advances in Neural Information Processing Systems 12: Proceedings of the 1999 Conference*. Cambridge, MA: MIT Press, 2000.

[30] S. Singhal and L. Wu, "Training multilayer perceptrons with the extended Kalman filter," in *Advances in Neural Information Processing Systems 1.* San Mateo, CA: Morgan Kauffman, 1989, pp. 133–140.

[31] G.V. Puskorius and L.A. Feldkamp, "Neural control of nonlinear dynamic systems with Kalman filter trained recurrent networks," *IEEE Transactions on Neural Networks*, **5** (1994).

[32] E.S. Plumer, "Training neural networks using sequential-update forms of the extended Kalman filter," Informal Report LA-UR-95-422, Los Alamos National Laboratory, January 1995.

[33] A.H. Sayed and T. Kailath, "A state-space approach to adaptive RLS filtering," *IEEE Signal Processing Magazine*, **11**(3), 18–60 (1994).

[34] S. Haykin, *Adaptive Filter Theory*, 3rd ed. Upper Saddle River, NJ: Prentice-Hall, 1996.

[35] R. Williams and D. Zipser, "A learning algorithm for continually running fully recurrent neural networks," *Neural Computation*, **1**, 270–280 (1989).

[36] F.L. Lewis, *Optimal Estimation*. New York: Wiley, 1986.

[37] R.K. Mehra, Identification of stochastic linear dynamic systems using Kalman filter representation. *AIAA Journal*, **9**, 28–31 (1971).

[38] B.D. Ripley, *Pattern Recognition and Neural Networks*. Cambridge University Press, 1996.

[39] T.M. Cover and J.A. Thomas, *Elements of Information Theory*. New York: Wiley, 1991.

[40] G.V. Puskorius and L.A. Feldkamp, "Extensions and enhancements of decoupled extended Kalman filter training, in *Proceedings of International Conference on Neural Networks, ICNN'97, IEEE, June 1997*, Vol. 3.

[41] N.N. Schraudolph, "Online local gain adaptation for multi-layer perceptrons," Technical report, IDSIA, Lugano, Switzerland, March 1998.

[42] P.J. Werbos, *Handbook of Intelligent Control*. New York: Van Nostrand Reinhold, 1992, pp. 283–356.

[43] FRED: Federal Reserve Economic Data. Available on the Internet at http://www.stls.frb.org/fred/. Accessed May 9, 2000.

[44] J. Moody, U. Levin, and S. Rehfuss, "Predicting the U.S. index of industrial production," *Neural Network World*, **3**, 791–794 (1993).

[45] J.H.L. Hansen, J.R. Deller, and J.G. Praokis, *Discrete-Time Processing of Speech Signals*. New York: Macmillan, 1993.

[46] E.A. Wan and A.T. Nelson, "Removal of noise from speech using the dual EKF algorithm," in *Proceedings of International Conference on Acoustics, Speech, and Signal Processing, IEEE, May 1998*.

[47] Center for Spoken Language Understanding, *Speech Enhancement Assessment Resource* (SpEAR). Available on the Internet at http://cslu.ece.ogi.edu/nsel/data/index.html. Accessed May 2000.

[48] Rice University, *Signal Processing Information Base* (SPIB). Available on the Internet at http://spib.ece.rice.edu/signal.html. Accessed September 15, 1999.

[49] J.H.L. Hansen and B.L. Pellom, "An effective quality evaluation protocol for speech enhancement algorithms," in *Proceedings of International Conference on Spoken Language Processing, ICSLP-98, Sidney, 1998*.

6

LEARNING NONLINEAR DYNAMICAL SYSTEMS USING THE EXPECTATION–MAXIMIZATION ALGORITHM

Sam Roweis and Zoubin Ghahramani

Gatsby Computational Neuroscience Unit, University College London, London U.K.
(zoubin@gatsby.ucl.ac.uk)

6.1 LEARNING STOCHASTIC NONLINEAR DYNAMICS

Since the advent of cybernetics, dynamical systems have been an important modeling tool in fields ranging from engineering to the physical and social sciences. Most realistic dynamical systems models have two essential features. First, they are stochastic – the observed outputs are a noisy function of the inputs, and the dynamics itself may be driven by some unobserved noise process. Second, they can be characterized by

Kalman Filtering and Neural Networks, Edited by Simon Haykin
ISBN 0-471-36998-5

some finite-dimensional internal state that, while not directly observable, summarizes at any time all information about the past behavior of the process relevant to predicting its future evolution.

From a modeling standpoint, stochasticity is essential to allow a model with a few fixed parameters to generate a rich variety of time-series outputs.[1] Explicitly modeling the internal state makes it possible to decouple the internal dynamics from the observation process. For example, to model a sequence of video images of a balloon floating in the wind, it would be computationally very costly to directly predict the array of camera pixel intensities from a sequence of arrays of previous pixel intensities. It seems much more sensible to attempt to infer the true state of the balloon (its position, velocity, and orientation) and decouple the process that governs the balloon dynamics from the observation process that maps the actual balloon state to an array of measured pixel intensities.

Often we are able to write down equations governing these dynamical systems directly, based on prior knowledge of the problem structure and the sources of noise – for example, from the physics of the situation. In such cases, we may want to infer the hidden state of the system from a sequence of observations of the system's inputs and outputs. Solving this *inference* or *state-estimation* problem is essential for tasks such as tracking or the design of state-feedback controllers, and there exist well-known algorithms for this.

However, in many cases, the exact parameter values, or even the gross structure of the dynamical system itself, may be unknown. In such cases, the dynamics of the system have to be *learned* or *identified* from sequences of observations only. Learning may be a necessary precursor if the ultimate goal is effective state inference. But learning nonlinear state-based models is also useful in its own right, even when we are not explicitly interested in the internal states of the model, for tasks such as prediction (extrapolation), time-series classification, outlier detection, and filling-in of missing observations (imputation). This chapter addresses the problem of learning time-series models when the internal state is hidden. Below, we briefly review the two fundamental algorithms that form the basis of our learning procedure. In section 6.2, we introduce our algorithm

[1]There are, of course, completely deterministic but *chaotic* systems with this property. If we separate the noise processes in our models from the deterministic portions of the dynamics and observations, we can think of the noises as another deterministic (but highly chaotic) system that depends on initial conditions and exogenous inputs that we do not know. Indeed, when we run simulations using a psuedo-random-number generator started with a particular seed, this is precisely what we are doing.

and derive its learning rules. Section 6.3 presents results of using the algorithm to identify nonlinear dynamical systems. Finally, we present some conclusions and potential extensions to the algorithm in Sections 6.4 and 6.5.

6.1.1 State Inference and Model Learning

Two remarkable algorithms from the 1960s – one developed in engineering and the other in statistics – form the basis of modern techniques in state estimation and model learning. The Kalman filter, introduced by Kalman and Bucy in 1961 [1], was developed in a setting where the physical model of the dynamical system of interest was readily available; its goal is optimal state estimation in systems with known parameters. The expectation–maximization (EM) algorithm, pioneered by Baum and colleagues [2] and later generalized and named by Dempster et al. [3], was developed to learn parameters of statistical models in the presence of incomplete data or hidden variables.

In this chapter, we bring together these two algorithms in order to learn the dynamics of stochastic nonlinear systems with hidden states. Our goal is twofold: both to develop a method for identifying the dynamics of nonlinear systems whose hidden states we wish to infer, and to develop a general nonlinear time-series modeling tool. We examine inference and learning in discrete-time[2] stochastic nonlinear dynamical systems with hidden states x_k, external inputs u_k, and noisy outputs y_k. (All lower-case characters (except indices) denote vectors. Matrices are represented by upper-case characters.) The systems are parametrized by a set of tunable matrices, vectors, and scalars, which we shall collectively denote as θ. The inputs, outputs, and states are related to each other by

$$x_{k+1} = f(x_k, u_k) + w_k, \tag{6.1a}$$

$$y_k = g(x_k, u_k) + v_k, \tag{6.1b}$$

[2]Continuous-time dynamical systems (in which *derivatives* are specified as functions of the current state and inputs) can be converted into discrete-time systems by sampling their outputs and using "zero-order holds" on their inputs. In particular, for a continuous-time linear system $\dot{x}(t) = A_c x(t) + B_c u(t)$ sampled at interval τ, the corresponding dynamics and input driving matrices so that $x_{k+1} = Ax_k + Bu_k$ are $A = \sum_{k=0}^{\infty} A_c^k \tau^k / k! = \exp(A_c \tau)$ and $B = A_c^{-1}(A - I)B_c$.

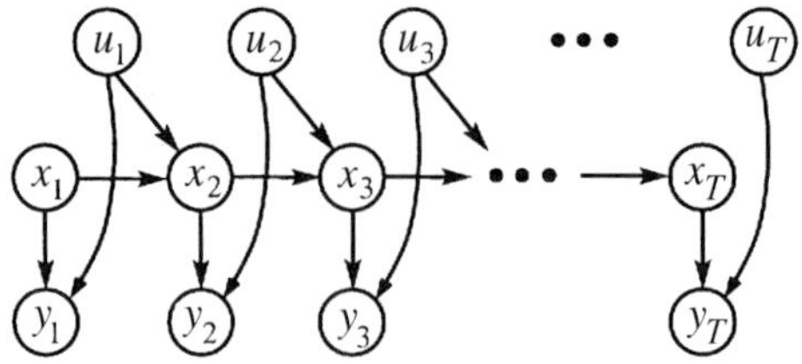

Figure 6.1 A probabilistic graphical model for stochastic dynamical systems with hidden states x_k, inputs u_k, and observables y_k.

where w_k and v_k are zero-mean Gaussian noise processes. The state vector x evolves according to a nonlinear but stationary Markov dynamics[3] driven by the inputs u and by the noise source w. The outputs y are nonlinear, noisy but stationary and instantaneous functions of the current state and current input. The vector-valued nonlinearities f and g are assumed to be differentiable, but otherwise arbitrary. The goal is to develop an algorithm that can be used to model the probability density of output sequences (or the conditional density of outputs given inputs) using only a finite number of example time series. The crux of the problem is that both the hidden state trajectory and the parameters are unknown.

Models of this kind have been examined for decades in systems and control engineering. They can also be viewed within the framework of *probabilistic graphical models*, which use graph theory to represent the conditional dependencies between a set of variables [4, 5]. A probabilistic graphical model has a node for each (possibly vector-valued) random variable, with directed arcs representing stochastic dependences. Absent connections indicate conditional independence. In particular, nodes are conditionally independent from their non-descendents, given their parents – where parents, children, descendents, etc, are defined with respect to the directionality of the arcs (i.e., arcs go from parent to child). We can capture the dependences in Eqs. (6.1*a*,*b*) compactly by drawing the graphical model shown in Figure 6.1.

One of the appealing features of probabilistic graphical models is that they explicitly diagram the mechanism that we assume generated the data. This *generative model* starts by picking randomly the values of the nodes that have no parents. It then picks randomly the values of their children

[3]Stationarity means here that neither f nor the covariance of the noise process w_k, depend on time; that is, the dynamics are *time-invariant*. Markov refers to the fact that given the current state, the next state does not depend on the past history of the states.

given the parents' values, and so on. The random choices for each child given its parents are made according to some assumed noise model. The combination of the graphical model and the assumed noise model at each node fully specify a probability distribution over all variables in the model.

Graphical models have helped clarify the relationship between dynamical systems and other probabilistic models such as hidden Markov models and factor analysis [6]. Graphical models have also made it possible to develop probabilistic inference algorithms that are vastly more general than the Kalman filter.

If we knew the parameters, the operation of interest would be to infer the hidden state sequence. The uncertainty in this sequence would be encoded by computing the posterior distributions of the hidden state variables given the sequence of observations. The Kalman filter (reviewed in Chapter 1) provides a solution to this problem in the case where f and g are linear. If, on the other hand, we had access to the hidden state trajectories as well as to the observables, then the problem would be one of model-fitting, i.e. estimating the parameters of f and g and the noise covariances. Given observations of the (no longer hidden) states and outputs, f and g can be obtained as the solution to a possibly nonlinear regression problem, and the noise covariances can be obtained from the residuals of the regression. How should we proceed when *both* the system model and the hidden states are unknown?

The classical approach to solving this problem is to treat the parameters θ as "extra" hidden variables, and to apply an extended Kalman filtering (EKF) algorithm (see Chapter 1) to the nonlinear system with the state vector augmented by the parameters [7, 8]. For stationary models, the dynamics of the parameter portion of this extended state vector are set to the identity function. The approach can be made inherently on-line, which may be important in certain applications. Furthermore, it provides an estimate of the covariance of the parameters at each time step. Finally, its objective, probabilistically speaking, is to find an optimum in the joint space of parameters and hidden state sequences.

In contrast, the algorithm we present is a batch algorithm (although, as we discuss in Section 6.4.2, online extensions are possible), and does not attempt to estimate the covariance of the parameters. Like other instances of the EM algorithm, which we describe below, its goal is to integrate over the uncertain estimates of the unknown hidden states and optimize the resulting marginal likelihood of the parameters given the observed data. An extended Kalman smoother (EKS) is used to estimate the approximate state distribution in the E-step, and a radial basis function (RBF) network [9, 10] is used for nonlinear regression in the M-step. It is important not to confuse this use of the extended Kalman algorithm, namely, to estimate just the hidden state as part of the E-step of EM, with the use that we described in the previous paragraph, namely to simultaneously estimate parameters and hidden states.

6.1.2 The Kalman Filter

Linear dynamical systems with additive white Gaussian noises are the most basic models to examine when considering the state-estimation problem, because they admit exact and efficient inference. (Here, and in what follows, we call a system linear if both the state evolution function and the state-to-output observation function are linear, and nonlinear otherwise.) The linear dynamics and observation processes correspond to matrix operations, which we denote by A, B and C, D, respectively, giving the classic state-space formulation of input-driven linear dynamical systems:

$$x_{k+1} = Ax_k + Bu_k + w_k, \tag{6.2a}$$

$$y_k = Cx_k + Du_k + v_k. \tag{6.2b}$$

The Gaussian noise vectors w and v have zero mean and covariances Q and R respectively. If the prior probability distribution $p(x_1)$ over initial states is taken to be Gaussian, then the joint probabilities of all states and outputs at future times are also Gaussian, since the Gaussian distribution is closed under the linear operations applied by state evolution and output mapping and under the convolution applied by additive Gaussian noise. Thus, all distributions over hidden state variables are fully described by their means and covariance matrices. The algorithm for exactly computing the posterior mean and covariance for x_k given some sequence of observations consists of two parts: a forward recursion, which uses the observations from y_1 to y_k, known as the *Kalman filter* [11], and a backward recursion, which uses the observations from y_T to y_{k+1}. The combined forward and backward recursions are known as the Kalman or Rauch–Tung–Streibel (RTS) *smoother* [12]. These algorithms are reviewed in detail in Chapter 1.

There are three key insights to understanding the Kalman filter. The first is that the Kalman filter is simply a method for implementing Bayes' rule. Consider the very general setting where we have a prior $p(x)$ on some state variable and an observation model $p(y|x)$ for the noisy outputs given the state. Bayes' rule gives us the state-inference procedure:

$$p(x|y) = \frac{p(y|x)p(x)}{p(y)} = \frac{p(y|x)p(x)}{Z}, \tag{6.3a}$$

$$Z = p(y) = \int_x p(y|x)p(x)\, dx, \tag{6.3b}$$

where the normalizer Z is the unconditional density of the observation. All we need to do in order to convert our prior on the state into a posterior is to multiply by the likelihood from the observation equation, and then renormalize.

The second insight is that there is no need to invert the output or dynamics functions, as long as we work with easily normalizable distributions over hidden states. We see this by applying Bayes' rule to the linear Gaussian case for a single time step.[4] We start with a Gaussian belief $\mathcal{N}(x_{k-1}, V_{k-1})$ on the current hidden state, use the dynamics to convert this to a prior $\mathcal{N}(x^+, V^+)$ on the next state, and then condition on the observation to convert this prior into a posterior $\mathcal{N}(x_k, V_k)$. This gives the classic Kalman filtering equations:

$$p(x_{k-1}) = \mathcal{N}(x^+, V^+), \tag{6.4a}$$

$$x^+ = Ax_{k-1}, \qquad V^+ = AV_{k-1}A^\top + Q, \tag{6.4b}$$

$$p(y_k|x_k) = \mathcal{N}(Cx_k, R), \tag{6.4c}$$

$$p(x_k|y_k) = \mathcal{N}(x_k, V_k), \tag{6.4d}$$

$$x_k = x^+ + K(y_k - Cx^+), \qquad V_k = (I - KC)V^+, \tag{6.4e}$$

$$K = V^+C^\top(CV^+C^\top + R)^{-1}. \tag{6.4f}$$

The posterior is again Gaussian and analytically tractable. Notice that neither the dynamics matrix A nor the observation matrix C needed to be inverted.

The third insight is that the state-estimation procedures can be implemented recursively. The posterior from the previous time step is run through the dynamics model and becomes our prior for the current time step. We then convert this prior into a new posterior by using the current observation.

[4]Some notation: A multivariate normal (Gaussian) distribution with mean μ and covariance matrix Σ is written as $\mathcal{N}(\mu, \Sigma)$. The same Gaussian evaluated at the point z is denoted by $\mathcal{N}(\mu, \Sigma)|_z$. The determinant of a matrix is denoted by $|A|$ and matrix inversion by A^{-1}. The symbol $\sim$ means "distributed according to."

For the general case of a nonlinear system with non-Gaussian noise, state estimation is much more complex. In particular, mapping through arbitrary nonlinearities f and g can result in arbitrary state distributions, and the integrals required for Bayes' rule can become intractable. Several methods have been proposed to overcome this intractability, each providing a distinct approximate solution to the inference problem. Assuming f and g are differentiable and the noise is Gaussian, one approach is to locally linearize the nonlinear system about the current state estimate so that applying the Kalman filter to the linearized system the approximate state distribution remains Gaussian. Such algorithms are known as *extended* Kalman filters (EKF) [13, 14]. The EKF has been used both in the classical setting of state estimation for nonlinear dynamical systems and also as a basis for on-line learning algorithms for feedforward neural networks [15] and radial basis function networks [16, 17]. For more details, see Chapter 2.

State inference in nonlinear systems can also be achieved by propagating a set of random samples in state space through f and g, while at each time step re-weighting them using the likelihood $p(y|x)$. We shall refer to algorithms that use this general strategy as *particle filters* [18], although variants of this sampling approach are known as sequential importance sampling, bootstrap filters [19], Monte Carlo filters [20], condensation [21], and dynamic mixture models [22, 23]. A recent survey of these methods is provided in [24]. A third approximate state-inference method, known as the *unscented* filter [25–27], deterministically chooses a set of balanced points and propagates them through the nonlinearities in order to recursively approximate a Gaussian state distribution; for more details, see Chapter 7. Finally, there are algorithms for approximate inference and learning based on mean field theory and variational methods [28, 29]. Although we have chosen to make local linearization (EKS) the basis of our algorithms below, it is possible to formulate the same learning algorithms using any approximate inference method (e.g., the unscented filter).

6.1.3 The EM Algorithm

The EM or *expectation–maximization* algorithm [3, 30] is a widely applicable iterative parameter re-estimation procedure. The objective of the EM algorithm is to maximize the likelihood of the observed data $P(Y|\theta)$ in the presence of hidden[5] variables X. (We shall denote the entire

[5]Hidden variables are often also called *latent variables*; we shall use both terms. They can also be thought of as missing data for the problem or as auxiliary parameters of the model.

sequence of observed data by $Y = \{y_1, \ldots, y_\tau\}$, observed inputs by $U = \{u_1, \ldots, u_T\}$, the sequence of hidden variables by $X = \{x_1, \ldots, x_\tau\}$, and the parameters of the model by θ.) Maximizing the likelihood as a function of θ is equivalent to maximizing the log-likelihood:

$$\mathcal{L}(\theta) = \log P(Y|U, \theta) = \log \int_X P(X, Y|U, \theta)\, dX. \tag{6.5}$$

Using *any* distribution $Q(X)$ over the hidden variables, we can obtain a lower bound on $\mathcal{L}$:

$$\log \int_X P(Y, X|U, \theta)\, dX = \log \int_X Q(X) \frac{P(X, Y|U, \theta)}{Q(X)}\, dX \tag{6.6a}$$

$$\geq \int_X Q(X) \log \frac{P(X, Y|U, \theta)}{Q(X)}\, dX \tag{6.6b}$$

$$= \int_X Q(X) \log P(X, Y|U, \theta)\, dX - \int_X Q(X) \log Q(X)\, dX \tag{6.6c}$$

$$= \mathcal{F}(Q, \theta), \tag{6.6d}$$

where the middle inequality (6.6*b*) is known as Jensen's inequality and can be proved using the concavity of the log function. If we define the *energy* of a global configuration (X, Y) to be $-\log P(X, Y|U, \theta)$, then the lower bound $\mathcal{F}(Q, \theta) \leq \mathcal{L}(\theta)$ is the negative of a quantity known in statistical physics as the *free energy*: the expected energy under Q minus the entropy of Q [31]. The EM algorithm alternates between maximizing $\mathcal{F}$ with respect to the *distribution* Q and the *parameters* θ, respectively, holding the other fixed. Starting from some initial parameters θ_0 we alternately apply:

$$\textbf{E-step:} \quad Q_{k+1} \leftarrow \arg\max_Q \mathcal{F}(Q, \theta_k), \tag{6.7a}$$

$$\textbf{M-step:} \quad \theta_{k+1} \leftarrow \arg\max_\theta \mathcal{F}(Q_{k+1}, \theta). \tag{6.7b}$$

It is easy to show that the maximum in the E-step results when Q is exactly the conditional distribution of X, $Q^*_{k+1}(X) = P(X|Y, U, \theta_k)$, at which point the bound becomes an equality: $\mathcal{F}(Q^*_{k+1}, \theta_k) = \mathcal{L}(\theta_k)$. The maximum in the M-step is obtained by maximizing the first term in (6.6*c*), since the entropy of Q does not depend on θ:

$$\textbf{M-step:} \quad \theta^*_{k+1} \leftarrow \arg\max_{\theta} \int_X P(X|Y, U, \theta_k) \log P(X, Y|U, \theta)\, dX. \tag{6.8}$$

This is the expression most often associated with the EM algorithm, but it obscures the elegant interpretation [31] of EM as coordinate ascent in $\mathcal{F}$ (see Fig. 6.2). Since $\mathcal{F} = \mathcal{L}$ at the beginning of each M-step, and since the E-step does not change θ, we are guaranteed not to decrease the likelihood after each combined EM step. (While this is obviously true of "complete" EM algorithms as described above, it may also be true for "incomplete" or "sparse" variants in which approximations are used during the E- and/or M-steps so long as $\mathcal{F}$ always goes up; see also the earlier work in [32].) For example, this can take the form of a gradient M- step algorithm (where we increase $P(Y|\theta)$ with respect to θ but do not strictly maximize it), or any E-step which improves the bound $\mathcal{F}$ without saturating it [31].)

In dynamical systems with hidden states, the E-step corresponds exactly to solving the smoothing problem: estimating the hidden state trajectory given both the observations/inputs and the parameter values. The M-step involves system identification using the state estimates from the smoother. Therefore, at the heart of the EM learning procedure is the following idea: use the solutions to the filtering/smoothing problem to estimate the unknown hidden states given the observations and the current

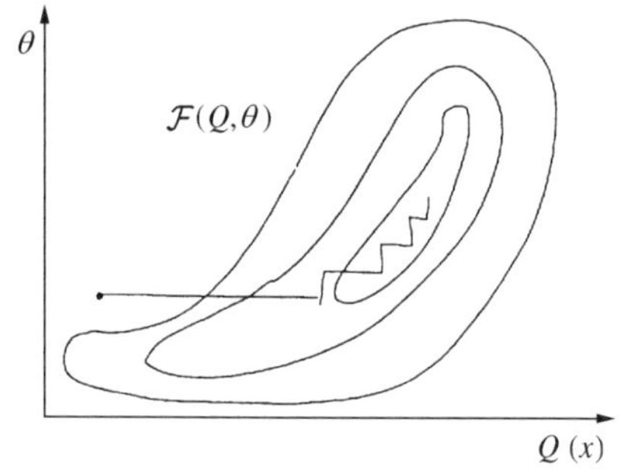

Figure 6.2 The EM algorithm can be thought of as coordinate ascent in the functional $\mathcal{F}(Q(X), \theta)$ (see text). The E-step maximizes $\mathcal{F}$ with respect to $Q(X)$ given fixed θ (horizontal moves), while the M-step maximizes $\mathcal{F}$ with respect to θ given fixed $Q(X)$ (vertical moves).

model parameters. Then use this fictitious complete data to solve for new model parameters. Given the estimated states obtained from the inference algorithm, it is usually easy to solve for new parameters. For example, when working with linear Gaussian models, this typically involves minimizing quadratic forms, which can be done with linear regression. This process is repeated, using these new model parameters to infer the hidden states again, and so on. Keep in mind that our goal is to maximize the log-likelihood (6.5) (or equivalently maximize the total likelihood) of the observed data with respect to the model parameters. This means integrating (or summing) over all the ways in which the model could have produced the data (i.e., hidden state sequences). As a consequence of using the EM algorithm to do this maximization, we find ourselves needing to compute (and maximize) the *expected* log-likelihood of the *joint* data (6.8), where the expectation is taken over the distribution of hidden values predicted by the current model parameters and the observations.

In the past, the EM algorithm has been applied to learning linear dynamical systems in specific cases, such as "multiple-indicator multiple-cause" (MIMC) models with a single latent variable [33] or state-space models with the observation matrix known [34]), as well as more generally [35]. This chapter applies the EM algorithm to learning nonlinear dynamical systems, and is an extension of our earlier work [36]. Since then, there has been similar work applying EM to nonlinear dynamical systems [37, 38]. Whereas other work uses sampling for the E-step and gradient M-steps, our algorithm uses the RBF networks to obtain a computationally efficient and exact M-step.

The EM algorithm has four important advantages over classical approaches. First, it provides a straightforward and principled method for handing missing inputs or outputs. (Indeed this was the original motivation for Shumway and Stoffer's application of the EM algorithm to learning partially unknown linear dynamical systems [34].) Second, EM generalizes readily to more complex models with combinations of discrete and real-valued hidden variables. For example, one can formulate EM for a *mixture* of nonlinear dynamical systems [39, 40]. Third, whereas it is often very difficult to prove or analyze stability within the classical on-line approach, the EM algorithm is always attempting to maximize the likelihood, which acts as a Lyapunov function for stable learning. Fourth, the EM framework facilitates Bayesian extensions to learning – for example, through the use of variational approximations [29].

6.2 COMBINING EKS AND EM

In the next sections, we shall describe the basic components of our EM learning algorithm. For the expectation step of the algorithm, we infer an approximate conditional distribution of the hidden states using Extended Kalman Smoothing (Section 6.2.1). For the maximization step, we first discuss the general case (Section 6.2.2), and then describe the particular case where the nonlinearities are represented using Gaussian radial basis function (RBF) networks (Section 6.2.3). Since, as with all EM or likelihood ascent algorithms, our algorithm is not guaranteed to find the globally optimum solutions, good initialization is a key factor in practical success. We typically use a variant of factor analysis followed by estimation of a purely linear dynamical system as the starting point for training our nonlinear models (Section 6.2.4).

6.2.1 Extended Kalman smoothing (E-step)

Given a system described by Eqs. (6.1*a*,*b*), the E-step of an EM learning algorithm needs to infer the hidden states from a history of observed inputs and outputs. The quantities at the heart of this inference problem are two conditional densities

$$P(x_k | u_1, \ldots, u_T, y_1, \ldots, y_T), \qquad 1 \leq k \leq T, \tag{6.9}$$

$$P(x_k, x_{k+1} | u_1, \ldots, u_T, y_1, \ldots, y_T), \qquad 1 \leq k \leq T-1. \tag{6.10}$$

For nonlinear systems, these conditional densities are in general non-Gaussian, and can in fact be quite complex. For all but a very few nonlinear systems, exact inference equations cannot be written down in closed form. Furthermore, for many nonlinear systems of interest, exact inference is intractable (even numerically), meaning that, in principle, the amount of computation required grows exponentially in the length of the time series observed. The intuition behind all extended Kalman algorithms is that they approximate a stationary nonlinear dynamical system with a *non-stationary* (time-varying) but linear system. In particular, extended Kalman smoothing (EKS) simply applies regular Kalman smoothing to a local linearization of the nonlinear system. At every point $\tilde{x}$ in x space, the derivatives of the vector-valued functions f and g define the matrices,

$$A_{\tilde{x}} \equiv \left.\frac{\partial f}{\partial x}\right|_{x=\tilde{x}} \quad \text{and} \quad C_{\tilde{x}} \equiv \left.\frac{\partial g}{\partial x}\right|_{x=\tilde{x}},$$

respectively. The dynamics are linearized about $\hat{x}_k$, the mean of the current filtered (not smoothed) state estimate at time t. The output equation can be similarly linearized. These linearizations yield

$$x_{k+1} \approx f(\hat{x}_k, u_k) + A_{\hat{x}_k}(x_k - \hat{x}_k) + w, \tag{6.11}$$

$$y_k \approx g(\hat{x}_k, u_k) + C_{\hat{x}_k}(x_k - \hat{x}_k) + v. \tag{6.12}$$

If the noise distributions and the prior distribution of the hidden state at $k = 1$ are Gaussian, then, in this progressively linearized system, the conditional distribution of the hidden state at any time k given the history of inputs and outputs will also be Gaussian. Thus, Kalman smoothing can be used on the linearized system to infer this conditional distribution; this is illustrated in Figure 6.3.

Notice that although the algorithm performs *smoothing* (in other words, it takes into account all observations, including future ones, when inferring the state at any time), the linearization is only done in the forward direction. Why not re-linearize about the backwards estimates during the RTS recursions? While, in principle, this approach might give better results, it is difficult to implement in practice because it requires the dynamics functions to be uniquely invertible, which it often is not true.

Unlike the normal (linear) Kalman smoother, in the EKS, the error covariances for the state estimates and the Kalman gain matrices *do*

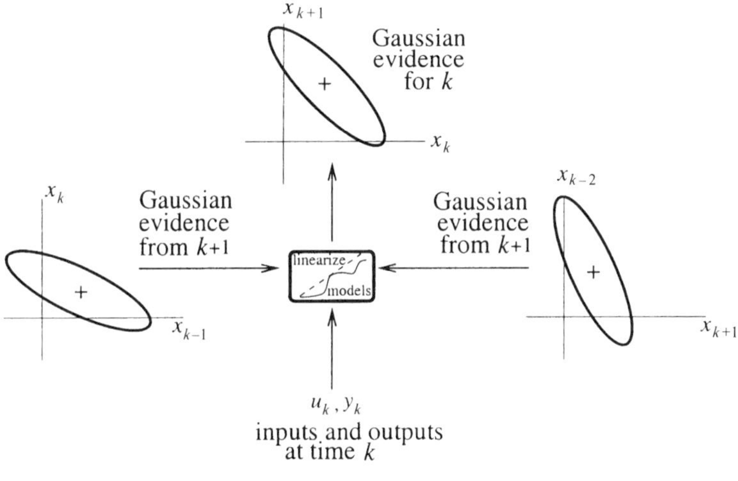

Figure 6.3 Illustration of the information used in extended Kalman smoothing (EKS), which infers the hidden state distribution during the E-step of our algorithm. The nonlinear model is linearized about the current state estimate at each time, and then Kalman smoothing is used on the linearized system to infer Gaussian state estimates.

depend on the observed data, not just on the time index t. Furthermore, it is no longer necessarily true that if the system is stationary, the Kalman gain will converge to a value that makes the smoother act as the optimal Wiener filter in the steady state.

6.2.2 Learning Model Parameters (M-step)

The M-step of our EM algorithm re-estimates the parameters of the model given the observed inputs, outputs, and the conditional distributions over the hidden states. For the model we have described, the parameters define the nonlinearities f and g, and the noise covariances Q and R (as well as the mean and covariance of the initial state, x_1).

Two complications can arise in the M-step. First, fully re-estimating f and g in each M-step may be computationally expensive. For example, if they are represented by neural network regressors, a single full M-step would be a lengthy training procedure using backpropagation, conjugate gradients, or some other optimization method. To avoid this, one could use partial M-steps that increase but do not maximize the expected log-likelihood (6.8) – for example, each consisting of one or a few gradient steps. However, this will in general make the fitting procedure much slower.

The second complication is that f and g have to be trained using the *uncertain* state-estimates output by the EKS algorithm. This makes it difficult to apply standard curve-fitting or regression techniques. Consider fitting f, which takes as inputs x_k and u_k and outputs x_{k+1}. For each t, the conditional density estimated by EKS is a full-covariance Gaussian in (x_k, x_{k+1}) space. So f has to be fit not to a set of data points but instead to a mixture of full-covariance Gaussians in input–output space (Gaussian "clouds" of data). Ideally, to follow the EM framework, this conditional density should be *integrated over* during the fitting process. Integrating over this type of data is nontrivial for almost any form of f. One simple but inefficient approach to bypass this problem is to draw a large sample from these Gaussian clouds of data and then fit f to these samples in the usual way. A similar situation occurs with the fitting of the output function g.

We present an alternative approach, which is to choose the form of the function approximator to make the integration easier. As we shall show, using Gaussian radial basis function (RBF) networks [9, 10] to model f and g allows us to do the integrals exactly and efficiently. With this choice of representation, both of the above complications vanish.

6.2.3 Fitting Radial Basis Functions to Gaussian Clouds

We shall present a general formulation of an RBF network from which it should be clear how to fit special forms for f and g. Consider the following nonlinear mapping from input vectors x and u to an output vector z:

$$z = \sum_{i=1}^{I} h_i \rho_i(x) + Ax + Bu + b + w, \tag{6.13}$$

where w is a zero-mean Gaussian noise variable with covariance Q, and ρ_i are scalar valved RBFs defined below. This general mapping can be used in several ways to represent dynamical systems, depending on which of the input to hidden to output mappings are assumed to be nonlinear. Three examples are: (1) representing f using (6.13) with the substitutions $x \leftarrow x_k$, $u \leftarrow u_k$, and $z \leftarrow x_{k+1}$; (2) representing f using $x \leftarrow (x_k, u_k)$, $u \leftarrow \emptyset$, and $z \leftarrow x_{k+1}$; and (3) representing g using the substitutions $x \leftarrow x_k$, $u \leftarrow u_k$, and $z \leftarrow y_k$. (Indeed, for different simulations, we shall use different forms.) The parameters are the I coefficients h_i of the RBFs; the matrices A and B multiplying inputs x and u, respectively; and an output bias vector b, and the noise covariance Q. Each RBF is assumed to be a Gaussian in x space, with center c_i and width given by the covariance matrix S_i:

$$\rho_i(x) = |2\pi S_i|^{-1/2} \exp[-\tfrac{1}{2}(x - c_i)^\top S_i^{-1}(x - c_i)], \tag{6.14}$$

where $|S_i|$ is the determinant of the matrix S_i. For now, we assume that the centers and widths of the RBFs are fixed, although we discuss learning their locations in Section 6.4.

The goal is to fit this RBF model to data (u, x, z). The complication is that the data set comes in the form of a mixture of Gaussian distributions. Here we show how to analytically integrate over this mixture distribution to fit the RBF model.

Assume the data set is

$$P(x, z, u) = \frac{1}{J} \sum_j \mathcal{N}_j(x, z)\delta(u - u_j). \tag{6.15}$$

That is, we observe samples from the u variables, each paired with a Gaussian "cloud" of data, $\mathcal{N}_j$, over (x, z). The Gaussian $\mathcal{N}_j$ has mean μ_j and covariance matrix C_j.

Let $\hat{z}_\theta(x, u) = \sum_{i=1}^{I} h_i \rho_i(x) + Ax + Bu + b$, where θ is the set of parameters. The log-likelihood of a single fully observed data point under the model would be

$$-\tfrac{1}{2}[z - \hat{z}_\theta(x, u)]^\top Q^{-1}[z - \hat{z}_\theta(x, u)] - \tfrac{1}{2}\ln|Q| + \text{const.}$$

Since the (x, z) values in the data set are uncertain, the maximum expected log-likelihood RBF fit to the mixture of Gaussian data is obtained by minimizing the following integrated quadratic form:

$$\min_{\theta, Q}\left\{\sum_j \int_x \int_z \mathcal{N}_j(x, z)[z - \hat{z}_\theta(x, u_j)]^\top Q^{-1}[z - \hat{z}_\theta(x, u_j)]\, dx\, dz + J\ln|Q|\right\}. \tag{6.16}$$

We rewrite this in a slightly different notation, using angular brackets $\langle\cdot\rangle_j$ to denote expectation over $\mathcal{N}_j$, and defining

$$\begin{aligned}\theta &\equiv [h_1, h_2, \ldots, h_I, A, B, b],\\ \Phi &\equiv [\rho_1(x), \rho_2(x), \ldots, \rho_I(x), x^\top, u^\top, 1]^\top.\end{aligned}$$

Then, the objective is written as

$$\min_{\theta, Q}\left\{\sum_j \langle (z - \theta\Phi)^\top Q^{-1}(z - \theta\Phi)\rangle_j + J\ln|Q|\right\}. \tag{6.17}$$

Taking derivatives with respect to θ, premultiplying by $-Q^{-1}$, and setting the result to zero gives the linear equations $\sum_j \langle (z - \theta\Phi)\Phi^T\rangle_j = 0$, which we can solve for θ and Q:

$$\hat{\theta} = \left(\sum_j \langle z\Phi^\top\rangle_j\right)\left(\sum_j \langle \Phi\Phi^\top\rangle_j\right)^{-1}, \qquad \hat{Q} = \frac{1}{J}\left(\sum_j \langle zz^\top\rangle_j - \hat{\theta}\sum_j \langle \Phi z^\top\rangle_j\right). \tag{6.18}$$

In other words, given the expectations in the angular brackets, the optimal parameters can be solved for via a set of linear equations. In the Appendix, we show that these expectations can be computed analytically and efficiently, which means that we can take full and exact M-steps. The derivation is somewhat laborious, but the intuition is very simple: the

Gaussian RBFs multiply the Gaussian densities $\mathcal{N}_j$ to form new unnormalized Gaussians in (x, y) space. Expectations under these new Gaussians are easy to compute. This fitting algorithm is illustrated in Figure 6.4.

Note that among the four advantages we mentioned previously for the EM algorithm – ability to handle missing observations, generalizability to extensions of the basic model, Bayesian approximations, and guaranteed stability through a Lyapunov function – we have had to forgo one. There is no guarantee that extended Kalman smoothing increases the lower bound on the true likelihood, and therefore stability cannot be assured. In practice, the algorithm is rarely found to become unstable, and the approximation works well: in our experiments, the likelihoods increased monotonically and good density models were learned. Nonetheless, it may be desirable to derive guaranteed-stable algorithms for certain special cases using lower-bound preserving variational approximations [29] or other approaches that can provide such proofs.

The ability to fully integrate over uncertain state estimates provides practical benefits as well as being theoretically pleasing. We have compared fitting our RBF networks using only the means of the state estimates with performing the full integration as derived above. When using only the means, we found it necessary to introduce a ridge

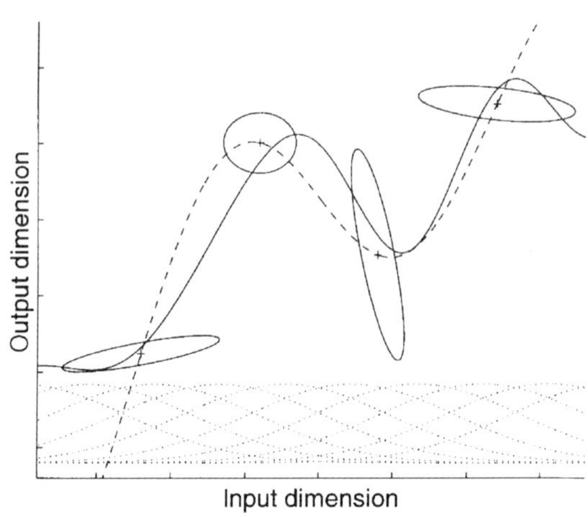

Figure 6.4 Illustration of the regression technique employed during the M-step. A fit to a mixture of Gaussian densities is required; if Gaussian RBF networks are used, then this fit can be solved analytically. The dashed line shows a regular RBF fit to the centers of the four Gaussian densities, while the solid line shows the analytical RBF fit using the covariance information. The dotted lines below show the support of the RBF kernels.

regression (weight decay) parameter in the M-step to penalize the very large coefficients that would otherwise occur based on precise cancellations between inputs. Since the model is linear in the parameters, this ridge regression regularizer is like adding white noise to the radial basis outputs $\rho_i(x)$ (i.e., *after* the RBF kernels have been applied).[6] By linearization, this is approximately equivalent to Gaussian noise at the inputs x with a covariance determined by the derivatives of the RBFs at the input locations. The uncertain state estimates provide exactly this sort of noise, and thus automatically regularize the RBF fit in the M-step. This naturally avoids the need to introduce a penalty on large coefficients, and improves generalization.

6.2.4 Initialization of Models and Choosing Locations for RBF Kernels

The practical success of our algorithm depends on two design choices that need to be made at the beginning of the training procedure. The first is to judiciously select the placement of the RBF kernels in the representation of the state dynamics and/or output function. The second is to sensibly initialize the parameters of the model so that iterative improvement with the EM algorithm (which finds only local maxima of the likelihood function) finds a good solution.

In models with low-dimensional hidden states, placement of RBF kernel centers can be done by gridding the state space and placing one kernel on each grid point. Since the scaling of the state variables is given by the covariance matrix of the state dynamics noise w_k in Eq. (6.1a) which, without loss of generality, we have set to I, it is possible to determine both a suitable size for the gridding region over the state space, and a suitable scaling of the RBF kernels themselves. However, the number of kernels in such a grid increases exponentially with the grid dimension, so, for more than three or four state variables, gridding the state space is impractical. In these cases, we first use a simple initialization, such as a linear dynamical system, to infer the hidden states, and then place RBF kernels on a randomly chosen subset of the inferred state means.[7] We set the widths (variances) of the RBF kernels once we have

[6]Consider a simple scalar linear regression example $y_j = \theta z_j$, which can be solved by minimizing $\sum_j(y_j - \theta z_j)^2$. If each z_j has mean $\bar{z}_j$ and variance λ, the expected value of this cost function is $\sum_j(y_j - \theta\bar{z}_j)^2 + J\lambda\theta^2$, which is exactly ridge regression with λ controlling the amount of regularization.

[7]In order to properly cover the portions of the state space that are most frequently used, we require a minimum distance between RBF kernel centers. Thus, in practice, we reject centers that fall too close together.

the spacing of their centers by attempting to make neighboring kernels cross when their outputs are half of their peak value. This ensures that, with all the coefficients set approximately equal, the RBF network will have an almost "flat" output across the space.[8]

These heuristics can be used both for fixed assignments of centers and widths, and as initialization to an adaptive RBF placement procedure. In Section 6.4.1, we discuss techniques for adapting both the positions of the RBF centers and their widths during training of the model.

For systems with nonlinear dynamics but approximately linear output functions, we initialize using maximum-likelihood factor analysis (FA) trained on the collection of output observations (or conditional factor analysis for models with inputs). Factor analysis is a very simple model, which assumes that the output variables are generated by linearly combining a small number of independent Gaussian hidden state variables and then adding independent Gaussian noise to each output variable [6]. One can think of factor analysis as a special case of linear dynamical systems with Gaussian noise where the states are not related in time (i.e., $A = 0$). We used the weight matrix (called the loading matrix) learned by factor analysis to initialize the observation matrix C in the dynamical system. By doing time-independent inference through the factor analysis model, we can also obtain approximate estimates for the state at each time. These estimates can be used to initialize the nonlinear RBF regressor by fitting the estimates at one time step as a function of those at the previous time step. (We also sometimes do a few iterations of training using a purely linear dynamical system before initializing the nonlinear RBF network.) Since such systems are nonlinear flows embedded in linear manifolds, this initialization estimates the embedding manifold using a linear statistical technique (FA) and the flow using a nonlinear regression based on projections into the estimated manifold.

If the output function is nonlinear but the dynamics are approximately linear, then a mixture of factor analyzers (MFA) can be trained on the output observations [41, 42]. A mixture of factor analyzers is a model that assumes that the data were generated from several Gaussian clusters with differing means, with the covariance within each cluster being modeled by a factor analyzer. Systems with nonlinear output function but linear dynamics capture linear flows in a nonlinear embedding manifold, and

[8]One way to see this is to consider Gaussian RBFs in an n-dimensional grid (i.e., a square lattice), all with heights 1. The RBF centers define a hypercube, the distance between neighboring RBFs being $2d$, where d is chosen such that $e^{-d^2/(2\sigma^2)} = \frac{1}{2}$. At the centers of the hypercubes, there are 2^n contributions from neighboring Gaussians, each of which is a distance $\sqrt{n}d$, and so contributes $(\frac{1}{2})^n$ to the height. Therefore, the height at the interiors is approximately equal to the height at the corners.

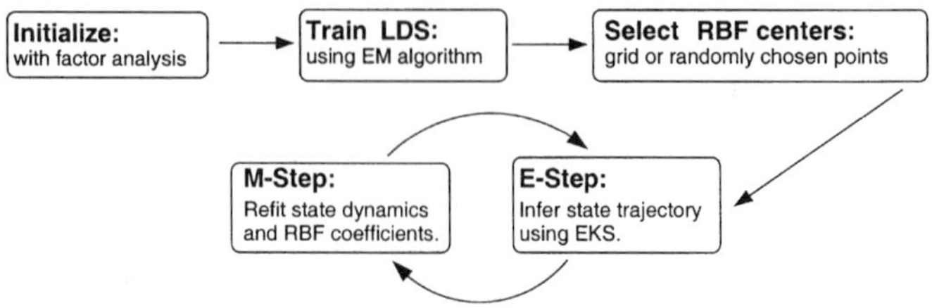

Figure 6.5 Summary of the main steps of the NLDS-EM algorithm.

the goal of the MFA initialization is to capture the nonlinear shape of the output manifold. Estimating the dynamics is difficult (since the hidden states of the individual analyzers in the mixture cannot be combined easily into a single internal state representation), but is still possible.[9] A summary of the algorithm including these initialization techniques is shown in Figure 6.5.

Ideally, Bayesian methods would be used to control the complexity of the model by estimating the internal state dimension and optimal number of RBF centers. However, in general, only approximate techniques such as cross-validation or variational approximations can be implemented in practice (see Section 6.4.4). Currently, we have set these complexity parameters either by hand or with cross-validation.

6.3 RESULTS

We tested how well our algorithm could learn the dynamics of a nonlinear system by observing only the system inputs and outputs. We investigated the behavior on simple one- and two-dimensional state-space problems whose nonlinear dynamics were known, as well as on a weather time-series problem involving real temperature data.

6.3.1 One- and Two-Dimensional Nonlinear State-Space Models

In order to be able to compare our algorithm's learned internal state representation with a ground truth state representation, we first tested it on

[9]As an approximate solution to the problem of getting a single hidden state from a MFA, we can use the following procedure: (1) Estimate the "similarity" between analyzer centers using average separation in time between data points for which they are active. (2) Use standard embedding techniques such as multidimensional scaling (MDS) [43] to place the MFA centers in a Euclidean space of dimension k. (3) Time-independent state inference for each observation now consists of the responsibility-weighted low-dimensional MFA centers, where the responsibilities are the posterior probabilities of each analyzer given the observation under the MFA.

synthetic data generated by nonlinear dynamics whose form was known. The systems we considered consisted of three inputs and four observables at each time, with either one or two hidden state variables. The relation of the state from one time step to the next was given by a variety of nonlinear functions followed by Gaussian noise. The outputs were a linear function of the state and inputs plus Gaussian noise. The inputs affected the state only through a linear driving function. The true and learned state transition functions for these systems, as well as sample outputs in response to Gaussian noise inputs and internal driving noise, are shown in Figures 6.6*c*,*d*, 6.7*c*, and 6.8*c*.

We initialized each nonlinear model with a linear dynamical model trained with EM, which, in turn, we initialized with a variant of factor analysis (see Section 6.2.4). The one-dimensional state-space models were given 11 RBFs in x space, which were uniformly spaced. (The range of maximum and minimum x values was automatically determined from the density of inferred points.) Two-dimensional state-space models were given 25 RBFs spaced in a 5×5 grid uniformly over the range of inferred

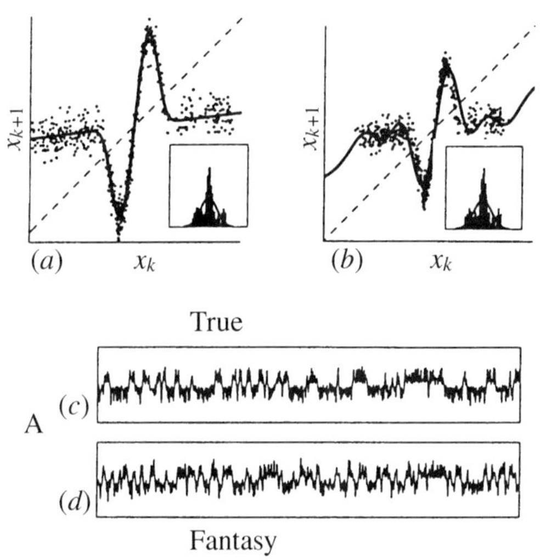

Figure 6.6 Example of fitting a system with nonlinear dynamics and linear observation function. The panels show the fitting of a nonlinear system with a one-dimensional hidden state and 4 noisy outputs driven by Gaussian noise inputs and internal state noise. (*a*) The true dynamics function (line) and states (dots) used to generate the training data (the inset is the histogram of internal states). (*b*) The learned dynamics function and states inferred on the training data (the inset is the histogram of inferred internal states). (*c*) The first component of the observable time series from the training data. (*d*) The first component of fantasy data generated from the learned model (on the same scale as *c*).

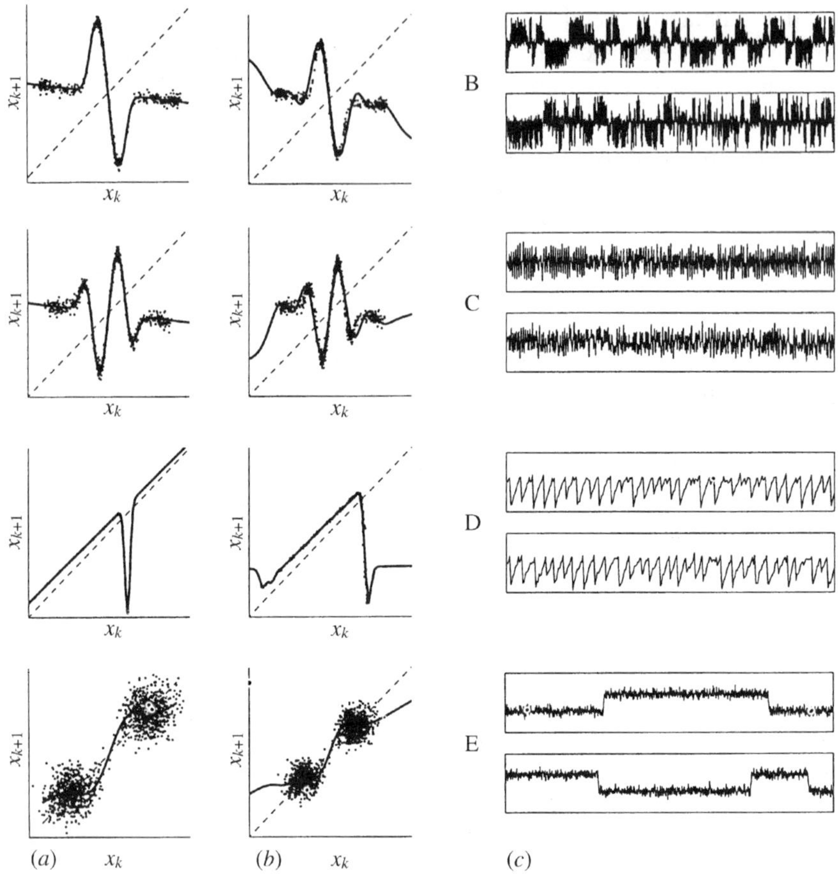

Figure 6.7 More examples of fitting systems with nonlinear dynamics and linear observation functions. Each of the five rows shows the fitting of a nonlinear system with a one-dimensional hidden state and four noisy outputs driven by Gaussian noise inputs and internal-state noise. (*a*) The true dynamics function (line) and states (dots) used to generate the training data. (*b*) The learned dynamics function and states inferred on the training data. (*c*) The first component of the observable time series: training data on the top and fantasy data generated from the learned model on the bottom. The nonlinear dynamics can produce quasi-periodic outputs in response to white driving noise.

states. After the initialization was over, the algorithm discovered the nonlinearities in the dynamics within less than 5 iterations of EM (see Figs. 6.6*a*,*b*, 6.7*a*,*b*, and 6.8*a*,*b*.

After training the models on input–output observations from the dynamics, we examined the learned internal state representation and

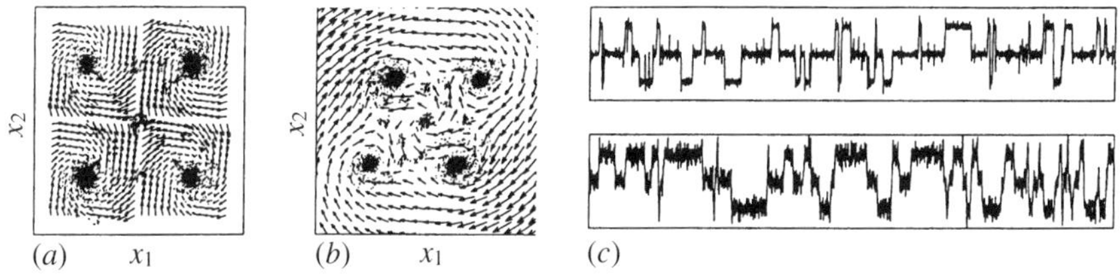

(a) (b) (c)

Figure 6.8 Multidimensional example of fitting a system with nonlinear dynamics and linear observation functions. The true system is piecewise-linear across the state space. The plots show the fitting of a nonlinear system with a two-dimensional hidden state and 4 noisy outputs driven by Gaussian noise inputs and internal state noise. (*a*) The true dynamics vector field (arrows) and states (dots) used to generate the training data. (*b*) The learned dynamics vector field and states inferred on the training data. (*c*) The first component of the observable time series: training data on the top and fantasy data generated from the learned model on the bottom.

compared it with the known structure of the generating system. As the figures show, the algorithm recovers the form of the nonlinear dynamics quite well. We are also able to generate "fantasy" data from the models once they have been learned by exciting them with Gaussian noise of similar variance to that applied during training. The resulting observation streams look qualitatively very similar to the time series from the true systems.

We can quantify this quality of fit by comparing the log-likelihood of the training sequences and novel test sequences under our nonlinear model with the likelihood under a basic linear dynamical system model or a static model such as factor analysis. Figure 6.9 presents this comparison. The nonlinear dynamical system had significantly superior likelihood on both training and test data for all the example systems. (Notice that for system **E**, the linear dynamical system is much better than factor analysis because of the strong hysteresis (mode-locking) in the system. Thus, the output at the previous time step is an excellent predictor of the current output.)

6.3.2 Weather Data

As an example of a real system with a nonlinear *output* function as well as important dynamics, we trained our model on records of the daily maximum and minimum temperatures in Melbourne, Australia, over the period 1981–1990.[10] We used a model with two internal state variables,

[10]This data is available on the world wide web from the Australian Bureau of Meteorology at http://www.bom.gov.au/climate.

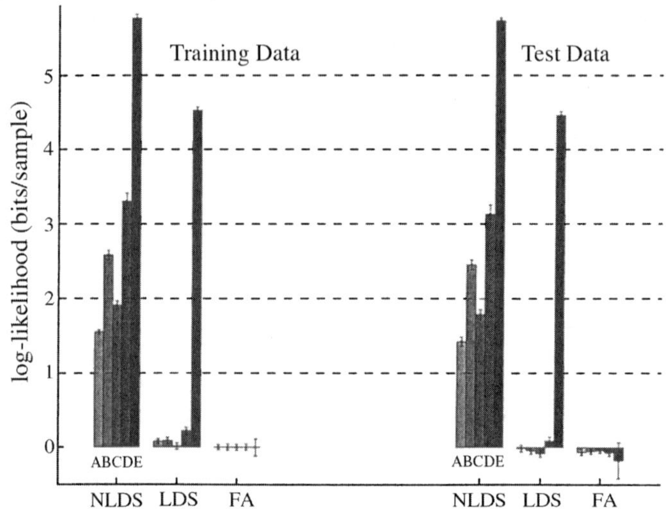

Figure 6.9 Differences in log-likelihood assigned by various models to training and test data from the systems in Figures 6.6 and 6.7. Each adjacent group of five bars shows the log-likelihood of the five examples (A–E) under factor analysis (FA), linear dynamical systems (LDS), and our nonlinear dynamical system model (NLDS). Results on training data appear on the left and results on test data on the right; taller bars represent better models. Log-likelihoods are offset so that FA on the training data is zero. Error bars represent the 68% quantile about the median across 100 repetitions of training or testing. For NLDS, the exact likelihood cannot be computed; what is shown is the pseudo-likelihood computed by EKS.

three outputs, and no inputs. During the training phase, the three outputs were the minimum and maximum daily temperature as well as a real valued output indicating the time of the year (month) in the range [0, 12]. The model was trained on 1500 days of temperature records, or just over four seasons. We tested on the remaining 2150 days by showing the model only the minimum and maximum daily temperatures and attempting to predict the time of year (month). The prediction was performed by using the EKS algorithm to do state inference given only the two available observation streams. Once state inference was performed, the learned output function of the model could be used to predict the time of year. This prediction problem inherently requires the use of information from previous and/or future times, since the static relationship between temperature and season is ambiguous during spring/fall. Figure 6.10 shows the results of this prediction after training; the algorithm has discovered a relationship between the hidden state and the observations that allows it to perform reasonable prediction for this task. Also shown

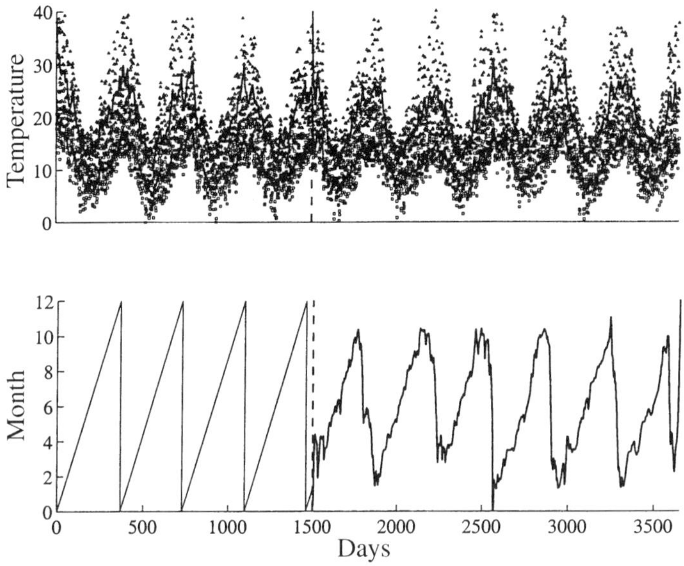

Figure 6.10 Model of maximum and minimum daily temperatures in Melbourne, Australia from 1981 to 1990. *Left of vertical line*: A system with two hidden states governed by linear dynamics and a non-linear output function was trained on observation vectors of a three-dimensional time series consisting of maximum and minimum temperatures for each day as well as the (real-valued) month of the year. Training points are shown as triangles (maximum temperature), squares (minimum temperature) and a solid line (sawtooth wave below). *Right of vertical line*: After training, the system can infer its internal state from only the temperature observations. Having inferred its internal state it can predict the month of the year as a missing output (line below). The solid lines in the upper plots show the model's prediction of minimum and maximum temperature given the inferred state at the time.

are the model predictions of minimum and maximum temperatures given the inferred state.

Although not explicitly part of the generative model, the learned system implicitly parameterizes a relationship between time of year and temperature. We can discover this relationship by evaluating the nonlinear output function at many points in the state space. Each evaluation yields a triple of month, minimum temperature and maximum temperature. These triples can then be plotted against each other as in Figure 6.11 to show that the model has discovered Melbourne's seasonal temperature variations.

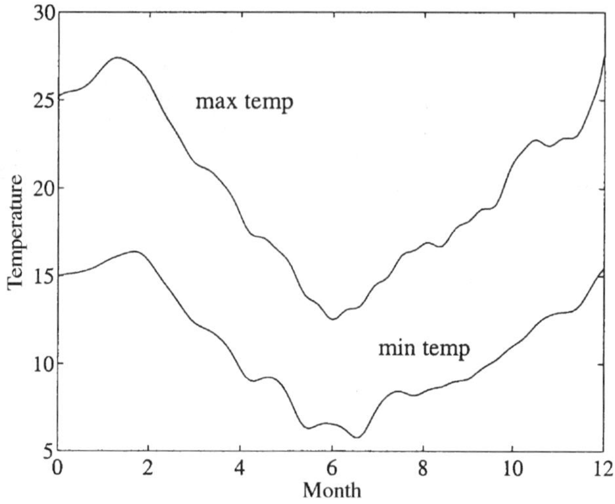

Figure 6.11 Prediction of maximum and minimum daily temperatures based on time of year. The model from Figure 6.10 implicitly learns a relationship between time of year and minimum/maximum temperature. This relationship is not directly invertible, but the temporal information used by extended Kalman smoothing correctly infers month given temperature as shown in Figure 6.10.

6.4 EXTENSIONS

6.4.1 Learning the Means and Widths of the RBFs

It is possible to relax the assumption that the Gaussian radial basis functions have fixed centers and widths, although this results in a somewhat more complicated and slower fitting algorithm. To derive learning rules for the RBF centers c_i and width matrices S_i, we need to consider how they play into the cost function (6.17) through the RBF kernel (6.14). We take derivatives with respect to the expectation of the cost function c, and exchange the order of the expectation and the derivative:

$$\left\langle \frac{\partial c}{\partial c_i} \right\rangle = \left\langle \frac{\partial c}{\partial \rho_i} \frac{\partial \rho_i}{\partial c_i} \right\rangle = 2\langle (\theta\Phi - z)^\top Q^{-1} h_i \rho_i(x) S_i^{-1}(x - c_i) \rangle. \qquad (6.19)$$

Recalling that $\Phi = [\rho_1(x) \quad \rho_2(x) \quad \dots \quad \rho_I(x) \quad x^\top \quad u^\top \quad 1]^\top$, it is clear that c_i figures nonlinearly in several places in this equation, and therefore it is not possible to solve for c_i in closed form. We can, however, use the above gradient to move the center c_i to decrease the cost, which corresponds to taking a *partial* M-step with respect to c_i. Equation (6.19) requires the computation of three third-order expectations in addition to the first- and second-order expectations needed to optimize θ and Q: $\langle \rho_i(x)\rho_k(x)x_l\rangle_j$, $\langle \rho_i(x)x_k x_l\rangle_j$, and $\langle \rho_i(x)z_k x_l\rangle_j$. Similarly, differentiating the cost with respect to S_i^{-1} gives

$$\left\langle \frac{\partial c}{\partial S_i^{-1}} \right\rangle = \left\langle \frac{\partial c}{\partial \rho_i} \frac{\partial \rho_i}{\partial S_i^{-1}} \right\rangle = \langle [(\theta\Phi - z)^\top Q^{-1} h_i] \rho_i(x)[S_i - (x - c_i)(x - c_i)^\top] \rangle. \tag{6.20}$$

We now need three *fourth*-order expectations as well: $\langle \rho_i(x)\rho_k(x)x_l x_m\rangle_j$, $\langle \rho_i(x)_k x_l x_m\rangle_j$, and $\langle \rho_i(x)z_k x_l x_m\rangle_j$.

These additional expectations increase both the storage and computation time of the algorithm – a cost that may not be compensated by the added advantage of moving of centers and widths by small gradient steps. One heuristic is to place centers and widths using unsupervised techniques such as the EM algorithm for Gaussian mixtures, which considers solely the input density and not the output nonlinearity. Alternatively, some of these higher-order expectations can be approximated using, for example, $\langle \rho_i(x)\rangle \approx \rho_i(\langle x\rangle)$.

6.4.2 On-line Learning

One of the major limitations of the algorithm that we have presented in this chapter is that it is a *batch* algorithm; that is, it assumes that we use the entire sequence of observations to estimate the model parameters. Fortunately, it is relatively straightforward to derive an on-line version of the algorithm, which updates parameters as it receives observations. This is achieved using the recursive least-squares (RLS) algorithm, which is in fact just a special case of the discrete Kalman filter (see, e.g., [8, 44]).

The key observation is that the cost minimized in the M-step of the algorithm (6.17) is a quadratic function of the parameters θ. RLS is simply a way of solving quadratic problems on-line. Using k to index time step, the resulting algorithm for scalar z is as follows:

$$\theta_k = \theta_{k-1} + (\langle z\Phi\rangle_k - \theta_{k-1}\langle \Phi\Phi^\top\rangle_k)P_k, \tag{6.21}$$

$$P_k = P_{k-1} - \frac{P_{k-1}\langle \Phi\Phi^\top\rangle_k P_{k-1}}{1 + \langle \Phi^\top P_{k-1}\Phi\rangle_k}, \tag{6.22}$$

$$Q_k = Q_{k-1} + \frac{1}{k}[\langle z^2\rangle_k - \theta_k\langle \Phi z\rangle_k - Q_{k-1}]. \tag{6.23}$$

Let us ignore the expectations for now. Initializing $\theta_0 = 0$, $Q_0 = \mathbf{I}$, and P_0 very large, it is easy to show that, after a few iterations, the estimates of θ_k will rapidly converge to the exact values obtained by the least-squares solution. The estimate of Q will converge to the correct values plus a bias incurred by the fact that the early estimates of Q were based on residuals from θ_k rather than $\lim_{k\to\infty}\theta_k$. P_k is a recursive estimate of $(\sum_{j=1}^{k}\langle\Phi\Phi\rangle_j)^{-1}$, obtained by using the matrix inversion lemma.

There is an important way in which this on-line algorithm is an approximation to the batch EM algorithm we have described for nonlinear state-space models. The expectations $\langle\cdot\rangle_k$ in the online algorithm are computed by running a single step of the extended Kalman filter using the previous parameters θ_{k-1}. In the batch EM algorithm, the expectations are computed by running an extended Kalman *smoother* over the entire sequence using the current parameter estimate. Moreover, these expectations are used to re-estimate the parameters, the smoother is then re-run, the parameters are re-re-estimated, and so on, to perform the usual iterations of EM. In general, we can expect that, unless the time series is nonstationary, the parameter estimates obtained by the batch algorithm after convergence will model the data better than those obtained by the on-line algorithm.

Interestingly, the updates for the RLS on-line algorithm described here are very similar to the parameter updates used a dual extended Kalman filter approach to system identification [45] (see Chapter 5 and Section 6.5.5). This similarity is not coincidental, since, as mentioned, the Kalman filter can be derived as a generalization of the RLS algorithm. In fact, this similarity can be exploited in an elegant manner to derive an on-line algorithm for parameter estimation for nonstationary nonlinear dynamical systems.

6.4.3 Nonstationarity

To handle nonstationary time series, we assume that the parameters can drift according to a Gaussian random walk with covariance Σ_θ:

$$\theta_k = \theta_{k-1} + \epsilon_k, \qquad \text{where} \qquad \epsilon_k \sim \mathcal{N}(0, \Sigma_\theta).$$

As before, we have the following function relating the z variables to the parameters θ and nonlinear kernels Φ:

$$z_k = \theta_k \Phi_k + w_k, \qquad \text{where} \qquad w_k \sim \mathcal{N}(0, Q),$$

which we can view as the observation model for a "state variable" θ_k with time-varying "output matrix" Φ_k. Since both the dynamics and observation models are linear in θ and the noise is Gaussian, we can apply the following Kalman filter to recursively compute the distribution of drifting parameters θ:

$$\hat{\theta}_k = \hat{\theta}_{k-1} + \frac{(\langle z\Phi \rangle_k - \hat{\theta}_{k-1}\langle \Phi\Phi^\top \rangle_k)P_{k|k-1}}{Q_{k-1} + \langle \Phi^\top P_{k|k-1}\Phi \rangle_k}, \tag{6.24}$$

$$P_{k|k-1} = P_{k-1} + \Sigma_\theta, \tag{6.25}$$

$$P_k = P_{k|k-1} - \frac{P_{k|k-1}\langle \Phi\Phi^\top \rangle_k P_{k|k-1}}{Q_{k-1} + \langle \Phi^\top P_{k|k-1}\Phi \rangle_k}, \tag{6.26}$$

$$Q_k = Q_{k-1} + \lambda(\langle z^2 \rangle_k - \hat{\theta}_k \langle \Phi z \rangle_k - Q_{k-1}). \tag{6.27}$$

There are two important things to note. First, these equations describe an ordinary Kalman filter, except that both the "output" z and "output matrix" Φ_k are jointly uncertain with a Gaussian distribution. Second, we have also assumed that the output noise covariance can drift by introducing a forgetting factor λ in its re-estimation equation. As before, the expectations are computed by running one step of the EKF over the hidden variables using θ_{k-1}.

While we derived this on-line algorithm starting from the batch EM algorithm, what we have ended up with appears almost identical to the dual extended Kalman filter (discussed in Chapter 5). Indeed, we have two Kalman filters – one extended and one ordinary – running in parallel, estimating the hidden states and parameters, respectively.

We can also view this on-line algorithm as an approximation to the Bayesian posterior over parameters and hidden variables. The true posterior would be some complicated distribution over the x, z, and θ parameters. Here, we have recursively approximated it with two independent Gaussians – one over (x, z) and one over θ. The approximated posterior for θ_k has mean $\hat{\theta}_k$ and covariance P_k.

6.4.4 Using Bayesian Methods for Model Selection and Complexity Control

Like any other maximum-likelihood procedure, the EM algorithm described in this chapter has the potential to overfit the data set – that is, to find spurious patterns in noise in the data, thereby generalizing poorly. In our implementation, we used some ridge regression, that is, a weight decay regularizer on the h_i parameters, which seemed to work well in practice but required some heuristics for setting regularization parameters. (Although, as mentioned previously, integrating over the hidden variables acts as a sort of modulated input noise, and so, in effect, performs ridge regression, which can eliminate the need for explicit regularization.)

A second closely related problem faced by maximum-likelihood methods is that there is no built-in procedure for doing model selection. That is, the value of the maximum of the likelihood is not a suitable way to choose between different model structures. For example, consider the problems of choosing the dimensionality of the state space x and choosing the number of basis functions I. Higher dimensions of x and more basis functions should always, in principle, result in higher maxima of the likelihood, which means that more complex models will always be preferred to simpler ones. But this, of course, leads to overfitting.

Bayesian methods provide a very general framework for simultaneously handling the overfitting and model selection problems in a consistent manner. The key idea of the Bayesian approach is to avoid maximization wherever possible. Instead, possible models, structures, parameters – in short, all settings of unknown quantities – should be weighted by their posterior probabilities, and predictions should be made according to this weighted posterior.

For our nonlinear dynamical system, we can, for example, treat the parameters θ as an unknown. Then the model's prediction of the output at time $k+1$ is

$$\begin{aligned} p(y_{k+1}|u_{1:k+1}, y_{1:k}) &= \int d\theta \, p(y_{k+1}|u_{k+1}, y_{1:k}, u_{1:k}, \theta) p(\theta | y_{1:k}, u_{1:k}) \\ &= \int d\theta \, p(\theta | y_{1:k}, u_{1:k}) \int dx_{k+1} p(y_{k+1}|u_{k+1}, x_{k+1}, \theta) \\ &\quad \times p(x_{k+1}|u_{1:k+1}, y_{1:k}, \theta), \end{aligned}$$

where the first integral on the last line is over the posterior distribution of the parameters and the second integral is over the posterior distribution of the hidden variables.

The posterior distribution over parameters can be obtained recursively from Bayes' rule:

$$p(\theta | y_{1:k}, u_{1:k}) = \frac{p(y_k|u_{1:k}, y_{1:k-1}, \theta) p(\theta | U_{1:k-1}, y_{1:k-1})}{p(y_k|u_{1:k}, y_{1:k-1})}.$$

The dual extended Kalman filter, the joint extended Kalman filter, and the nonstationary on-line algorithm from Section 6.4.3 are all coarse approximations of these Bayesian recursions.

The above equations are all implicitly conditioned on some choice of model structure s_m, that is, the dimension of x and the number of basis functions. Although the Bayesian modeling philosophy advocates averaging predictions of different model structures, if necessary it is also possible to use Bayes' rule to *choose* between model structures according to their probabilities:

$$P(s_m | y_{1:k}, u_{1:k}) = \frac{p(y_{1:k}|u_{1:k}, s_m)P(s_m)}{\sum_n p(y_{1:k}|u_{1:k}, s_n)P(s_n)}.$$

Tractable approximations to the required integrals can be obtained in several ways. We highlight three ideas, without going into much detail; an adequate solution to this problem for nonlinear dynamical systems requires further research. The first idea is the use of Markov-chain Monte Carlo (MCMC) techniques to sample over both parameters and hidden variables. Sampling can be an efficient way of computing high-dimensional integrals if the samples are concentrated in regions where parameters and states have high probability. MCMC methods such as Gibbs sampling have been used for linear dynamical systems [46, 47], while a promising method for nonlinear systems is particle filtering [18, 24], in which samples ("particles") can be used to represent the joint distribution over parameters and hidden states at each time step. The second idea is the use of so-called "automatic relevance determination" (ARD [48, 49]). This consists of using a zero-mean Gaussian prior on each parameter with tunable variances. Since these variances are parameters that control the prior distribution of the model parameters, they are referred to as *hyperparameters*. Optimizing these variance hyperparameters while integrating over the parameters results in "irrelevant" parameters being eliminated from the model. This occurs when the variance controlling a particular parameter goes to zero. ARD for RBF networks with a center on each data point has been used by Tipping [50] successfully for nonlinear regression, and given the name "relevance vector machine" in analogy to support vector machines. The third idea is the use of variational methods to lower-bound the model structure posterior probabilities. In exactly the same way that EM can be thought of as forming a lower bound on the likelihood using a distribution over the hidden variables, variational Bayesian methods lower-bound the evidence using a distribution over both hidden variables and parameters. Variational Bayesian methods have been used in [51] to infer the structure of linear dynamical systems, although the generalization to nonlinear systems of the kind described in this chapter is not straightforward.

Of course, in principle, the Bayesian approach would advocate averaging over all possible choices of c_i, S_i, I, Q, etc. It is easy to see how this can rapidly get very unwieldy.

6.5 DISCUSSION

6.5.1 Identifiability and Expressive Power

As we saw from the experiments described above, the algorithm that we have presented is capable of learning good density models for a variety of nonlinear time series. Specifying the class of nonlinear systems that our algorithm can model well defines its *expressive power*. A related question is: What is the ability of this model, in principle, to recover the actual parameters of specific nonlinear systems? This is the question of *model identifiability*. These two questions are intimately tied, since they both describe the mapping between actual nonlinear systems and model parameter settings.

There are three trivial degeneracies that make our model technically unidentifiable, but should not concern us. First, it is always possible to permute the dimensions in the state space and, by permuting the domain of the output mapping and dynamics in the corresponding fashion, obtain an exactly equivalent model. Second, the state variables can be rescaled or, in fact, transformed by any invertible linear mapping. This transformation can be absorbed by the output and dynamics functions, yielding a model with identical input–output behavior. Without loss of generality, we always set the covariance of the state evolution noise to be the identity matrix, which both sets the scale of the state space and disallows certain state transformations without reducing the expressive power of the model. Third, we take the observation noise to be uncorrelated with the state noise and both noises to be zero-mean, since, again without loss of generality, these can be absorbed into the f and g functions.[11]

There exist other forms of unidentifiability that are more difficult to overcome. For example, if both f and g are nonlinear, then (at least in the noise-free case), for any arbitrary invertible transformation of the state,

[11]Imagine that the joint noise covariance was nonzero: $\langle w_k v_k^\top \rangle = S$. Replacing A with $A' = A - SR^{-1}C$ gives a new noise process w' with covariance $Q' = Q - SR^{-1}S^\top$ that is uncorrelated with v, leaving the input–output behavior invariant. Similarly, any nonzero noise means can be absorbed into the b terms in the functions f and g.

there exist transformations of f and g that result in identical input–output behavior. In this case, it would he very hard to detect that the recovered model is indeed a faithful model of the actual system, since the estimated and actual states would appear to be unrelated.

Clearly, not all systems can be modeled by assuming that f is linear and g is nonlinear. Similarly, not all systems can be modeled by assuming that f is nonlinear and g is linear. For example, consider the case where the observations y_k and y_{k+n} are statistically independent, but each observation lies on a curved low-dimensional manifold in a high-dimensional space. Modeling this would require a nonlinear g as in nonlinear factor analysis, but an $f = 0$. Therefore, choosing either f or g to be linear restricts the expressive power of the model.

Unlike the state noise covariance Q, assuming that the observation noise covariance R is diagonal *does* restrict the expressive power of the model. This is easy to see for the case where the dimension of the state space is small and the dimension of the observation vector is large. A full covariance R can capture all correlations between observations at a single time step, while a diagonal R model cannot.

For nonlinear dynamical systems, the Gaussian-noise assumption is not as restrictive as it may initially appear. This is because the nonlinearity can be used to turn Gaussian noise into non-Gaussian noise [6].

Of course, we have restricted our expressive power by using an RBF network, especially one in which the means and centers of the RBFs are fixed. One could try to appeal to universal approximation theorems to make the claim that one could, in principle, model any nonlinear dynamical system. But this would be misleading in the light of the noise assumptions and the fact that only a finite and usually small number of RBFs are going to be used in practice.

6.5.2 Embedded Flows

There are two ways to think about the dynamical models we have investigated, shown in Figure 6.12. One is as a nonlinear Markov process (flow) x_k that has been embedded (or potentially projected) into a manifold y_k. From this perspective, the function f controls the evolution of the stochastic process, and the function g specifies the nonlinear embedding (or projection) operation.[12]

[12]To simplify presentation, we shall neglect driving inputs u_k in this section, although the arguments extend as well to systems with inputs.

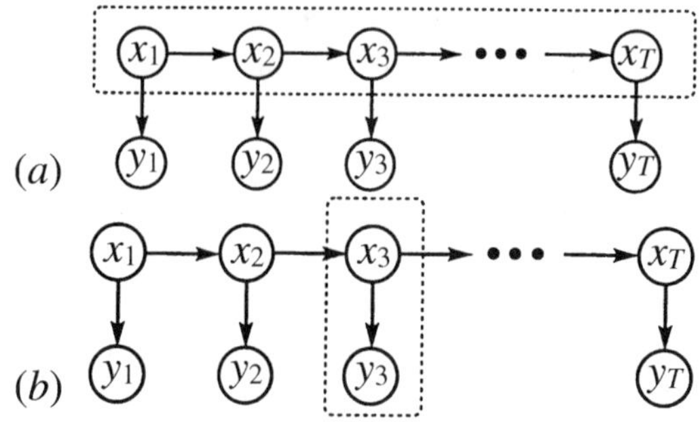

Figure 6.12 Two interpretations of the graphical model for stochastic (non)linear dynamical systems (see text). (*a*) A Markov process embedded in a manifold. (*b*) Nonlinear factor analysis through time.

Another way to think of the same model is as a nonlinear version of a latent-variable model such as factor analysis (but possibly with external inputs as well) in which the latent variables or factors evolve through time rather than being drawn independently for each observation. The nonlinear factor analysis model is represented by g and the time evolution of the latent variables by f.

If the state space is of lower dimension than the observation space and the observation noise is additive, then a useful geometrical intuition applies. In such cases, we have observed a flow inside an embedded manifold. The observation function g specifies the structure (shape) of the manifold, while the dynamics f specifies the flow within the manifold. Armed with this intuition, the learning problem looks as if it might be decoupled into two separate stages: first find the manifold by doing some sort of density modeling on the collection of observed outputs (ignoring their time order); second, find the flow (dynamics) by projecting the observations into the manifold and doing nonlinear regression from one time step to the next. This intuition is partly true, and indeed provides the basis for many of the practical and effective initialization schemes that we have tried. However, the crucial point as far as the design of learning algorithms is concerned is that the two learning problems interact in a way that makes the problem *easier*. Once we know something about the dynamics, this information gives some prior knowledge when trying to learn the manifold shape. For example, if the dynamics suggest that the next state will be near a certain point, we can use this information to do better than naive projection when we locate a noisy observation on the manifold. Conversely, knowing something about the manifold allows us to estimate the dynamics more effectively.

We discuss separately two special cases of flows in manifolds: systems with linear output functions but nonlinear dynamics, and systems with linear dynamics but nonlinear output function.

When the output function g is linear and the dynamics f is nonlinear (Fig. 6.13), the observed sequence forms a nonlinear flow in a linear subspace of the observation space. The manifold estimation is made easier, even with high levels of observation noise, by the fact that its shape is known to be a hyperplane. All that is required is to find its orientation and the character of the output noise. Time-invariant analysis of the observations by algorithms such as factor analysis is an excellent way to initialize estimates of the hyperplane and noises. However, during learning, we may have cause to tilt the hyperplane to make the dynamics fit better, or conversely cause to modify the dynamics to make the hyperplane model better.

This setting is actually more expressive than it might seem initially. Consider a nonlinear output function $g(x)$ that is "invertible" in the sense that it be written in the form $g(x) = C\tilde{g}(x)$ for invertible $\tilde{g}$ and non-square matrix C. Any such nonlinear output function can be made strictly linear if we transform to a new state variable $\tilde{x}$:

$$\tilde{x} = \tilde{g}(x) \Rightarrow \tilde{x}_{k+1} = \tilde{f}(\tilde{x}_k, w_k) = \tilde{g}(f(\tilde{g}^{-1}(\tilde{x})) + w_k), \qquad (6.28a)$$

$$y_k = C\tilde{x}_k + v_k = g(x_k) + v_k, \qquad (6.28b)$$

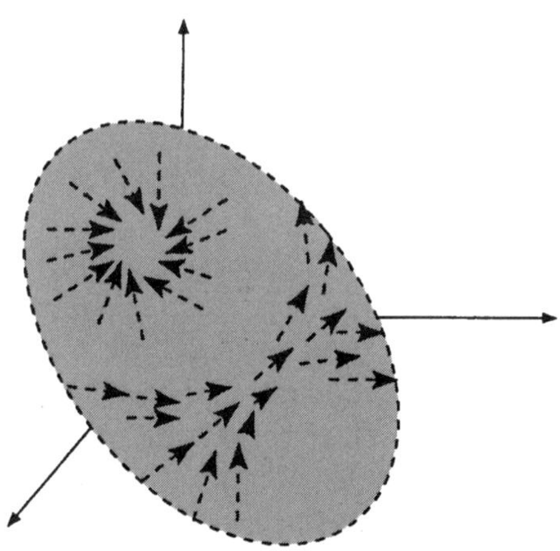

Figure 6.13 Linear and nonlinear dynamical systems represent flow fields embedded in manifolds. For systems with linear output functions, such as the one illustrated, the manifold is a hyperplane while the dynamics may be complex. For systems with nonlinear output functions the shape of the embedding manifold is also curved.

which gives an equivalent model but with a purely linear output process, and potentially nonadditive dynamics noise.

For nonlinear output functions g paired with linear dynamics f, the observation sequence forms a matrix (linear) flow in a nonlinear manifold:

$$x_{k+1} = Ax_k + w_k, \tag{6.29a}$$

$$y_k = g(x_k) + v_k. \tag{6.29b}$$

The manifold learning is harder now, because we must estimate a thin, curved subspace of the observation space in the presence of noise. However, once we have learned this manifold approximately, we project the observations into it and learn only linear dynamics. The win comes from the following fact: in the locations where the projected dynamics do not look linear, we know that we should bend the manifold to make the dynamics more linear. Thus, not only the shape of the outputs (ignoring time) but also the linearity of the dynamics give us clues to learning the manifold.

6.5.3 Stability

Stability is a key issue in the study of any dynamical system. Here we have to consider stability at two levels: the stability of the learning procedure, and the stability of the learned nonlinear dynamical system.

Since every step of the EM algorithm is guaranteed to increase the log-likelihood until convergence, it has a built-in Lyapunov function for stable learning. However, as we have pointed out, our use of extended Kalman smoothing in the E-step of the algorithm represents an approximation to the exact E-step, and therefore we have to forego any guarantees of stability of learning. While we rarely had problems with stability of learning, this is sure to be problem-specific, depending both on the quality of the EKS approximation and on how close the true system dynamics is to the boundary of stability. In contrast to the EKS approximations, certain variational approximations [29] transform the intractable Lyapunov function into a tractable one, and therefore preserve stability of learning. It is not clear how to apply these variational approximations to nonlinear dynamics, although this would clearly be an interesting area of research.

Stability of the learned nonlinear dynamical system can be analyzed by making use of some linear systems theory. We know that, for discrete-time linear dynamical systems, if all eigenvalues of the A matrix lie inside the unit circle, then the system is globally stable. The nonlinear dynamics of our RBF network f can be decomposed into two parts (cf. Eq. (6.13)): a linear component given by A, and a nonlinear component given by $\sum_i h_i \rho_i(x)$. Clearly, for the system to be globally stable, A has to satisfy the eigenvalue criterion for linear systems. Moreover, if the RBF coefficients for both f and g have bounded norm (i.e., $\max_i |h_i| < \bar{h}$) and the RBF is bounded, with $\min_i \det(S_i) > s_{\min} > 0$ and $\max_{ij} |c_i - c_j| < \bar{c}$, then the nonlinear system is stable in the following sense. The conditions on S_i and h mean that

$$\left| \sum_i h_i \rho_i(x) \right| < \frac{I\bar{h}}{(2\pi)^{d/2} \sqrt{s_{\min}}} \equiv \kappa.$$

Therefore, the noise-free nonlinear component of the dynamics alone will always maintain the state within a sphere of radius κ around $\bar{c}$. So, if the linear component is stable, then for any sequence of bounded inputs, the output sequence of the noise-free system will be bounded. Intuitively, although unstable behavior might occur in the region of RBF support, once x leaves this region it is drawn back in by A.

For the on-line EM learning algorithm, the hidden state dynamics and the parameter re-estimation dynamics will interact, and therefore a stability analysis would be quite challenging. However, since there is no stability guarantee for the batch EKS-EM algorithm, it seems very unlikely that a simple form of the on-line algorithm could be provably stable.

6.5.4 Takens' Theorem and Hidden States

It has long been known that for linear systems, there is an equivalence between so called state-space formulations involving hidden variables and direct vector autoregressive models of the time series. In 1980, Takens proved a remarkable theorem [52] that tells us that, for almost any deterministic nonlinear dynamical system with a d-dimensional state space, the state can be effectively reconstructed by observing $2d + 1$ time lags of any one of its outputs. In particular, Takens showed that such a lag vector will be a smooth embedding (diffeomorphism) of the true state, if one exists. This notion of finding an "embedding" for the state has been used to justify a nonlinear regression approach to learning nonlinear dynamical systems. That is, if you suspect that the system is nonlinear and that it has d state dimensions, then instead of building a state-space model, you can do away with representing states and just build an autoregressive (AR) model directly on the observations that nonlinearly relates previous outputs and the current output. (Chapter 4 discusses the case of chaotic dynamics.) This view begs the question: Do we need our models to have hidden states at all?

While no constructive realization for Takens' theorem exists in general, there are very strong results for linear systems. For purely linear systems, we can appeal to the Cayley–Hamilton theorem[13] to show that the hidden state can always be eliminated to obtain an equivalent vector autoregressive model by taking only d time lags of the output. Furthermore, there is a construction that allows this conversion to be performed explicitly.[14] Takens' theorem offers us a similar guarantee for elimination of hidden states in nonlinear dynamical systems, as long as we take $2d + 1$ output lags. (However, no similar recipe exists for explicitly converting to an autoregressive form). These results appear to make hidden states unnecessary.

The problem with this view is that it does not generalize well to many realistic high-dimensional and noisy scenarios. Consider the example mentioned in the introduction. While it is mathematically true that the pixels in the video frame of a balloon floating in the wind are a (highly nonlinear) function of the pixels in the previous video frames, it would be ludicrous from the modeling perspective to build an AR model of the video images. This would require a number of parameters of the order of the number of pixels squared. Furthermore, unlike the noise-free case of Takens' theorem, when the dynamics are noisy, the optimal prediction of the observation would have to depend on the entire history of past observations. Any truncation of this history throws away potentially valuable information about the unobserved state. The state-space formulation of nonlinear dynamical systems allows us to overcome both of these limitations of nonlinear autoregressive models. That is, it allows us to have compact representations of dynamics, and to integrate uncertain information over time. The price paid for this is that it requires inference over the hidden state.

[13]Any square matrix A of size n satisfies its own characteristic equation. Equivalently, any matrix power A^m for $m \geq n$ can be written as a linear combination of lower matrix powers $I, A, A^2, \ldots, A^{n-1}$.

[14]Start with the system $x_{k+1} = Ax_k + w_k$, $y_k = Cx_k + v_k$. Create a d-dimensionl lag vector $z_k = [y_k; y_{k+1}; \ldots; y_{k+d-1}]$ that holds the current and $d - 1$ future outputs. Write $z_k = Gx_k + n_k$ for $G = [CI; CA; CA^2; \ldots; CA^{d-1}]$ and Gaussian noise n (although with nondiagonal covariance). The Cayley–Hamilton theorem assures us that G is full rank, and thus we need not take any more lags. Given the lag vector z_k, we can solve the system $z_k = Gx_k$ for x_k; write this solution as G^+z_k. Using the original observation equation d times, to solve for $y_k, \ldots, y_{k+d-1}$ in terms of z_k, we can write an autoregression for z_k as $z_{k+1} = G^+AGz_k + m_k$ for Gaussian noise m.

6.5.5 Should Parameters and Hidden States be Treated Differently?

The maximum-likelihood framework on which the EM algorithm is based makes a distinction between parameters and hidden variables: it attempts to *integrate* over hidden variables to *maximize* the likelihood as a function of parameters. This leads to the two-step approach, which computes sufficient statistics over the hidden variables in the E-step and optimizes parameters in the M-step. In contrast, a fully Bayesian approach to learning nonlinear dynamical state-space models would treat both hidden variables and parameters as unknown and attempt to compute or approximate the joint posterior distribution over them – in effect integrating over both.

It is important to compare these approaches to system identification with more traditional ones. We highlight two such approaches: *joint EKF* approaches and *dual EKF* approaches.

In joint EKF approaches [7, 8], an augmented hidden state space is constructed that comprises the original hidden state space and the parameters. Since parameters and hidden states interact, even for linear dynamical systems this approach results in nonlinear dynamics over the augmented hidden states. Initializing a Gaussian prior distribution both over parameters and over states, an extended Kalman filter is then used to recursively update the joint distribution over states and parameters based on the observations, $p(X, \theta | Y)$. This approach has the advantage that it can model uncertainties in the parameters and correlations between parameters and hidden variables. In fact, this approach treats parameters and state variables completely symmetrically, and can be thought of as iteratively implementing a Gaussian approximation to the recursive Bayes' rule computations. Nonstationarity can be easily built in by giving the parameters (e.g., random-walk) dynamics. Although it has some very appealing properties, this approach is known to suffer from instability problems, which is the reason why dual EKF approaches have been proposed.

In dual EKF approaches (see Chapter 5), two interacting but distinct extended Kalman filters run simultaneously. One computes a Gaussian approximation of the state posterior given a parameter estimate and the observations: $p(X | \hat{\theta}_{\text{old}}, Y)$, while the other computes a Gaussian approximation of the parameter posterior given the estimated states $p(\theta | \hat{X}_{\text{old}}, Y)$. The two EKFs interact by each feeding its estimate (i.e., the posterior means $\hat{X}$ and $\hat{\theta}$) into the other. One can think of the dual EKF as performing approximate coordinate ascent in $p(X, \theta | Y)$ by iteratively maximizing $p(X|\hat{\theta}_{\text{old}}, Y)$ and $p(\theta|\hat{X}_{\text{old}}, Y)$ under the assumption that each conditional is Gaussian. Since the only interaction between parameters and hidden variables occurs through their respective means, the procedure has the flavor of mean-field methods in physics and neural networks [53]. Like these methods, it is also likely to suffer from the overconfidence problem – namely, since the parameter estimate does not take into account the uncertainty in the states, the parameter covariance will be overly narrow, and likewise for the states.

For large systems, both joint and dual EKF methods suffer from the fact that the parameter covariance matrix is quadratic in the number of parameters. This problem is more pronounced for the joint EKF, since it considers the concatenated state space. Furthermore, both joint and dual EKF methods rely on Gaussian approximations to parameter distributions. This can sometimes be problematic – for example, consider retaining positive-definiteness of a noise covariance matrix under the assumption that its parameters are Gaussian-distributed.

6.6 CONCLUSIONS

This chapter has brought together two classic algorithms – one from statistics and another from systems engineering – to address the learning of stochastic nonlinear dynamical systems. We have shown that by pairing the extended Kalman smoothing algorithm for approximate state estimation in the E-step with a radial basis function learning model that permits exact analytic solution of the M-step, the EM algorithm is capable of learning a nonlinear dynamical model from data. As a side-effect we have derived an algorithm for training a radial basis function network to fit data in the form of a mixture of Gaussians. We have also derived an on-line version of the algorithm and a version for dealing with nonstationary time series.

We have demonstrated the algorithm on a series of synthetic and realistic nonlinear dynamical systems, and have shown that it is able to learn accurate models from only observations of inputs and outputs. Initialization of model parameters and placement of the radial basis kernels are important to the practical success of the algorithm. We have discussed techniques for making these choices, and have provided gradient rules for adapting the centers and widths of the basis functions.

The main strength of our algorithm is that by making a specific choice of nonlinear estimator (Gaussian radial basis networks), we are able to exactly account for the uncertain state estimates generated during inference. Furthermore, the parameter-update procedures still only require the solution of systems of linear equations. However, we rely on the standard, but potentially inaccurate, extended Kalman smoother for approximate inference. For certain problems where local linearization is an extremely poor approximation, greater accuracy may be achieved using other approximate inference techniques such as the unscented filter (see Chapter 7). Another area worthy of further investigation is how to initialize the parameters more effectively when the data lie on a nonlinear manifold; in these cases, factor analysis provides an inadequate static model.

The belief network literature has recently been dominated by two methods for approximate inference: Markov-chain Monte Carlo [54] and variational approximations [29]. To the best of our knowledge, [36] and [45] were the first instances where extended Kalman smoothing was used to perform approximate inference in the E-step of EM. While EKS does not have the theoretical guarantees of variational methods (which are also approximate, but monotonically optimize a computable objective function during learning), its simplicity has gained it wide acceptance in the estimation and control literatures as a method for doing inference in nonlinear dynamical systems. Our practical success in modeling a variety of nonlinear time series suggests that the combination of extended Kalman algorithms and the EM algorithm can provide powerful tools for learning nonlinear dynamical systems.

ACKNOWLEDGMENTS

The authors acknowledge support from the Gatsby Charitable Fund, and thank Simon Haykin for helpful comments that helped to clarify the presentation. S.R. is supported in part by a National Science Foundation PDF Fellowship and a grant from the Natural Sciences and Engineering Research Council of Canada.

APPENDIX: EXPECTATIONS REQUIRED TO FIT THE RBFs

The expectations that we need to compute for Eq. (6.78) are $\langle x\rangle_j$, $\langle z\rangle_j$, $\langle xx^\top\rangle_j$, $\langle zz^\top\rangle_j$, $\langle xz^\top\rangle_j$, $\langle\rho_i(x)\rangle_j$, $\langle x\rho_i(x)\rangle_j$, $\langle z\rho_i(x)\rangle_j$, and $\langle\rho_i(x)\rho_l(x)\rangle_j$. Starting with some of the easier ones that do not depend on the RBF kernel ρ, we have

$$\begin{aligned} \langle x\rangle_j &= \mu_j^x, & \langle z\rangle_j &= \mu_j^z, \\ \langle xx^\top\rangle_j &= \mu_j^x\mu_j^{x,\top} + C_j^{xx}, & \langle zz^\top\rangle_j &= \mu_j^z\mu_n^{z,\top} + C_j^{zz}, \\ \langle xz^\top\rangle_j &= \mu_j^x\mu_j^{z,\top} + C_j^{xz}. \end{aligned}$$

Observe that when we multiply the Gaussian RBF kernel $\rho_i(x)$ (Eq. (6.14)) and $\mathcal{N}_j$, we get a Gaussian density over (x, z) with mean and covariance

$$\mu_{ij} = C_{ij}\left(C_j^{-1}\mu_j + \begin{bmatrix} S_i^{-1}c_i \\ 0 \end{bmatrix}\right), \quad C_{ij} = \left(C_j^{-1} + \begin{bmatrix} S_i^{-1} & 0 \\ 0 & 0 \end{bmatrix}\right)^{-1},$$

and an extra constant (due to lack of normalization),

$$\beta_{ij} = (2\pi)^{-d_x/2}|S_i|^{-1/2}|C_j|^{-1/2}|C_{ij}|^{1/2}\exp(-\tfrac{1}{2}\delta_{ij}),$$

where

$$\delta_{ij} = c_i^\top S_i^{-1}c_i + \mu_j^\top C_j^{-1}\mu_j - \mu_{ij}^\top C_{ij}^{-1}\mu_{ij}.$$

Using β_{ij} and μ_{ij}, we can evaluate the other expectations:

$$\langle\rho_i(x)\rangle_j = \beta_{ij}, \qquad \langle x\rho_i(x)\rangle_j = \beta_{ij}\mu_{ij}^x, \qquad \langle z\rho_i(x)\rangle_j = \beta_{ij}\mu_{ij}^z.$$

Finally,

$$\langle\rho_i(x)\rho_l(x)\rangle_j = (2\pi)^{-d_x}|C_j|^{-1/2}|S_i|^{-1/2}|S_l|^{-1/2}|C_{ilj}|^{1/2}\exp(\tfrac{1}{2}\gamma_{ilj}),$$

where

$$C_{ilj} = \left(C_j^{-1} + \begin{bmatrix} S_i^{-1} + S_l^{-1} & 0 \\ 0 & 0 \end{bmatrix}\right)^{-1},$$

$$\mu_{ilj} = C_{ilj}\left(C_j^{-1}\mu_j + \begin{bmatrix} S_i^{-1}c_i + S_l^{-1}c_l \\ 0 \end{bmatrix}\right),$$

$$\gamma_{ilj} = c_i^\top S_i^{-1}c_i + c_l^\top S_l^{-1}c_l + \mu_j^\top C_j^{-1}\mu_j - \mu_{ilj}^\top C_{ilj}^{-1}\mu_{ilj}.$$

REFERENCES

[1] R.E. Kalman and R.S. Bucy, "New results in linear filtering and prediction," *Transactions of the ASME Ser. D, Journal of Basic Engineering*, **83**, 95–108 (1961).

[2] L.E. Baum and J.A. Eagon, "An equality with applications to statistical estimation for probabilistic functions of markov processes and to a model for ecology," *Bulletin of the American Mathematical Society*, **73**, 360–363 (1967).

[3] A.P. Dempster, N.M. Laird, and D.B. Rubin, "Maximum likelihood from incomplete data via the EM algorithm," *Journal of the Royal Statistical Society, Ser. B*, **39**, 1–38 (1977).

[4] J. Pearl, *Probabilistic Reasoning in Intelligent Systems: Networks of Plausible Inference*. San Mateo, CA: Morgan Kaufmann, 1988.

[5] S.L. Lauritzen and D.J. Spiegelhalter, "Local computations with probabilities on graphical structures and their application to expert systems," *Journal of the Royal Statistical Society, Ser. B*, **50**, 157–224 (1988).

[6] S. Roweis and Z. Ghahramani, "A unifying review of linear Gaussian models," *Neural Computation*, **11**, 305–345 (1999).

[7] L. Ljung and T. Söderström, *Theory and Practice of Recursive Identification*. Cambridge, MA: MIT Press, 1983.

[8] G.C. Goodwin and K.S. Sin, *Adaptive Filtering, Prediction and Control*. Englewood Cliffs, NJ: Prentice-Hall, 1984.

[9] J. Moody and C. Darken, "Fast learning in networks of locally-tuned processing units," *Neural Computation*, **1**, 281–294 (1989).

[10] D.S. Broomhead and D. Lowe, "Multivariable functional interpolation and adaptive networks," *Complex Systems*, **2**, 321–355 (1988).

[11] R.E. Kalman, "A new approach to linear filtering and prediction problems," *Journal of Basic Engineering*, **82**, 35–45 (1960).

[12] H.E. Rauch, "Solutions to the linear smoothing problem," *IEEE Transactions on Automatic Control*, **8**, 371–372 (1963).

[13] R.E. Kopp and R.J. Orford, "Linear regression applied to system identification and adaptive control systems," *AIEE Journal*, **1**, 2300–2306 (1963).

[14] H. Cox, "On the estimation of state variables and parameters for noisy dynamic systems." *IEEE Transactions on Automatic Control*, **9**, 5–12 (1964).

[15] S. Singhal and L. Wu, "Training multiplayer perceptrons with the extended Kalman algorithm," in *Advances in Neural Information Processing Systems*, Vol. 1, San Mateo, CA: Morgan Kaufmann, 1989, 133–140.

[16] V. Kadirkamanathan and M. Niranjan, "A functional estimation approach to sequential learning with neural networks," *Neural Computation*, **5**, 954–975 (1993).

[17] I.T. Nabney, A. McLachlan, and D. Lowe, "Practical methods of tracking of nonstationary time series applied to real-world data," in S. K. Rogers and D. W. Ruck, Eds. *AeroSense '96: Applications and Science of Artificial Neural Networks II*. SPIE Proceedings, 1996, pp. 152–163.

[18] J.E. Handschin and D.Q. Mayne, "Monte Carlo techniques to estimate the conditional expectation in multi-stage non-linear filtering," *International Journal of Control*, **9**, 547–559 (1969).

[19] N.J. Gordon, D.J. Salmond, and A.F.M. Smith, "A novel approach to nonlinear/non-Gaussian Bayesian state space estimation," *IEE Proceedings F: Radar and Signal Processing*, **140**, 107–113 (1993).

[20] G. Kitagawa, "Monte Carlo filter and smoother for non-Gaussian nonlinear state space models," *Journal of Computer Graphics and Graphical Statistics*, **5**, 1–25 (1996).

[21] M. Israd and A. Blake, "CONDENSATION – conditional density propagation for visual tracking," *International Journal of Computer Vision*, **29**, 5–28 (1998).

[22] M. West, "Approximating posterior distributions by mixtures," *Journal of the Royal Statistical Society, Ser, B*, **54**, 553–568 (1993).

[23] M. West, "Mixture models, monte carlo, Bayesian updating and dynamic models," *Computing Science and Statistics*, **24**, 325–333 (1993).

[24] A. Doucet, J.F.G. de Freitas, and N.J. Gordon, *Sequential Monte Carlo Methods in Practice*. New York: Springer-Verlag, 2000.

[25] S.J. Julier, J.K. Uhlmann, and H.F. Durrant-Whyte, "A new approach for filtering nonlinear systems," in *Proceedings of the 1995 American Control Conference, Seattle, 1995*, pp. 1628–1632.

[26] S.J. Julier and J.K. Uhlmann, "A new extension of the Kalman filter to nonlinear systems," in *Proceedings of AeroSense: The 11th International Symposium on Aerospace/Defense Sensing, Simulation and Controls, Orlando, FL, 1997*.

[27] E.A. Wan, R. van der Merwe, and A.T. Nelson, "Dual estimation and the unscented transformation," in *Advances in Neural Information Processing Systems*, Vol. 12, Cambridge, MA: MIT Press, 1999.

[28] Z. Ghahramani and G.E. Hinton, "Variational learning for switching state-space models," *Neural Computation*, **12**, 831–864 (2000).

[29] M.I. Jordan, Z. Ghahramani, T.S. Jaakkola, and L.K. Saul, "An introduction to variational methods in graphical models," *Machine Learning*, **37**, 183–233 (1999).

[30] L.E. Baum and T. Petrie, "Statistical inference for probabilistic functions of finite state Markov chains," *Annals of Mathematical Statistics*, **37**, 1554–1563 (1966).

[31] R.M. Neal and G.E. Hinton, "A view of the EM algorithm that justifies incremental sparse and other variants," in M.I. Jordan, ed., *Learning in Graphical Models*. Dordrecht: Kluwer Academic, 1998, pp. 355–368.

[32] I. Csiszár and G. Tusnády, "Information geometry and alternating minimization procedures," *Statistics & Decisions*, Supplement Issue 1, pp. 205–237, 1984.

[33] C.F. Chen, "The EM approach to the multiple indicators and multiple causes model via the estimation of the latent variables." *Journal of the American Statistical Association*, **76**, 704–708 (1981).

[34] R.H. Shumway and D.S. Stoffer, "An approach to time series smoothing and forecasting using the EM algorithm," *Journal of Time Series Analysis*, **3**, 253–264 (1982).

[35] Z. Ghahramani and G. Hinton, "Parameter estimation for linear dynamical systems," Technical Report CRG-TR-96-2, Department of Computer Science, University of Toronto, February 1996.

[36] Z. Ghahramani and S. Roweis, "Learning nonlinear dynamical systems using an EM algorithm," in *Advances in Neural Information Processing Systems*, Vol. 11. Cambridge, MA: MIT Press, 1999, pp. 431–437.

[37] J.F.G. de Freitas, M. Niranjan, and A.H. Gee, "Nonlinear state space estimation with neural networks and the EM algorithm," Technical Report, Cambridge University Engineering Department, 1999.

[38] T. Briegel and V. Tresp, "Fisher scoring and a mixture of modes approach for approximate inference and learning in nonlinear state space models," in *Advances in Neural Information Processing Systems*, Vol. 11. Cambridge, MA, MIT Press, 1999.

[39] Z. Ghahramani and G. Hinton "Switching state-space models," Technical Report CRG-TR-96-3, Department of Computer Science, University of Toronto, July 1996.

[40] K. Murphy, "Switching Kalman filters," Technical Report, Department of Computer Science, University of California, Berkeley, August, 1998.

[41] G.E. Hinton, P. Dayan, and M. Revow, "Modeling the manifolds of Images of handwritten digits," *IEEE Transactions on Neural Networks*, **8**, 65–74 (1997).

[42] Z. Ghahramani and G. Hinton, "The EM algorithm for mixtures of factor analyzers," Technical Report CRG-TR-96-1, Department of Computer Science, University of Toronto, May 1996 (revised February 1997).

[43] W.S. Torgerson, "Multidimensional scaling I. Theory and method," *Psychometrika*, **17**, 401–419 (1952).

[44] S. Haykin, *Adaptive Filter Theory*, 3rd ed. Upper Saddle River, NJ: Prentice-Hall, 1996.

[45] E.A. Wan and A.T. Nelson, "Dual Kalman filtering methods for nonlinear prediction," in *Advances in Neural Information Processing Systems*, Vol. 9. Cambridge, MA: MIT Press, 1997.

[46] C.K. Carter and R. Kohn, "On Gibbs sampling for state space models," *Biometrika*, **81**, 541–553 (1994).

[47] S. Früwirth-Schnatter, "Bayesian model discrimination and Bayes factors for linear Gaussian state space models." *Journal of the Royal Statistical Society, Ser. B*, **57**, 237–246 (1995).

[48] D.J.C. MacKay, "Bayesian non-linear modelling for the prediction competition," *ASHRAE Transcations*, **100**, 1053–1062 (1994).

[49] R.M. Neal, "Assessing relevance determination methods using DELVE," in C.M. Bishop, Ed. *Neural Networks and Machine Learning*. New York: Springer-Verlag, 1998, pp. 97–129.

[50] M.E. Tipping, "The relevance vector machine," in *Advances in Neural Information Processiing Systems*, Vol. 12. Cambridge, MA: MIT Press, 2000, pp. 652–658.

[51] Z. Ghahramani and M.J. Beal, "Propagation algorithms for variational Bayesian learning." in *Advances in Neural Information Processing Systems*, Vol. 13. Cambridge, MA: MIT Press, 2001.

[52] F. Takens, "Detecting strange attractors in turbulence," in D.A. Rand and L.-S. Young, Eds., *Dynamical Systems and Turbulence, Warwick 1980*, Lecture Notes in Mathematics, Vol. 898. Berlin: Springer-Verlag, 1981, pp. 365–381.

[53] J. Hertz, A. Krogh, and R.G. Palmer, *Introduction to the Theory of Neural Computation*. Redwood City, CA: Addison-Wesley, 1991.

[54] R.M. Neal, "Probablistic inference using Markov chain Monte Carlo methods," Technical Report CRG-TR-93-1, Department of Computer Science, University of Toronto, 1993.

7

THE UNSCENTED KALMAN FILTER

Eric A. Wan and Rudolph van der Merwe

Department of Electrical and Computer Engineering, Oregon Graduate Institute of Science and Technology, Beaverton, Oregon, U.S.A.

7.1 INTRODUCTION

In this book, the *extended Kalman filter* (EKF) has been used as the standard technique for performing recursive nonlinear estimation. The EKF algorithm, however, provides only an *approximation* to optimal nonlinear estimation. In this chapter, we point out the underlying assumptions and flaws in the EKF, and present an alternative filter with performance superior to that of the EKF. This algorithm, referred to as the *unscented Kalman filter* (UKF), was first proposed by Julier et al. [1–3], and further developed by Wan and van der Merwe [4–7].

The basic difference between the EKF and UKF stems from the manner in which Gaussian random variables (GRV) are represented for propagating through system dynamics. In the EKF, the state distribution is

Kalman Filtering and Neural Networks, Edited by Simon Haykin
ISBN 0-471-36998-5

approximated by a GRV, which is then propagated analytically through the first-order linearization of the nonlinear system. This can introduce large errors in the true posterior mean and covariance of the transformed GRV, which may lead to suboptimal performance and sometimes divergence of the filter. The UKF address this problem by using a deterministic sampling approach. The state distribution is again approximated by a GRV, but is now represented using a minimal set of carefully chosen sample points. These sample points completely capture the true mean and covariance of the GRV, and, when propagated through the *true* nonlinear system, captures the posterior mean and covariance accurately to second order (Taylor series expansion) for *any* nonlinearity. The EKF, in contrast, only achieves first-order accuracy. No explicit Jacobian or Hessian calculations are necessary for the UKF. Remarkably, the computational complexity of the UKF is the same order as that of the EKF.

Julier and Uhlman demonstrated the substantial performance gains of the UKF in the context of state estimation for nonlinear control. A number of theoretical results were also derived. This chapter reviews this work, and presents extensions to a broader class of nonlinear estimation problems, including nonlinear system identification, training of neural networks, and dual estimation problems. Additional material includes the development of an unscented Kalman *smoother* (UKS), specification of efficient recursive *square-root* implementations, and a novel use of the UKF to improve *particle filters* [6].

In presenting the UKF, we shall cover a number of application areas of nonlinear estimation in which the EKF has been applied. General application areas may be divided into *state estimation*, *parameter estimation* (e.g., learning the weights of a neural network), and *dual estimation* (e.g., the expectation–maximization (EM) algorithm). Each of these areas place specific requirements on the UKF or EKF, and will be developed in turn. An overview of the framework for these areas is briefly reviewed next.

State Estimation The basic framework for the EKF involves estimation of the state of a discrete-time nonlinear dynamical system,

$$\mathbf{x}_{k+1} = \mathbf{F}(\mathbf{x}_k, \mathbf{u}_k, \mathbf{v}_k), \tag{7.1}$$

$$\mathbf{y}_k = \mathbf{H}(\mathbf{x}_k, \mathbf{n}_k), \tag{7.2}$$

where $\mathbf{x}_k$ represents the unobserved state of the system, $\mathbf{u}_k$ is a known exogeneous input, and $\mathbf{y}_k$ is the observed measurement signal. The *process noise* $\mathbf{v}_k$ drives the dynamic system, and the *observation noise* is given by $\mathbf{n}_k$. Note that we are not assuming additivity of the noise sources. The system dynamical model $\mathbf{F}$ and $\mathbf{H}$ are assumed known. A simple block diagram of this system is shown in Figure 7.1. In state estimation, the EKF is the standard method of choice to achieve a recursive (approximate) maximum-likelihood estimate of the state $\mathbf{x}_k$. For completeness, we shall review the EKF and its underlying assumptions in Section 7.2 to help motivate the presentation of the UKF for state estimation in Section 7.3.

Parameter Estimation Parameter estimation, sometimes referred to as system identification or machine learning, involves determining a nonlinear mapping

$$\mathbf{y}_k = \mathbf{G}(\mathbf{x}_k, \mathbf{w}), \tag{7.3}$$

where $\mathbf{x}_k$ is the input, $\mathbf{y}_k$ is the output, and the nonlinear map $\mathbf{G}(\cdot)$ is parameterized by the vector $\mathbf{w}$. The nonlinear map, for example, may be a feedforward or recurrent neural network ($\mathbf{w}$ are the weights), with numerous applications in regression, classification, and dynamic modeling. Learning corresponds to estimating the parameters $\mathbf{w}$. Typically, a training set is provided with sample pairs consisting of known input and desired outputs, $\{\mathbf{x}_k, \mathbf{d}_k\}$. The error of the machine is defined as $\mathbf{e}_k = \mathbf{d}_k - \mathbf{G}(\mathbf{x}_k, \mathbf{w})$, and the goal of learning involves solving for the parameters $\mathbf{w}$ in order to minimize the expectation of some given function of the error.

While a number of optimization approaches exist (e.g., gradient descent using backpropagation), the EKF may be used to estimate the parameters by writing a new state-space representation,

$$\mathbf{w}_{k+1} = \mathbf{w}_k + \mathbf{r}_k, \tag{7.4}$$

$$\mathbf{d}_k = \mathbf{G}(\mathbf{x}_k, \mathbf{w}_k) + \mathbf{e}_k, \tag{7.5}$$

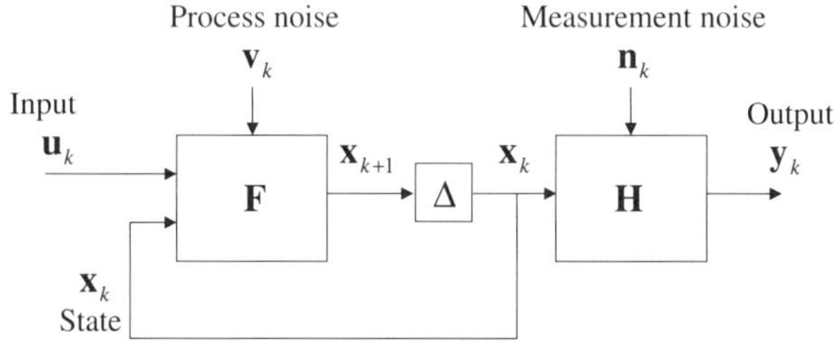

Figure 7.1 Discrete-time nonlinear dynamical system.

where the parameters $\mathbf{w}_k$ correspond to a stationary process with identity state transition matrix, driven by process noise $\mathbf{r}_k$ (the choice of variance determines convergence and tracking performance and will be discussed in further detail in Section 7.4). The output $\mathbf{d}_k$ corresponds to a nonlinear observation on $\mathbf{w}_k$. The EKF can then be applied directly as an efficient "second-order" technique for learning the parameters. The use of the EKF for training neural networks has been developed by Singhal and Wu [8] and Puskorious and Feldkamp [9], and is covered in Chapter 2 of this book. The use of the UKF in this role is developed in Section 7.4.

Dual Estimation A special case of machine learning arises when the input $\mathbf{x}_k$ is unobserved, and requires coupling both state estimation and parameter estimation. For these *dual estimation* problems, we again consider a discrete-time nonlinear dynamical system,

$$\mathbf{x}_{k+1} = \mathbf{F}(\mathbf{x}_k, \mathbf{u}_k, \mathbf{v}_k, \mathbf{w}), \tag{7.6}$$

$$\mathbf{y}_k = \mathbf{H}(\mathbf{x}_k, \mathbf{n}_k, \mathbf{w}), \tag{7.7}$$

where both the system states $\mathbf{x}_k$ and the set of model parameters $\mathbf{w}$ for the dynamical system must be simultaneously estimated from only the observed noisy signal $\mathbf{y}_k$. Example applications include adaptive nonlinear control, noise reduction (e.g., speech or image enhancement), determining the underlying price of financial time series, etc. A general theoretical and algorithmic framework for dual Kalman-based estimation has been presented in Chapter 5. An expectation–maximization approach has also been covered in Chapter 6. Approaches to dual estimation utilizing the UKF are developed in Section 7.5.

In the next section, we review optimal estimation to explain the basic assumptions and flaws with the EKF. This will motivate the use of the UKF as a method to amend these flaws. A detailed development of the UKF is given in Section 7.3. The remainder of the chapter will then be divided based on the application areas reviewed above. We conclude the chapter in Section 7.6 with the *unscented particle filter*, in which the UKF is used to improve sequential Monte-Carlo-based filtering methods. Appendix A provides a derivation of the accuracy of the UKF. Appendix B details an efficient *square-root* implementation of the UKF.

7.2 OPTIMAL RECURSIVE ESTIMATION AND THE EKF

Given observations $\mathbf{y}_k$, the goal is to estimate the state $\mathbf{x}_k$. We make no assumptions about the nature of the system dynamics at this point. The optimal estimate in the minimum mean-squared error (MMSE) sense is given by the conditional mean:

$$\hat{\mathbf{x}}_k = \mathbb{E}[\mathbf{x}_k | \mathbf{Y}_0^k], \tag{7.8}$$

where $\mathbf{Y}_0^k$ is the sequence of observations up to time k. Evaluation of this expectation requires knowledge of the *a posteriori* density $p(\mathbf{x}_k | \mathbf{Y}_0^k)$.[1] Given this density, we can determine not only the MMSE estimator, but any "best" estimator under a specified performance criterion. The problem of determining the *a posteriori* density is in general referred to as the Bayesian approach, and can be evaluated recursively according to the following relations:

$$p(\mathbf{x}_k | \mathbf{Y}_0^k) = \frac{p(\mathbf{x}_k | \mathbf{Y}_0^{k-1}) p(\mathbf{y}_k | \mathbf{x}_k)}{p(\mathbf{y}_k | \mathbf{Y}_0^{k-1})}, \tag{7.9}$$

where

$$p(\mathbf{x}_k | \mathbf{Y}_0^{k-1}) = \int p(\mathbf{x}_k | \mathbf{x}_{k-1}) p(\mathbf{x}_{k-1} | \mathbf{Y}_0^{k-1})\, d\mathbf{x}_{k-1}, \tag{7.10}$$

and the normalizing constant $p(\mathbf{y}_k | \mathbf{Y}_0^k)$ is given by

$$p(\mathbf{y}_k | \mathbf{Y}_0^{k-1}) = \int p(\mathbf{x}_k | \mathbf{Y}_0^{k-1}) p(\mathbf{y}_k | \mathbf{x}_k)\, d\mathbf{x}_k. \tag{7.11}$$

This recursion specifies the current state density as a function of the previous density and the most recent measurement data. The state-space model comes into play by specifying the state transition probability $p(\mathbf{x}_k | \mathbf{x}_{k-1})$ and measurement probability or likelihood, $p(\mathbf{y}_k | \mathbf{x}_x)$. Specifically, $p(\mathbf{x}_k | \mathbf{x}_{k-1})$ is determined by the process noise density $p(\mathbf{v}_k)$ with the state-update equation

$$\mathbf{x}_{k+1} = \mathbf{F}(\mathbf{x}_k, \mathbf{u}_k, \mathbf{v}_k). \tag{7.12}$$

For example, given an additive noise model with Gaussian density, $p(\mathbf{v}_k) = \mathcal{N}(0, \mathbf{R}^\mathbf{v})$, then $p(\mathbf{x}_k | \mathbf{x}_{k-1}) = \mathcal{N}(\mathbf{F}(\mathbf{x}_{k-1}, \mathbf{u}_{k-1}), \mathbf{R}^\mathbf{v})$. Similarly,

[1] Note that we do not write the implicit dependence on the observed input $\mathbf{u}_k$, since it is not a random variable.

$p(\mathbf{y}_k|\mathbf{x}_x)$ is determined by the observation noise density $p(\mathbf{n}_k)$ and the measurement equation

$$\mathbf{y}_k = \mathbf{H}(\mathbf{x}_k, \mathbf{n}_k). \tag{7.13}$$

In principle, knowledge of these densities and the initial condition $p(\mathbf{x}_0|\mathbf{y}_0) = p(\mathbf{y}_0|\mathbf{x}_0)p(\mathbf{x}_0)/p(\mathbf{y}_0)$ determines $p(\mathbf{x}_k|\mathbf{Y}_0^k)$ for all k. Unfortunately, the multidimensional integration indicated by Eqs. (7.9)–(7.11) makes a closed-form solution intractable for most systems. The only general approach is to apply Monte Carlo sampling techniques that essentially convert integrals to finite sums, which converge to the true solution in the limit. The *particle filter* discussed in the last section of this chapter is an example of such an approach.

If we make the basic assumption that all densities remain Gaussian, then the Bayesian recursion can be greatly simplified. In this case, only the conditional mean $\hat{\mathbf{x}}_k = \mathbb{E}[\mathbf{x}_k|\mathbf{Y}_0^k]$ and covariance $\mathbf{P}_{\mathbf{x}_k}$ need to be evaluated. It is straightforward to show that this leads to the recursive estimation

$$\hat{\mathbf{x}}_k = (\text{prediction of } \mathbf{x}_k) + \mathcal{K}_k[\mathbf{y}_k - (\text{prediction of } \mathbf{y}_k)], \tag{7.14}$$

$$\mathbf{P}_{\mathbf{x}_k} = \mathbf{P}_{\mathbf{x}_k}^- - \mathcal{K}_k \mathbf{P}_{\tilde{\mathbf{y}}_k} \mathcal{K}_k^T. \tag{7.15}$$

While this is a *linear* recursion, we have not assumed linearity of the model. The optimal terms in this recursion are given by

$$\hat{\mathbf{x}}_k^- = \mathbb{E}[\mathbf{F}(\mathbf{x}_{k-1}, \mathbf{u}_{k-1}, \mathbf{v}_{k-1})], \tag{7.16}$$

$$\mathcal{K}_k = \mathbf{P}_{\mathbf{x}_k \mathbf{y}_k} \mathbf{P}_{\tilde{\mathbf{y}}_k \tilde{\mathbf{y}}_k}^{-1}, \tag{7.17}$$

$$\hat{\mathbf{y}}_k^- = \mathbb{E}[\mathbf{H}(\mathbf{x}_k^-, \mathbf{n}_k)], \tag{7.18}$$

where the optimal prediction (i.e., prior mean) of $\mathbf{x}_k$ is written as $\hat{\mathbf{x}}_k^-$, and corresponds to the expectation of a nonlinear function of the random variables $\mathbf{x}_{k-1}$ and $\mathbf{v}_{k-1}$ (with a similar interpretation for the optimal prediction $\hat{\mathbf{y}}_k^-$). The optimal gain term $\mathcal{K}_k$ is expressed as a function of posterior covariance matrices (with $\tilde{\mathbf{y}}_k = \mathbf{y}_k - \hat{\mathbf{y}}_k^-$). Note that evaluation of the covariance terms also require taking expectations of a nonlinear function of the prior state variable. $\mathbf{P}_{\mathbf{x}_k}^-$ is the prediction of the covariance of $\mathbf{x}_k$, and $\mathbf{P}_{\tilde{\mathbf{y}}_k}$ is the covariance of $\tilde{\mathbf{y}}_k$.

The celebrated Kalman filter [10] calculates all terms in these equations exactly in the linear case, and can be viewed as an efficient method for analytically propagating a GRV through linear system dynamics. For nonlinear models, however, the EKF *approximates* the optimal terms as

$$\hat{\mathbf{x}}_k^- \approx \mathbf{F}(\hat{\mathbf{x}}_{k-1}, \mathbf{u}_{k-1}, \bar{\mathbf{v}}), \tag{7.19}$$

$$\mathcal{K}_k \approx \hat{\mathbf{P}}_{\mathbf{x}_k\mathbf{y}_k} \hat{\mathbf{P}}_{\tilde{\mathbf{y}}_k\tilde{\mathbf{y}}_k}^{-1}, \tag{7.20}$$

$$\hat{\mathbf{y}}_k^- \approx \mathbf{H}(\hat{\mathbf{x}}_k^-, \bar{\mathbf{n}}), \tag{7.21}$$

where predictions are approximated simply as functions of the prior *mean* value (no expectation taken).[2] The covariances are determined by linearizing the dynamical equations ($\mathbf{x}_{k+1} \approx \mathbf{A}\mathbf{x}_k + \mathbf{B}_u\mathbf{u}_k + \mathbf{B}\mathbf{v}_k$, $\mathbf{y}_k \approx \mathbf{C}\mathbf{x}_\mathbf{k} + \mathbf{D}\mathbf{n}_k$), and then determining the posterior covariance matrices analytically for the linear system. In other words, in the EKF, the state distribution is approximated by a GRV, which is then propagated analytically through the "first-order" linearization of the nonlinear system. The explicit equations for the EKF are given in Table 7.1. As such, the EKF

Table 7.1 Extended Kalman filter (EKF) equations

Initialize with

$$\hat{\mathbf{x}}_0 = \mathbb{E}[\mathbf{x}_0], \tag{7.22}$$

$$\mathbf{P}_{\mathbf{x}_0} = \mathbb{E}[(\mathbf{x}_0 - \hat{\mathbf{x}}_0)(\mathbf{x}_0 - \hat{\mathbf{x}}_0)^T]. \tag{7.23}$$

For $k \in \{1, \ldots, \infty\}$, the time-update equations of the extended Kalman filter are

$$\hat{\mathbf{x}}_k^- = \mathbf{F}(\hat{\mathbf{x}}_{k-1}, \mathbf{u}_{k-1}, \bar{\mathbf{v}}), \tag{7.24}$$

$$\mathbf{P}_{\mathbf{x}_k}^- = \mathbf{A}_{k-1}\mathbf{P}_{\mathbf{x}_{k-1}}\mathbf{A}_{k-1}^T + \mathbf{B}_k\mathbf{R}^\mathbf{v}\mathbf{B}_k^T, \tag{7.25}$$

and the measurement-update equations are

$$\mathcal{K}_k = \mathbf{P}_{\mathbf{x}_k}^-\mathbf{C}_k^T(\mathbf{C}_k\mathbf{P}_{\mathbf{x}_k}^-\mathbf{C}_k^T + \mathbf{D}_k\mathbf{R}^\mathbf{n}\mathbf{D}_k^T)^{-1}, \tag{7.26}$$

$$\hat{\mathbf{x}}_k = \hat{\mathbf{x}}_k^- + \mathcal{K}_k[\mathbf{y}_k - \mathbf{H}(\hat{\mathbf{x}}_k^-, \bar{\mathbf{n}})], \tag{7.27}$$

$$\mathbf{P}_{\mathbf{x}_k} = (\mathbf{I} - \mathcal{K}_k\mathbf{C}_k)\mathbf{P}_{\mathbf{x}_k}^-, \tag{7.28}$$

where

$$\begin{aligned} \mathbf{A}_k &\triangleq \left.\frac{\partial \mathbf{F}(\mathbf{x}, \mathbf{u}_k, \bar{\mathbf{v}})}{\partial \mathbf{x}}\right|_{\hat{\mathbf{x}}_k}, &\quad \mathbf{B}_k &\triangleq \left.\frac{\partial \mathbf{F}(\hat{\mathbf{x}}_k^-, \mathbf{u}_k, \mathbf{v})}{\partial \mathbf{v}}\right|_{\bar{\mathbf{v}}}, \\ \mathbf{C}_k &\triangleq \left.\frac{\partial \mathbf{H}(\mathbf{x}, \bar{\mathbf{n}})}{\partial \mathbf{x}}\right|_{\hat{\mathbf{x}}_k}, &\quad \mathbf{D}_k &\triangleq \left.\frac{\partial \mathbf{H}(\hat{\mathbf{x}}_k^-, \mathbf{n})}{\partial \mathbf{n}}\right|_{\bar{\mathbf{n}}}, \end{aligned} \tag{7.29}$$

and where $\mathbf{R}^\mathbf{v}$ and $\mathbf{R}^\mathbf{n}$ are the covariances of $\mathbf{v}_k$ and $\mathbf{n}_k$, respectively.

[2]The noise means are denoted by $\bar{\mathbf{n}} = \mathbb{E}[\mathbf{n}]$ and $\bar{\mathbf{v}} = \mathbb{E}[\mathbf{v}]$, and are usually assumed to equal zero.

can be viewed as providing "first-order" approximations to the optimal terms.[3] These approximations, however, can introduce large errors in the true posterior mean and covariance of the transformed (Gaussian) random variable, which may lead to suboptimal performance and sometimes divergence of the filter.[4] It is these "flaws" that will be addressed in the next section using the UKF.

7.3 THE UNSCENTED KALMAN FILTER

The UKF addresses the approximation issues of the EKF. The state distribution is again represented by a GRV, but is now specified using a minimal set of carefully chosen sample points. These sample points completely capture the true mean and covariance of the GRV, and when propagated through the *true* nonlinear system, capture the posterior mean and covariance accurately to the second order (Taylor series expansion) for *any* nonlinearity. To elaborate on this, we begin by explaining the *unscented transformation*.

Unscented Transformation The unscented transformation (UT) is a method for calculating the statistics of a random variable which undergoes a nonlinear transformation [3]. Consider propagating a random variable $\mathbf{x}$ (dimension L) through a nonlinear function, $\mathbf{y} = f(\mathbf{x})$. Assume $\mathbf{x}$ has mean $\bar{\mathbf{x}}$ and covariance $\mathbf{P}_{\mathbf{x}}$. To calculate the statistics of $\mathbf{y}$, we form a matrix $\boldsymbol{\mathcal{X}}$ of $2L+1$ *sigma* vectors $\mathcal{X}_i$ according to the following:

$$
\begin{aligned}
\mathcal{X}_0 &= \bar{\mathbf{x}}, \\
\mathcal{X}_i &= \bar{\mathbf{x}} + \left(\sqrt{(L+\lambda)\mathbf{P}_{\mathbf{x}}}\right)_i, \qquad i = 1, \ldots, L, \\
\mathcal{X}_i &= \bar{\mathbf{x}} - \left(\sqrt{(L+\lambda)\mathbf{P}_{\mathbf{x}}}\right)_{i-L}, \qquad i = L+1, \ldots, 2L,
\end{aligned}
\tag{7.30}
$$

[3]While "second-order" versions of the EKF exist, their increased implementation and computational complexity tend to prohibit their use.

[4]A popular technique to improve the "first-order" approach is the *iterated EKF*, which effectively iterates the EKF equations at the current time step by redefining the nominal state estimate and re-linearizing the measurement equations. It is capable of providing better performance than the basic EKF, especially in the case of significant nonlinearity in the measurement function [11]. We have not performed a comparison to the UKF at this time, though a similar procedure may also be adapted to iterate the UKF.

where $\lambda = \alpha^2(L + \kappa) - L$ is a scaling parameter. The constant α determines the spread of the sigma points around $\bar{\mathbf{x}}$, and is usually set to a small positive value (e.g., $1 \le \alpha \le 10^{-4}$). The constant κ is a secondary scaling parameter, which is usually set to $3 - L$ (see [1] for details), and β is used to incorporate prior knowledge of the distribution of $\mathbf{x}$ (for Gaussian distributions, $\beta = 2$ is optimal). $(\sqrt{(L+\lambda)\mathbf{P}_{\mathbf{x}}})_i$ is the ith column of the matrix square root (e.g., lower-triangular Cholesky factorization). These sigma vectors are propagated through the nonlinear function

$$\mathcal{Y}_i = f(\mathcal{X}_i), \qquad i = 0, \ldots, 2L, \tag{7.31}$$

and the mean and covariance for $\mathbf{y}$ are approximated using a weighted sample mean and covariance of the posterior sigma points,

$$\bar{\mathbf{y}} \approx \sum_{i=0}^{2L} W_i^{(m)} \mathcal{Y}_i, \tag{7.32}$$

$$\mathbf{P}_{\mathbf{y}} \approx \sum_{i=0}^{2L} W_i^{(c)} (\mathcal{Y}_i - \bar{\mathbf{y}})(\mathcal{Y}_i - \bar{\mathbf{y}})^T, \tag{7.33}$$

with weights W_i given by

$$\begin{aligned} W_0^{(m)} &= \frac{\lambda}{L+\lambda}, \\ W_0^{(c)} &= \frac{\lambda}{L+\lambda} + 1 - \alpha^2 + \beta \\ W_i^{(m)} &= W_i^{(c)} = \frac{1}{2(L+\lambda)}, \qquad i = 1, \ldots, 2L. \end{aligned} \tag{7.34}$$

A block diagram illustrating the steps in performing the UT is shown in Figure 7.2. Note that this method differs substantially from general Monte Carlo sampling methods, which require orders of magnitude more sample points in an attempt to propagate an accurate (possibly non-Gaussian) distribution of the state. The deceptively simple approach taken with the UT results in approximations that are accurate to the third order for Gaussian inputs for all nonlinearities. For non-Gaussian inputs, approximations are accurate to at least the second order, with the accuracy of third- and higher-order moments being determined by the choice of α and β. The proof of this is provided in Appendix A. Valuable insight into the UT can also be gained by relating it to a numerical technique called

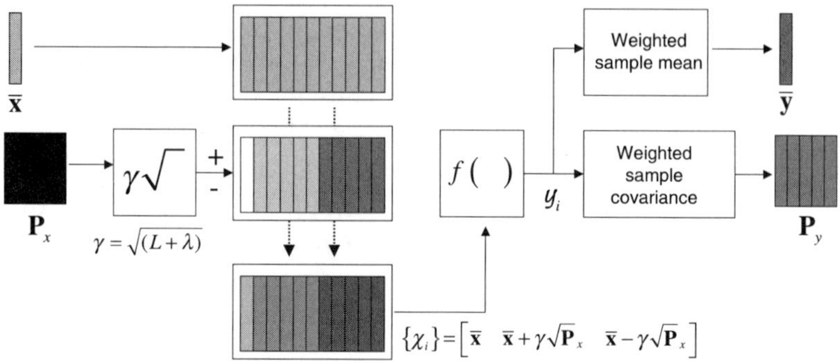

Figure 7.2 Block diagram of the UT.

Gaussian quadrature numerical evaluation of integrals. Ito and Xiong [12] recently showed the relation between the UT and the *Gauss–Hermite* quadrature rule[5] in the context of state estimation. A close similarity also exists between the UT and the *central difference interpolation* filtering (CDF) techniques developed separately by Ito and Xiong [12] and Nørgaard, Poulsen, and Ravn [13]. In [7] van der Merwe and Wan show how the UKF and CDF can be unified in a general family of *derivative-free* Kalman filters for nonlinear estimation.

A simple example is shown in Figure 7.3 for a two-dimensional system: Figure 7.3*a* shows the true mean and covariance propagation using Monte Carlo sampling; Figure 7.3*b* shows the results using a linearization approach as would be done in the EKF; Figure 7.3*c* shows the performance of the UT (note that only five sigma points are required). The superior performance of the UT is clear.

Unscented Kalman Filter The *unscented Kalman filter* (UKF) is a straightforward extension of the UT to the recursive estimation in Eq. (7.14), where the state RV is redefined as the concatenation of the original state and noise variables: $\mathbf{x}_k^a = [\mathbf{x}_k^T \quad \mathbf{v}_k^T \quad \mathbf{n}_k^T]^T$. The UT sigma point selection scheme, Eq. (7.30), is applied to this new augmented state RV to calculate the corresponding sigma matrix, $\mathcal{X}_k^a$. The UKF equations are

[5]In the scalar case, the Gauss–Hermite rule is given by $\int_{-\infty}^{\infty} f(x)(2\pi)^{-1/2} e^{-x^2}\,dx = \sum_{i=1}^{m} w_i f(x_i)$, where the equality holds for all polynomials, $f(\cdot)$, of degree up to $2m-1$ and the quadrature points x_i and weights w_i are determined according to the rule type (see [12] for details). For higher dimensions, the Gauss–Hermite rule requires on the order of m^L functional evaluations, where L is the dimension of the state. For the scalar case, the UT with $\alpha = 1$, $\beta = 0$, and $\kappa = 2$ coincides with the three-point Gauss–Hermite quadrature rule.

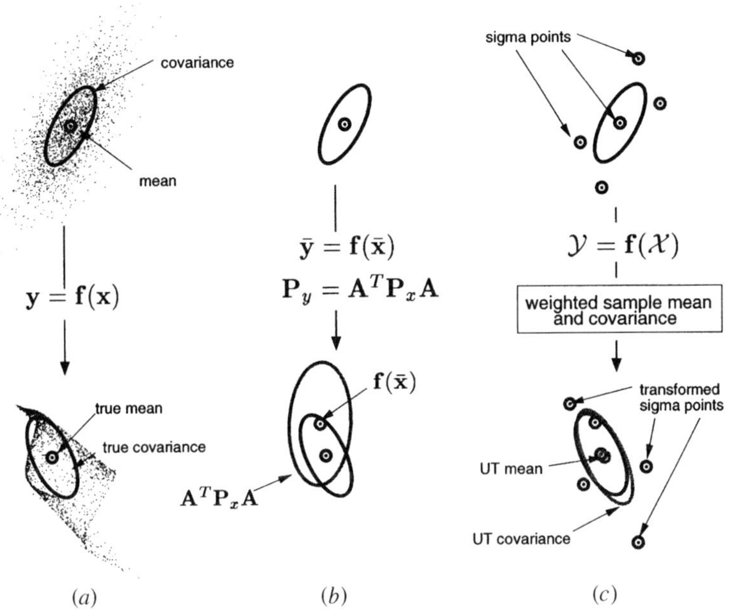

Figure 7.3 Example of the UT for mean and covariance propagation: (*a*) actual; (*b*) first-order linearization (EFK); (*c*) UT.

given in Table 7.2. Note that no explicit calculations of Jacobians or Hessians are necessary to implement this algorithm. Furthermore, the overall number of computations is of the same order as the EKF.

Implementation Variations For the special (but often encountered) case where the process and measurement noise are purely additive, the computational complexity of the UKF can be reduced. In such a case, the system state need not be augmented with the noise RVs. This reduces the dimension of the sigma points as well as the total number of sigma points used. The covariances of the noise source are then incorporated into the state covariance using a simple additive procedure. This implementation is given in Table 7.3. The complexity of the algorithm is of order L^3, where L is the dimension of the state. This is the same complexity as the EKF. The most costly operation is in forming the sample prior covariance matrix $\mathbf{P}_k^-$. Depending on the form of $\mathbf{F}$, this may be simplified; for example, for univariate time series or with parameter estimation (see Section 7.4), the complexity reduces to order L^2.

Table 7.2 Unscented Kalman filter (UKF) equations

Initialize with

$$\hat{\mathbf{x}}_0 = \mathbb{E}[\mathbf{x}_0], \tag{7.35}$$

$$\mathbf{P}_0 = \mathbb{E}[(\mathbf{x}_0 - \hat{\mathbf{x}}_0)(\mathbf{x}_0 - \hat{\mathbf{x}}_0)^T], \tag{7.36}$$

$$\hat{\mathbf{x}}_0^a = \mathbb{E}[\mathbf{x}^a] = [\hat{\mathbf{x}}_0^T \quad \mathbf{0} \quad \mathbf{0}]^T, \tag{7.37}$$

$$\mathbf{P}_0^a = \mathbb{E}[(\mathbf{x}_0^a - \hat{\mathbf{x}}_0^a)(\mathbf{x}_0^a - \hat{\mathbf{x}}_0^a)^T] = \begin{bmatrix} \mathbf{P}_0 & \mathbf{0} & \mathbf{0} \\ \mathbf{0} & \mathbf{R}^{\mathrm{v}} & \mathbf{0} \\ \mathbf{0} & \mathbf{0} & \mathbf{R}^{\mathrm{n}} \end{bmatrix}. \tag{7.38}$$

For $k \in \{1, \ldots, \infty\}$,
calculate the sigma points:

$$\mathcal{X}_{k-1}^a = [\hat{\mathbf{x}}_{k-1}^a \quad \hat{\mathbf{x}}_{k-1}^a + \gamma\sqrt{\mathbf{P}_{k-1}^a} \quad \hat{\mathbf{x}}_{k-1}^a - \gamma\sqrt{\mathbf{P}_{k-1}^a}]. \tag{7.39}$$

The time-update equations are

$$\mathcal{X}_{k|k-1}^x = \mathbf{F}(\mathcal{X}_{k-1}^x, \mathbf{u}_{k-1}, (\mathcal{X}_{k-1}^v), \tag{7.40}$$

$$\hat{\mathbf{x}}_k^- = \sum_{i=0}^{2L} W_i^{(m)} \mathcal{X}_{i,k|k-1}^x, \tag{7.41}$$

$$\mathbf{P}_k^- = \sum_{i=0}^{2L} W_i^{(c)} (\mathcal{X}_{i,k|k-1}^x - \hat{\mathbf{x}}_k^-)(\mathcal{X}_{i,k|k-1}^x - \hat{\mathbf{x}}_k^-)^T, \tag{7.42}$$

$$\mathcal{Y}_{k|k-1} = \mathbf{H}(\mathcal{X}_{k|k-1}^x, \mathcal{X}_{k-1}^n), \tag{7.43}$$

$$\hat{\mathbf{y}}_k^- = \sum_{i=0}^{2L} W_i^{(m)} \mathcal{Y}_{i,k|k-1}, \tag{7.44}$$

and the measurement-update equations are

$$\mathbf{P}_{\tilde{\mathbf{y}}_k \tilde{\mathbf{y}}_k} = \sum_{i=0}^{2L} W_i^{(c)} (\mathcal{Y}_{i,k|k-1} - \hat{\mathbf{y}}_k^-)(\mathcal{Y}_{i,k|k-1} - \hat{\mathbf{y}}_k^-)^T, \tag{7.45}$$

$$\mathbf{P}_{\mathbf{x}_k \mathbf{y}_k} = \sum_{i=0}^{2L} W_i^{(c)} (\mathcal{X}_{i,k|k-1} - \hat{\mathbf{x}}_k^-)(\mathcal{Y}_{i,k|k-1} - \hat{\mathbf{y}}_k^-)^T, \tag{7.46}$$

$$\mathcal{K}_k = \mathbf{P}_{\mathbf{x}_k \mathbf{y}_k} \mathbf{P}_{\tilde{\mathbf{y}}_k \tilde{\mathbf{y}}_k}^{-1}, \tag{7.47}$$

$$\hat{\mathbf{x}}_k = \hat{\mathbf{x}}_k^- + \mathcal{K}_k (\mathbf{y}_k - \hat{\mathbf{y}}_k^-), \tag{7.48}$$

$$\mathbf{P}_k = \mathbf{P}_k^- - \mathcal{K}_k \mathbf{P}_{\tilde{\mathbf{y}}_k \tilde{\mathbf{y}}_k} \mathcal{K}_k^T, \tag{7.49}$$

where

$$\mathbf{x}^a = [\mathbf{x}^T \quad \mathbf{v}^T \quad \mathbf{n}^T]^T, \qquad \mathcal{X}^a = [(\mathcal{X}^x)^T \quad (\mathcal{X}^v)^T \quad (\mathcal{X}^n)^T]^T, \qquad \gamma = \sqrt{L + \lambda},$$

λ is the composite scaling parameter, L is the dimension of the augmented state, $\mathbf{R}^{\mathrm{v}}$ is the process-noise covariance, $\mathbf{R}^{\mathrm{n}}$ is the measurement-noise covariance, and W_i are the weights as calculated in Eq. (7.34).

Table 7.3 UKF – additive (zero mean) noise case

Initialize with

$$\hat{\mathbf{x}}_0 = \mathbb{E}[\mathbf{x}_0], \tag{7.50}$$

$$\mathbf{P}_0 = \mathbb{E}[(\mathbf{x}_0 - \hat{\mathbf{x}}_0)(\mathbf{x}_0 - \hat{\mathbf{x}}_0)^T]. \tag{7.51}$$

For $k \in \{1, \ldots, \infty\}$,
calculate the sigma points:

$$\mathcal{X}_{k-1} = [\hat{\mathbf{x}}_{k-1} \quad \hat{\mathbf{x}}_{k-1} + \gamma\sqrt{\mathbf{P}_{k-1}} \quad \hat{\mathbf{x}}_{k-1} - \gamma\sqrt{\mathbf{P}_{k-1}}]. \tag{7.52}$$

The time-update equations are

$$\mathcal{X}^*_{k|k-1} = \mathbf{F}(\mathcal{X}_{k-1}, \mathbf{u}_{k-1}), \tag{7.53}$$

$$\hat{\mathbf{x}}_k^- = \sum_{i=0}^{2L} W_i^{(m)} \mathcal{X}^*_{i,k|k-1} \tag{7.54}$$

$$\mathbf{P}_k^- = \sum_{i=0}^{2L} W_i^{(c)} (\mathcal{X}^*_{i,k|k-1} - \hat{\mathbf{x}}_k^-)(\mathcal{X}^*_{i,k|k-1} - \hat{\mathbf{x}}_k^-)^T + \mathbf{R}^{\mathbf{v}}, \tag{7.55}$$

(augment sigma points)[6]

$$\mathcal{X}_{k|k-1} = [\mathcal{X}^*_{k|k-1} \quad \mathcal{X}^*_{0,k|k-1} + \gamma\sqrt{\mathbf{R}^{\mathbf{v}}} \quad \mathcal{X}^*_{0,k|k-1} - \gamma\sqrt{\mathbf{R}^{\mathbf{v}}}] \tag{7.56}$$

$$\mathcal{Y}_{k|k-1} = \mathbf{H}(\mathcal{X}_{k|k-1}), \tag{7.57}$$

$$\hat{\mathbf{y}}_k^- = \sum_{i=0}^{2L} W_i^{(m)} \mathcal{Y}_{i,k|k-1}, \tag{7.58}$$

and the measurement-update equations are

$$\mathbf{P}_{\tilde{\mathbf{y}}_k\tilde{\mathbf{y}}_k} = \sum_{i=0}^{2L} W_i^{(c)} (\mathcal{Y}_{i,k|k-1} - \hat{\mathbf{y}}_k^-)(\mathcal{Y}_{i,k|k-1} - \hat{\mathbf{y}}_k^-)^T + \mathbf{R}^{\mathbf{n}}, \tag{7.59}$$

$$\mathbf{P}_{\mathbf{x}_k\mathbf{y}_k} = \sum_{i=0}^{2L} W_i^{(c)} (\mathcal{X}_{i,k|k-1} - \hat{\mathbf{x}}_k^-)(\mathcal{Y}_{i,k|k-1} - \hat{\mathbf{y}}_k^-)^T \tag{7.60}$$

$$\mathcal{K}_k = \mathbf{P}_{\mathbf{x}_k\mathbf{y}_k} \mathbf{P}_{\tilde{\mathbf{y}}_k\tilde{\mathbf{y}}_k}^{-1} \tag{7.61}$$

$$\hat{\mathbf{x}}_k = \hat{\mathbf{x}}_k^- + \mathcal{K}_k(\mathbf{y}_k - \hat{\mathbf{y}}_k^-) \tag{7.62}$$

$$\mathbf{P}_k = \mathbf{P}_k^- - \mathcal{K}_k \mathbf{P}_{\tilde{\mathbf{y}}_k\tilde{\mathbf{y}}_k} \mathcal{K}_k^T, \tag{7.63}$$

where $\gamma = \sqrt{L + \lambda}$, λ is the composite scaling parameter, L is the dimension of the state, $\mathbf{R}^{\mathbf{v}}$ is the process-noise covariance, $\mathbf{R}^{\mathbf{n}}$ is the measurement-noise covariance and W_i are the weights as calculated in Eq. (7.34).

[6]Here we augment the sigma points with additional points derived from the matrix square root of the process noise covariance. This requires setting $L \to 2L$ and recalculating the various weights W_i accordingly. Alternatively, we may *redraw* a complete new set of sigma points, i.e., $\mathcal{X}_{k|k-1} = [\hat{\mathbf{x}}_k^- \quad \hat{\mathbf{x}}_k^- + \gamma\sqrt{\mathbf{P}_k^-} \quad \hat{\mathbf{x}}_k^- - \gamma\sqrt{\mathbf{P}_k^-}]$. This alternative approach results in fewer sigma points being used, but also discards any odd-moments information captured by the original propagated sigma points.

A number of variations for numerical purposes are also possible. For example, the matrix square root, which can be implemented directly using a Cholesky factorization, is in general of order $\frac{1}{6}L^3$. However, the covariance matrices are expressed recursively, and thus the square root can be computed in only order $M \times L^2$ (where M is the dimension of the output $\mathbf{y}_k$) by performing a recursive update to the Cholesky factorization. Details of an efficient recursive square-root UKF implementation are given in Appendix B.

7.3.1 State-Estimation Examples

The UKF was originally designed for state estimation applied to nonlinear control applications requiring full-state feedback [1–3]. We provide an example for a double inverted pendulum control system. In addition, we provide a new application example corresponding to noisy time-series estimation with neural networks.

Double Inverted Pendulum A double inverted pendulum (see Fig. 7.4) has states corresponding to cart position and velocity, and top and bottom pendulum angle and angular velocity, $\mathbf{x} = [x, \dot{x}, \theta_1, \dot{\theta}_1, \theta_2, \dot{\theta}_2]$. The system parameters correspond to the length and mass of each pendulum, and the cart mass, $\mathbf{w} = [l_1, l_2, m_1, m_2, M]$. The dynamical equations are

$$\begin{aligned}(M + m_1 + m_2)\ddot{x} - (m_1 + 2m_2)l_1\ddot{\theta}_1 \cos\theta_1 - m_2 l_2 \ddot{\theta}_2 \cos\theta_2 \\ = u + (m_1 + 2m_2)l_1(\dot{\theta}_1)^2 \sin\theta_1 + m_2 l_2 (\dot{\theta}_2)^2 \sin\theta_2, \end{aligned} \quad (7.64)$$

$$\begin{aligned}-(m_1 + 2m_2)l_1\ddot{x}\cos\theta_1 + 4(\tfrac{1}{3}m_1 + m_2)(l_1)^2\ddot{\theta}_1 + 2m_2 l_1 l_2 \ddot{\theta}_2 \cos(\theta_2 - \theta_1) \\ = (m_1 + 2m_2)g l_1 \sin\theta_1 + 2m_2 l_1 l_2 (\dot{\theta}_2)^2 \sin(\theta_2 - \theta_1), \end{aligned} \quad (7.65)$$

$$\begin{aligned}-m_2\ddot{x}l_2\cos\theta_2 + 2m_2 l_1 l_2 \ddot{\theta}_1 \cos(\theta_2 - \theta_1) + \tfrac{4}{3}m_2(l_2)^2\ddot{\theta}_2 \\ = m_2 g l_2 \sin\theta_2 - 2m_2 l_1 l_2 (\dot{\theta}_1)^2 \sin(\theta_2 - \theta_1). \end{aligned} \quad (7.66)$$

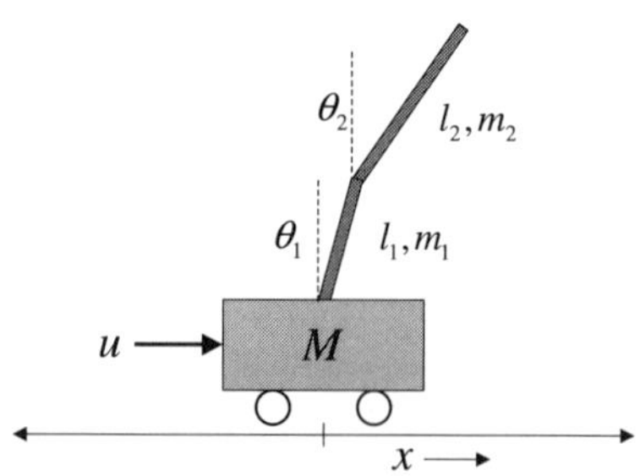

Figure 7.4 Double inverted pendulum.

These continuous-time dynamics are discretized with a sampling period of 0.02 seconds. The pendulum is stabilized by applying a control force u to the cart. In this case, we use a state-dependent Ricatti equation (SDRE) controller to stabilize the system.[7] A state estimator is run outside the control loop in order to compare the EKF with the UKF (i.e., the estimated states are not used in the feedback control for evaluation purposes). The observation corresponds to noisy measurements of the cart position, cart velocity, and angle of the top pendulum. This is a challenging problem, since no measurements are made for the bottom pendulum, nor for the angular velocity of the top pendulum. For this experiment, the pendulum is initialized in a jack-knife position ($+25^\circ/-25^\circ$), with a cart offset of 0.5 meters. The resulting state estimates are shown in Figure 7.5. Clearly, the UKF is better able to track the unobserved states.[8] If the estimated states are used for feedback in the control loop, the UKF system is still able to stabilize the pendulum, while the EKF system crashes. We shall return to the double inverted pendulum problem later in this chapter for both model estimation and dual estimation.

Noisy Time-Series Estimation In this example, the UKF is used to estimate an underlying clean time series corrupted by additive Gaussian white noise. The time-series used is the Mackey–Glass-30 chaotic series [15, 16]. The clean time-series is first modeled as a nonlinear autoregression

$$x_k = f(x_{k-1}, \ldots x_{k-M}, \mathbf{w}) + v_k, \tag{7.67}$$

where the model f (parameterised by $\mathbf{w}$) was approximated by training a feedforward neural network on the clean sequence. The residual error after convergence was taken to be the process-noise variance.

Next, white Gaussian noise was added to the clean Mackey–Glass series to generate a noisy time series $y_k = x_k + n_k$. The corresponding

[7]An SDRE controller [11] is designed by formulating the dynamical equations as $\mathbf{x}_{k+1} = \mathbf{A}(\mathbf{x}_k)\mathbf{x}_k + \mathbf{B}(\mathbf{x}_k)\mathbf{u}_k$. Note, this representation is *not* a linearization, but rather a reformulation of the nonlinear dynamics into a pseudo-linear form. Based on this state-space representation, we design an optimal LQR controller, $\mathbf{u}_k = -\mathbf{R}^{-1}\mathbf{B}^T(\mathbf{x}_k)\mathbf{P}(\mathbf{x}_k)\mathbf{x}_k \equiv \mathbf{K}(\mathbf{x}_k)\mathbf{x}_k$, where $\mathbf{P}(\mathbf{x}_k)$ is a solution of the standard Ricatti equations using state-dependent matrices $\mathbf{A}(\mathbf{x}_k)$ and $\mathbf{B}(\mathbf{x}_k)$. The procedure is repeated at every time step at the current state $\mathbf{x}_k$, and provides local asymptotic stability of the plant [14]. The approach has been found to be far more robust than LQR controllers based on standard linearization techniques.

[8]Note that if all six states are observed with noise, then the performances of the EKF and UKF are comparable.

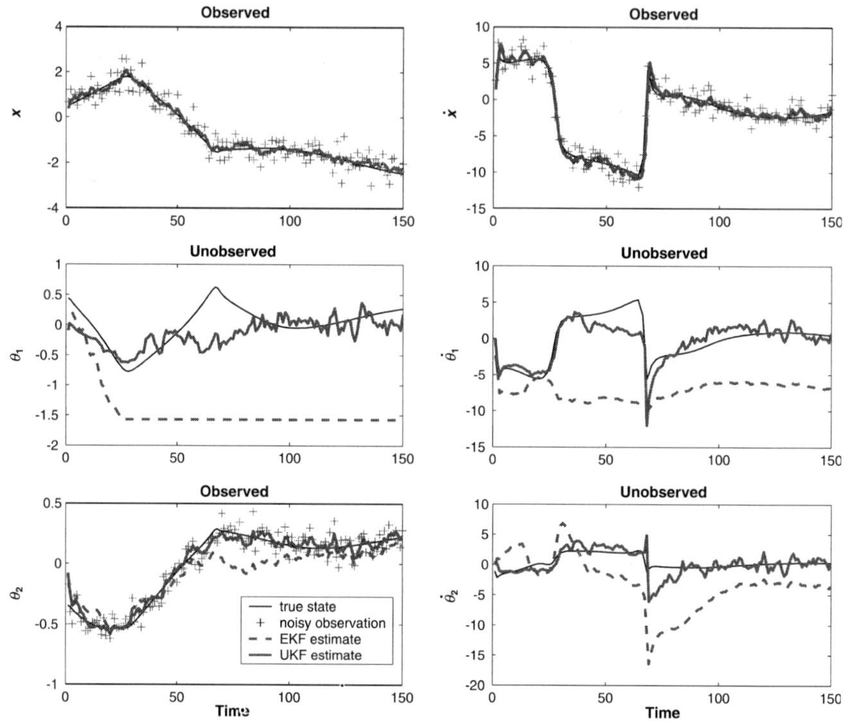

Figure 7.5 State estimation for the double inverted pendulum problem. Only three noisy states are observed: cart position, cart velocity, and the angle of the top pendulum. (10 dB SNR; $\alpha = 1$, $\beta = 0$, $\kappa = 0$.)

state-space representation is given by

$$\mathbf{x}_{k+1} = \mathbf{F}(\mathbf{x}_k, \mathbf{w}) + \mathbf{B}v_k,$$

$$\begin{bmatrix} x_{k+1} \\ x_k \\ \vdots \\ x_{k-M} \end{bmatrix} = \begin{bmatrix} f(x_k, \ldots, x_{k-M+1}, \mathbf{w}) \\ \begin{bmatrix} 1 & 0 & 0 & 0 \\ 0 & \ddots & 0 & \vdots \\ 0 & 0 & 1 & 0 \end{bmatrix} \begin{bmatrix} x_k \\ \vdots \\ x_{k-M+1} \end{bmatrix} \end{bmatrix} + \begin{bmatrix} 1 \\ 0 \\ \vdots \\ 0 \end{bmatrix} v_k,$$

$$y_k = [1 \quad 0 \quad \ldots \quad 0]\mathbf{x}_k + n_k. \tag{7.68}$$

In the estimation problem, the noisy time-series y_k is the only observed input to either the EKF or UKF algorithms (both utilize the known neural network model). Figure 7.6 shows a subsegment of the estimates generated by both the EKF and the UKF (the original noisy time series has a 3 dB SNR). The superior performance of the UKF is clearly visible.

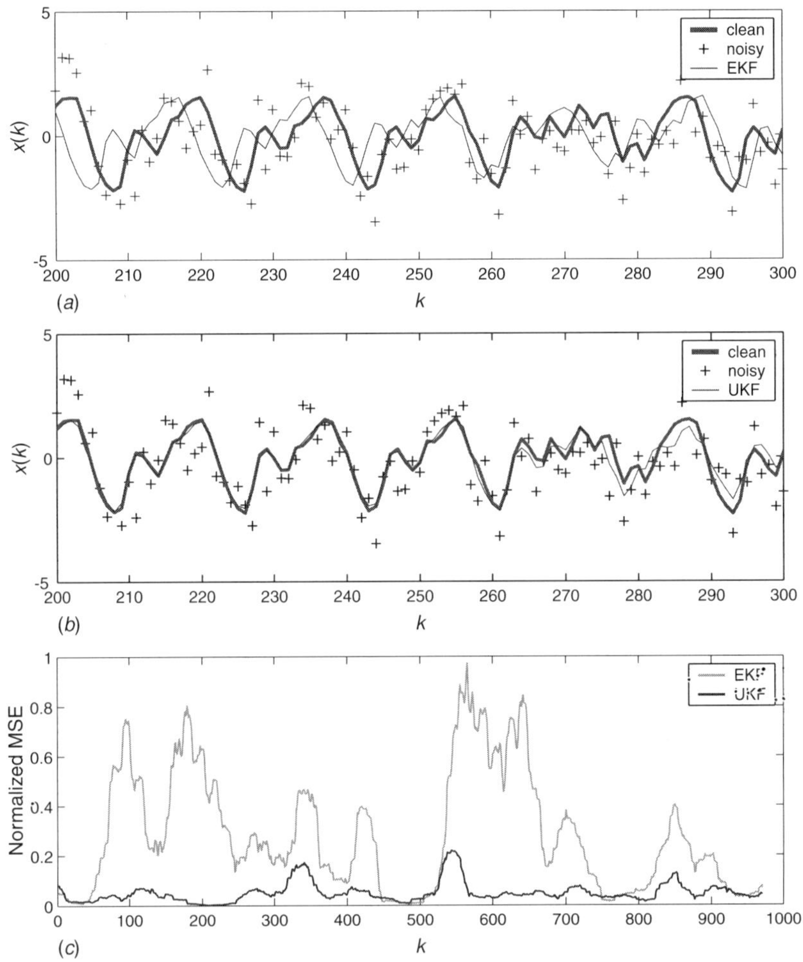

Figure 7.6 Estimation of Mackey–Glass time series using a known model: (*a*) with the EKF; (*b*) with the UKF. (*c*) shows a comparison of estimation errors for the complete sequence.

7.3.2 The Unscented Kalman Smoother

As has been discussed, the Kalman filter is a recursive algorithm providing the conditional expectation of the state $\mathbf{x}_k$ given all observations $\mathbf{Y}_0^k$ up to the current time k. In contrast, the Kalman smoother estimates the state given all observations past and future, $\mathbf{Y}_0^N$, where N is the final time. Kalman smoothers are commonly used for applications such as trajectory planning, noncausal noise reduction, and the *E-step* in the EM algorithm

[17, 18]. A thorough treatment of the Kalman smoother in the linear case is given in [19]. The basic idea is to run a Kalman filter forward in time to estimate the mean and covariance $(\hat{\mathbf{x}}_k^f, \mathbf{P}_k^f)$ of the state, given *past* data. A second Kalman filter is then run backward in time to produce a backward-time predicted mean and covariance $(\hat{\mathbf{x}}_k^{-b}, \mathbf{P}_k^{-b})$, given the *future* data. These two estimates are then combined, producing the following smoothed statistics, given *all* the data:

$$(\mathbf{P}_k^s)^{-1} = (\mathbf{P}_k^f)^{-1} + (\mathbf{P}_k^{-b})^{-1}, \tag{7.69}$$

$$\hat{\mathbf{x}}_k^s = \mathbf{P}_k^s[(\mathbf{P}_k^{-b})^{-1}\hat{\mathbf{x}}_k^{-b} + (\mathbf{P}_k^f)^{-1}\hat{\mathbf{x}}_k^f]. \tag{7.70}$$

For the nonlinear case, the EKF replaces the Kalman filter. The use of the EKF for the forward filter is straightforward. However, implementation of the backward filter is achieved by using the following linearized backward-time system:

$$\mathbf{x}_{k-1} = \mathbf{A}^{-1}\mathbf{x}_k + \mathbf{A}^{-1}\mathbf{B}\mathbf{v}_k \tag{7.71}$$

that is, the forward nonlinear dynamics are linearized, and then inverted for the backward model. A linear Kalman filter is then applied.

Our proposed unscented Kalman smoother (UKS) replaces the EKF with the UKF. In addition, we consider using a nonlinear backward model as well, either derived from first principles or by training a backward predictor using a neural network model, as illustrated for the time-series case in Figure 7.7. The nonlinear backward model allows us to take full advantage of the UKF, which requires no linearization step.

To illustrate performance, we reconsider the noisy Mackey–Glass time-series problem of the previous section, as well as a second time series generated using a chaotic autoregressive neural network. Table 7.4 compares *smoother* performance. In this case, the network models are trained on the clean time series, and then tested on the noisy data using the standard extended Kalman smoother with linearized backward model

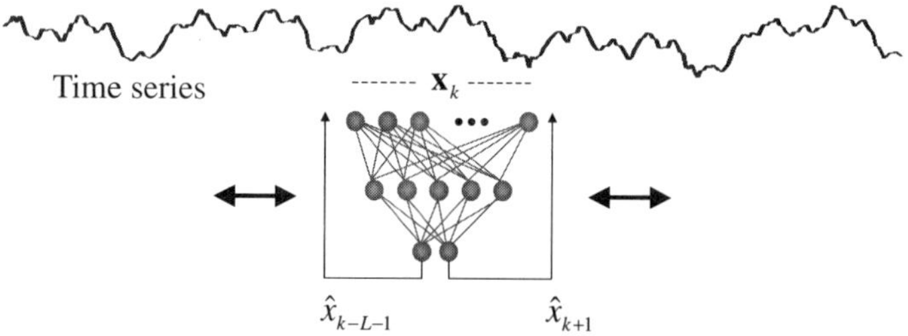

Figure 7.7 Forward/backward neural network prediction training.

Table 7.4 Comparison of smoother performance

Mackey–Glass			
	Normalized MSE		
Algorithm	F	B	S
EKS1	0.20	0.70	0.27
EKS2	0.20	0.31	0.19
UKS	0.10	0.24	0.08

Chaotic AR–NN			
	Normalized MSE		
Algorithm	F	B	S
EKS1	0.35	0.32	0.28
EKS2	0.35	0.22	0.23
UKS	0.23	0.21	0.16

(EKS1), an extended Kalman smoother with a second nonlinear backward model (EKS2), and the unscented Kalman smoother (UKS). The forward (F), backward (B), and smoothed (S) estimation errors are reported. Again, the performance benefits of the unscented approach are clear.

7.4 UKF PARAMETER ESTIMATION

Recall that parameter estimation involves learning a nonlinear mapping $\mathbf{y}_k = \mathbf{G}(\mathbf{x}_k, \mathbf{w})$, where $\mathbf{w}$ corresponds to the set of unknown parameters. $\mathbf{G}(\cdot)$ may be a neural network or another parameterized function. The EKF may be used to estimate the parameters by writing a new state-space representation

$$\mathbf{w}_{k+1} = \mathbf{w}_k + \mathbf{r}_k, \tag{7.73}$$

$$\mathbf{d}_k = \mathbf{G}(\mathbf{x}_k, \mathbf{w}_k) + \mathbf{e}_k, \tag{7.74}$$

where $\mathbf{w}_k$ corresponds to a stationary process with identity state transition matrix, driven by process noise $\mathbf{r}_k$. The desired output $\mathbf{d}_k$ corresponds to a nonlinear observation on $\mathbf{w}_k$. In the linear case, the relationship between the Kalman Filter (KF) and the popular recursive least-squares (RLS) is given in [20] and [25]. In the nonlinear case, the EKF training corresponds to a modified-Newton method [22] (see also Chapter 2).

From an optimization perspective, the following *prediction error* cost is minimized:

$$J(\mathbf{w}) = \sum_{t=1}^{k} [\mathbf{d}_t - \mathbf{G}(\mathbf{x}_t, \mathbf{w})]^{\mathrm{T}} (\mathbf{R}^{\mathrm{e}})^{-1} [\mathbf{d}_t - \mathbf{G}(\mathbf{x}_t, \mathbf{w})]. \qquad (7.75)$$

Thus, if the "noise" covariance $\mathbf{R}^{\mathrm{e}}$ is a constant diagonal matrix, then, in fact, it cancels out of the algorithm (this can be shown explicitly), and hence can be set arbitrarily (e.g., $\mathbf{R}^{\mathrm{e}} = 0.5\mathbf{I}$). Alternatively, $\mathbf{R}^{\mathrm{e}}$ can be set to specify a weighted MSE cost. The innovations covariance $\mathbb{E}[\mathbf{r}_k \mathbf{r}_k^T] = \mathbf{R}_k^{\mathbf{r}}$, on the other hand, affects the convergence rate and tracking performance. Roughly speaking, the larger the covariance, the more quickly older data is discarded. There are several options on how to choose $\mathbf{R}_k^{\mathbf{r}}$.

- Set $\mathbf{R}_k^{\mathbf{r}}$ to an arbitrary "fixed" diagonal value, which may then be "annealed" towards zero as training continues.
- Set $\mathbf{R}_k^{\mathbf{r}} = (\lambda_{RLS}^{-1} - 1)\mathbf{P}_{\mathbf{w}_k}$, where $\lambda_{RLS} \in (0, 1]$ is often referred to as the "forgetting factor," as defined in the recursive least-squares (RLS) algorithm [21]. This provides for an approximate exponentially decaying weighting on past data, and is described more fully in [22]. Note that λ_{RLS} should not be confused with λ used for sigma-point calculation.
- Set

$$\mathbf{R}_k^{\mathbf{r}} = (1 - \alpha_{\mathrm{RM}})\mathbf{R}_{k-1}^{\mathbf{r}} + \alpha_{\mathrm{RM}} \mathbf{K}_k^{\mathbf{w}} [\mathbf{d}_k - \mathbf{G}(\mathbf{x}_k, \hat{\mathbf{w}})] \times [\mathbf{d}_k - \mathbf{G}(\mathbf{x}_k, \hat{\mathbf{w}})]^T (\mathbf{K}_k^{\mathbf{w}})^T,$$

  which is a Robbins–Monro stochastic approximation scheme for estimating the innovations [23]. The method assumes that the covariance of the Kalman update model is consistent with the actual update model. Typically, $\mathbf{R}_k^{\mathbf{r}}$ is also constrained to be a diagonal matrix, which implies an independence assumption on the parameters. Note that a similar update may also be used for $\mathbf{R}_k^{\mathbf{e}}$.

Our experience indicates that the "Robbins–Monro" method provides the fastest rate of absolute convergence and lowest final MMSE values (see the experiments in the next section). The "fixed" $\mathbf{R}_k^{\mathbf{r}}$ in combination with annealing can also achieve good final MMSE performance, but requires more monitoring and a greater prior knowledge of the noise levels. For problems where the MMSE is zero, the covariance should be lower-bounded to prevent the algorithm from stalling and potential numerical problems. The "forgetting-factor" and "fixed" $\mathbf{R}_k^{\mathbf{r}}$ methods are most appropriate for on-line learning problems in which tracking of time-varying parameters is necessary. In this case, the parameter covariance stays lower-bounded, allowing the most recent data to be emphasized. This leads to some misadjustment, but also keeps the Kalman gain sufficiently large to maintain good tracking. In general, study of the various trade-offs between these different approaches is still an area of open research.

The UKF represents an alternative to the EKF for parameter estimation. However, as the state transition function is linear, the advantage of the UKF may not be as obvious. Note that the observation function is still nonlinear. Furthermore, the EKF essentially builds up an approximation to the expected Hessian by taking outer products of the gradient. The UKF, however, may provide a more accurate estimate through direct approximation of the expectation of the Hessian. While both the EKF and UKF can be expected to achieve similar final MMSE performance, their covergence properties may differ. In addition, a distinct advantage of the UKF occurs when either the architecture or error metric is such that differentiation with respect to the parameters is not easily derived, as is necessary in the EKF. The UKF effectively evaluates both the Jacobian and Hessian precisely through its sigma-point propagation, without the need to perform any analytical differentiation.

Specific equations for UKF parameter estimation are given in Table 7.5. Simplifications have been made relative to the state UKF, accounting for the specific form of the state transition function. In Table 7.5, we have provided two options on how the function output $\hat{\mathbf{d}}_k$ is achieved. In the first option, the output is given as

$$\hat{\mathbf{d}}_k = \sum_{i=0}^{2L} W_i^{(m)} \mathcal{D}_{i,k|k-1} \approx \mathbb{E}[\mathbf{G}(\mathbf{x}_k, \mathbf{w}_k)], \tag{7.89}$$

corresponding to the direct interpretation of the UKF equations. The output is the expected value (mean) of a function of the random variable $\mathbf{w}_k$. In the second option, we have

$$\hat{\mathbf{d}}_k = \mathbf{G}(\mathbf{x}_k, \hat{\mathbf{w}}_k^-), \tag{7.90}$$

corresponding to the typical interpretation, in which the output is the function with the current "best" set of parameters. This option yields convergence performance that is indistinguishable from the EKF. The first option, however, has different convergence characteristics, and requires

Table 7.5 UKF parameter estimation

Initialize with

$$\hat{\mathbf{w}}_0 = E[\mathbf{w}], \tag{7.76}$$

$$\mathbf{P}_{\mathbf{w}_0} = E[(\mathbf{w} - \hat{\mathbf{w}}_0)(\mathbf{w} - \hat{\mathbf{w}}_0)^T]. \tag{7.77}$$

For $k \in \{1, \ldots, \infty\}$,
The time update and sigma-point calculation are given by

$$\hat{\mathbf{w}}_k^- = \hat{\mathbf{w}}_{k-1}, \tag{7.78}$$

$$\mathbf{P}_{\mathbf{w}_k}^- = \mathbf{P}_{\mathbf{w}_{k-1}} + \mathbf{R}_{k-1}^{\mathbf{r}}, \tag{7.79}$$

$$\mathcal{W}_{k|k-1} = [\hat{\mathbf{w}}_k^- \quad \hat{\mathbf{w}}_k^- + \gamma\sqrt{\mathbf{P}_{\mathbf{w}_k}^-} \quad \hat{\mathbf{w}}_k^- - \gamma\sqrt{\mathbf{P}_{\mathbf{w}_k}^-}], \tag{7.80}$$

$$\mathcal{D}_{k|k-1} = \mathbf{G}(\mathbf{x}_k, \mathcal{W}_{k|k-1}), \tag{7.81}$$

$$\text{option 1:} \quad \hat{\mathbf{d}}_k = \sum_{i=0}^{2L} W_i^{(m)} \mathcal{D}_{i,k|k-1}, \tag{7.82}$$

$$\text{option 2:} \quad \hat{\mathbf{d}}_k = \mathbf{G}(\mathbf{x}_k, \hat{\mathbf{w}}_k^-). \tag{7.83}$$

and the measurement-update equations are

$$\mathbf{P}_{\tilde{\mathbf{d}}_k\tilde{\mathbf{d}}_k} = \sum_{i=0}^{2L} W_i^{(c)} (\mathcal{D}_{i,k|k-1} - \hat{\mathbf{d}}_k)(\mathcal{D}_{i,k|k-1} - \hat{\mathbf{d}}_k)^T + \mathbf{R}_k^{\mathbf{e}}, \tag{7.84}$$

$$\mathbf{P}_{\mathbf{w}_k\mathbf{d}_k} = \sum_{i=0}^{2L} W_i^{(c)} (\mathcal{W}_{i,k|k-1} - \hat{\mathbf{w}}_k^-)(\mathcal{D}_{i,k|k-1} - \hat{\mathbf{d}}_k)^T, \tag{7.85}$$

$$\mathcal{K}_k = \mathbf{P}_{\mathbf{w}_k\mathbf{d}_k} \mathbf{P}_{\tilde{\mathbf{d}}_k\tilde{\mathbf{d}}_k}^{-1}, \tag{7.86}$$

$$\hat{\mathbf{w}}_k = \hat{\mathbf{w}}_k^- + \mathcal{K}_k(\mathbf{d}_k - \hat{\mathbf{d}}_k), \tag{7.87}$$

$$\mathbf{P}_{\mathbf{w}_k} = \mathbf{P}_{\mathbf{w}_k}^- - \mathcal{K}_k \mathbf{P}_{\tilde{\mathbf{d}}_k\tilde{\mathbf{d}}_k} \mathcal{K}_k^T, \tag{7.88}$$

where $\gamma = \sqrt{L + \lambda}$, λ is the composite scaling parameter, L is the dimension of the state, $\mathbf{R}^{\mathbf{r}}$ is the process-noise covariance, $\mathbf{R}^{\mathbf{e}}$ is the measurement-noise covariance, and W_i are the weights as calculated in Eq. (7.34).

---

further explanation. In the state-space approach to parameter estimation, absolute convergence is achieved when the parameter covariance $\mathbf{P}_{\mathbf{w}_k}$ goes to zero (this also forces the Kalman gain to zero). At this point, the output for either option is identical. However, prior to this, the finite covariance provides a form of averaging on the output of the function, which in turn prevents the parameters from going to the minimum of the error surface. Thus, the method may help avoid falling into a local minimum. Furthermore, it provides a form of built-in regularization for short or noisy data sets that are prone to overfitting (exact specification of the level of regularization requires further study).

Note that the complexity of the UKF algorithm is still of order L^3 (L is the number of parameters), owing to the need to compute a matrix square root at each time step. An order L^2 complexity (same as the EKF) can be achieved by using a recursive square-root formulation as given in Appendix B.

7.4.1 Parameter Estimation Examples

We have performed a number of experiments to illustrate the performance of the UKF parameter-estimation approach. The first set of experiments corresponds to benchmark problems for neural network training, and serve to illustrate some of the differences between the EKF and UKF, as well as the different options discussed above. Two parametric optimization problems are also included, corresponding to model estimation of the double pendulum, and the benchmark "Rosenbrock's Banana" optimization problem.

Benchmark NN Regression and Time-Series Problems The *Mackay robot-arm* dataset [24, 25] and the *Ikeda* chaotic time series [26] are used as benchmark problems to compare neural network training. Figure 7.8 illustrates the differences in learning curves for the EKF versus UKF (option 1). Note the slightly lower final MSE performance of the UKF weight training. If option 2 for the UKF output is used (see Eq. (7.82), then the learning curves for the EKF and UKF are indistinguishable; this has been found to be consistent with all experiments; therefore, we shall not show explicit learning curves for the UKF with option 2.

Figure 7.9 illustrates performance differences based on the choice of processing noise covariance $\mathbf{R}_k^{\mathbf{r}}$. The Mackey–Glass and Ikeda time series are used. The plots show only comparisons for the UKF (differences are similar for the EKF). In general, the Robbins–Monro method is the most robust approach, with the fastest rate of convergence. In some examples, we have seen faster convergence with the "annealed" approach; however, this also requires additional insight and heuristic methods to monitor the learning. We should reiterate that the "fixed" and "lambda" approaches are more appropriate for on-line tracking problems.

Four-Regions Classification In the next example, we consider a benchmark pattern classification problem having four interlocking regions

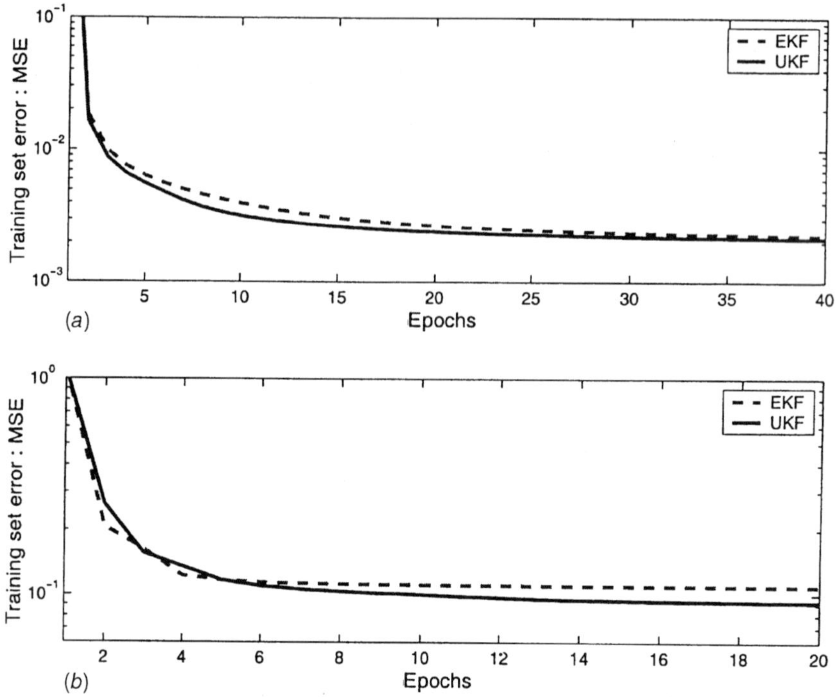

Figure 7.8 (*a*) MacKay robot-arm problem: comparison of learning curves for the EKF and UKF training, 2-12-2 MLP, annealing noise estimation. (*b*) Ikeda chaotic time series: comparison of learning curves for the EKF and UKF training, 10-7-1 MLP, Robbins–Monro noise estimation.

[8]. A three-layer feedforward network (MLP) with 2-10-10-4 nodes is trained using inputs randomly drawn within the pattern space, $S = [-1, -1] \times [1, 1]$, with the desired output value of $+0.8$ if the pattern fell within the assigned region and -0.8 otherwise. Figure 7.10 illustrates the classification task, learning curves for the UKF and EKF, and the final classification regions. For the learning curve, each epoch represents 100 randomly drawn input samples. The test set evaluated on each epoch corresponds to a uniform grid of 10,000 points. Again, we see the superior performance of the UKF.

Double Inverted Pendulum Returning to the double inverted pendulum (Section 7.3.1), we consider learning the system parameters, $\mathbf{w} = [l_1, l_2, m_1, m_2, M]$. These parameter values are treated as unknown (all initialized to 1.0). The full state, $\mathbf{x} = [x, \dot{x}, \theta_1, \dot{\theta}_1, \theta_2, \dot{\theta}_2]$, is observed.

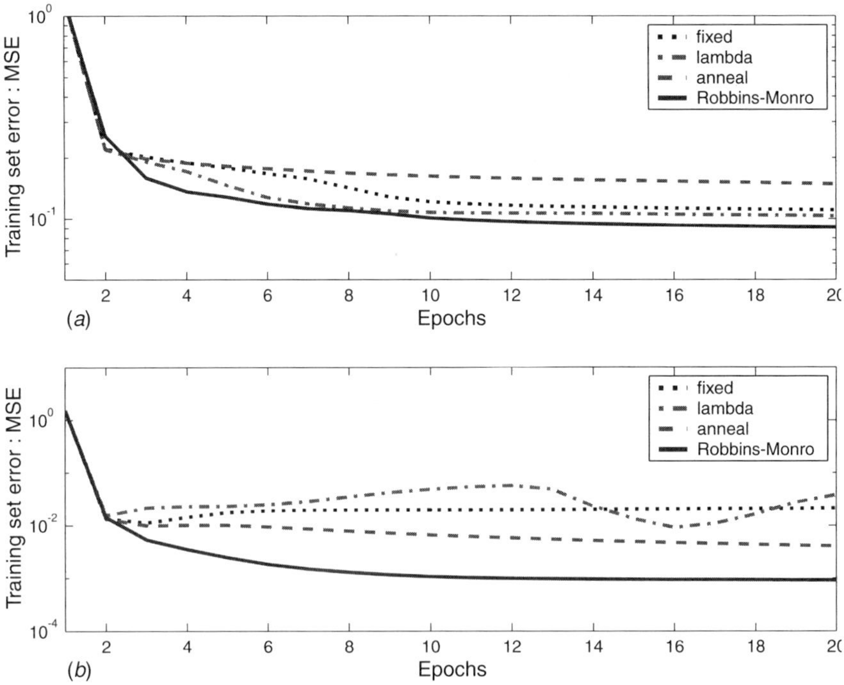

Figure 7.9 Neural network parameter estimation using different methods for noise estimation. (*a*) Ikeda chaotic time series. (*b*) Mackey–Glass chaotic time series. (UKF settings: $\alpha = 10^{-4}$, $\beta = 2$, $\kappa = 3 - L$, where L is the state dimension.)

Figure 7.11 shows the total model MSE versus iteration comparing EKF with UKF. Each iteration represents a pendulum crash with different initial conditions for the state (no control is applied). The final converged parameter estimates are as follows:

	l_1	l_2	m_1	m_2	M
True model	0.50	0.75	0.75	0.50	1.50
UKF estimate	0.50	0.75	0.75	0.50	1.49
EKF estimate	0.50	0.75	0.68	0.45	1.35

In this case, the EKF has converged to a biased solution, possibly corresponding to a local minimum in the error surface.

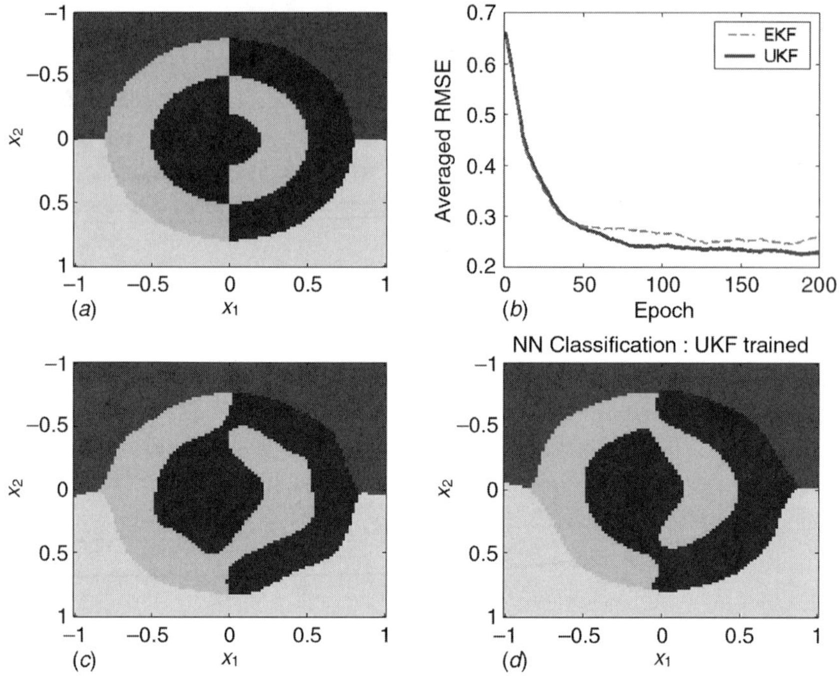

Figure 7.10 Singhal and Wu's four-region classification problem. (*a*) True mapping. (*b*) Learning curves on the test set. (*c*) NN classification: EKF-trained. (*d*) NN classification: UKF-trained. (UKF settings: $\alpha = 10^{-4}$, $\beta = 2$, $\kappa = 3 - L$, where L is the state dimension; 2-10-10-4 MLP; Robbins–Monro; 1 epoch = 100 random examples.)

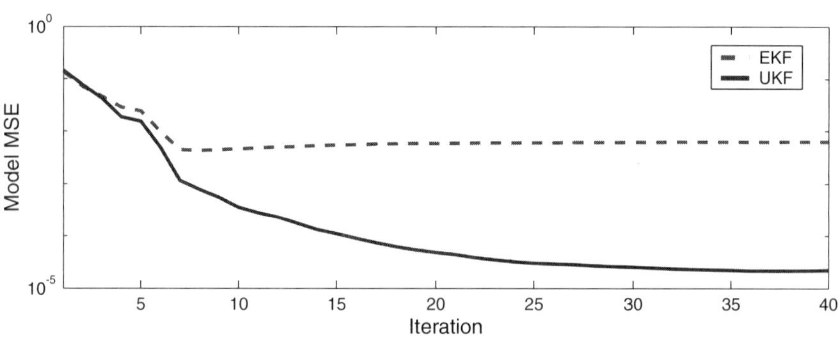

Figure 7.11 Inverted double pendulum parameter estimation. (UKF settings: $\alpha = 10^{-4}$, $\beta = 2$, $\kappa = 3 - L$, where L is the state dimension; Robbins–Monro.)

Rosenbrock's Banana Function For the last parameter estimation example, we turn to a pure optimization problem. The Banana function [27] can be thought of as a two-dimensional surface with a saddle-like curvature that bends around the origin. Specifically, we wish to find the values of x_1 and x_2 that minimize the function

$$f(x_1, x_2) = 100(x_2 - x_1^2)^2 + (1 - x_1)^2. \tag{7.91}$$

The true minimum is at $x_1 = 1$ and $x_2 = 1$. The Banana function is a well-known test problem used to compare the convergence rates of competing minimization techniques.

In order to use the UKF or EKF, the basic parameter estimation equations need to be reformulated to minimize a non-MSE cost function. To do this we write the state-space equations in observed error form [28]:

$$\mathbf{w}_k = \mathbf{w}_{k-1} + \mathbf{r}_k, \tag{7.92}$$

$$0 = -\mathbf{e}_k + \mathbf{e}_k, \tag{7.93}$$

where the target "observation" is fixed at zero, and $\mathbf{e}_k$ is an error term resulting in the optimization of the sum of instantaneous costs $J_k = \mathbf{e}_k^T\mathbf{e}_k$. The MSE cost is optimized by setting $\mathbf{e}_k = \mathbf{d}_k - \mathbf{G}(\mathbf{x}_k, \mathbf{w}_k)$. However, arbitrary costs (e.g., cross-entropy) can also be minimized simply by specifying $\mathbf{e}_k$ appropriately. Further discussion of this approach has been given in Chapter 5. Reformulation of the UKF equations requires changing only the effective output to $\mathbf{e}_k$, and setting the desired response to zero.

For the example at hand, we set $\mathbf{e}_k = [10(x_2 - x_1) \quad 1 - x_1]^T$. Furthermore, since this optimization problem is a special case of "noiseless" parameter estimation where the actual error can be minimized to zero, we make use of Eq. (7.89) (option 2) to calculate the output of the UKF algorithm. This will allow the UKF to reach the true minimum of the error surface more rapidly.[9] We also set the scaling parameter α to a small value, which we have found to be appropriate again for zero MSE problems. Under these circumstances, the performances of the UKF and EKF are indistinguishable, as illustrated in Figure 7.12. Overall, the performances

[9]Note that the use of option 1, where the expected value of the function is used as the output, essentially involves averaging of the output based on the current parameter covariance. This shows convergence in the case where zero MSE is possible, since convergence of the state covariance to zero would also be necessary through proper annealing of the state noise innovations $\mathbf{R}^{\mathbf{r}}$.

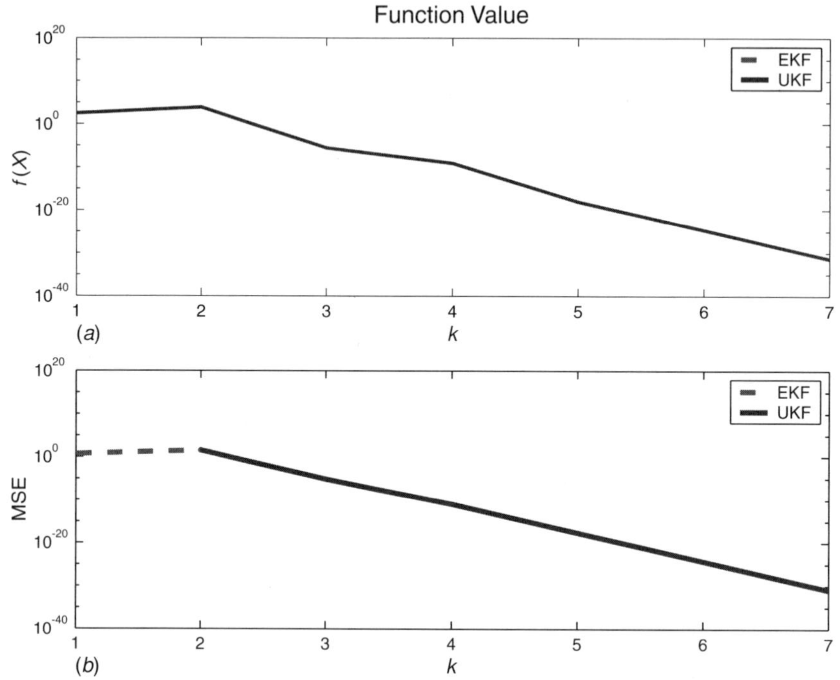

Figure 7.12 Rosenbrock's "Banana" optimization problem. (*a*) Function value. (*b*) Model error. (UKF settings: $\alpha = 10^{-4}$, $\beta = 2$, $\kappa = 3 - L$, where L is the state dimension; Fixed.)

of the two filters are comparable or superior to those of a number of alternative optimization approaches (e.g., Davidson–Fletcher–Powell, Levenburg–Marquardt, etc. See "optdemo" in *Matlab*). The main purpose of this example was to illustrate the versatility of the UKF to general optimization problems.

7.5 UKF DUAL ESTIMATION

Recall that the dual estimation problem consists of simultaneously estimating the clean state $\mathbf{x}_k$ and the model parameters $\mathbf{w}$ from the noisy data y_k (see Eq. (7.7)). A number of algorithmic approaches exist for this problem, including joint and dual EKF methods (recursive prediction error and maximum-likelihood versions), and expectation–maximization (EM) approaches. A thorough coverage of these algorithms is given in Chapters 5 and 6. In this section, we present results for the dual UKF (prediction error) and joint UKF methods.

In the *dual extended Kalman filter* [29], a separate state-space representation is used for the signal and the weights. Two EKFs are run simultaneously for signal and weight estimation. At every time step, the current estimate of the weights is used in the signal filter, and the current estimate of the signal state is used in the weight filter. In the *dual UKF* algorithm, both state and weight estimation are done with the UKF.

In the *joint extended Kalman filter* [30], the signal-state and weight vectors are concatenated into a single, *joint* state vector: $[\mathbf{x}_k^T \ \mathbf{w}_k^T]^T$. Estimation is done recursively by writing the state-space equations for the joint state as

$$\begin{bmatrix} \mathbf{x}_{k+1} \\ \mathbf{w}_{k+1} \end{bmatrix} = \begin{bmatrix} \mathbf{F}(\mathbf{x}_k, \mathbf{u}_k, \mathbf{w}_k) \\ \mathbf{I}\mathbf{w}_k \end{bmatrix} + \begin{bmatrix} \mathbf{B}v_k \\ \mathbf{r}_k \end{bmatrix}. \tag{7.94}$$

$$y_k = [1 \quad 0 \quad \ldots \quad 0] \begin{bmatrix} \mathbf{x}_k \\ \mathbf{w}_k \end{bmatrix} + n_k, \tag{7.95}$$

and running an EKF on the joint state space to produce simultaneous estimates of the states $\mathbf{x}_k$ and $\mathbf{w}$. Again, our approach is to use the UKF instead of the EKF.

7.5.1 Dual Estimation Experiments

Noisy Time-Series We present results on two time-series to provide a clear illustration of the use of the UKF over the EKF. The first series is again the Mackey–Glass-30 chaotic series with additive noise (SNR $\approx$ 3 dB). The second time series (also chaotic) comes from an autoregressive neural network with random weights driven by Gaussian process noise and also corrupted by additive white Gaussian noise (SNR $\approx$ 3 dB). A standard 6-10-1 MLP with tanh hidden activation functions and a linear output layer was used for all the filters in the Mackey–Glass problem. A 5-3-1 MLP was used for the second problem. The process- and measurement-noise variances associated with the state were assumed to be known. Note that, in contrast to the state estimation example in the previous section, only the noisy time series is observed. A clean reference is never provided for training.

Example training curves for the different dual and joint Kalman-based estimation methods are shown in Figure 7.13. A final estimate for the Mackey–Glass series is also shown for the dual UKF. The superior performance of the UKF-based algorithms is clear.

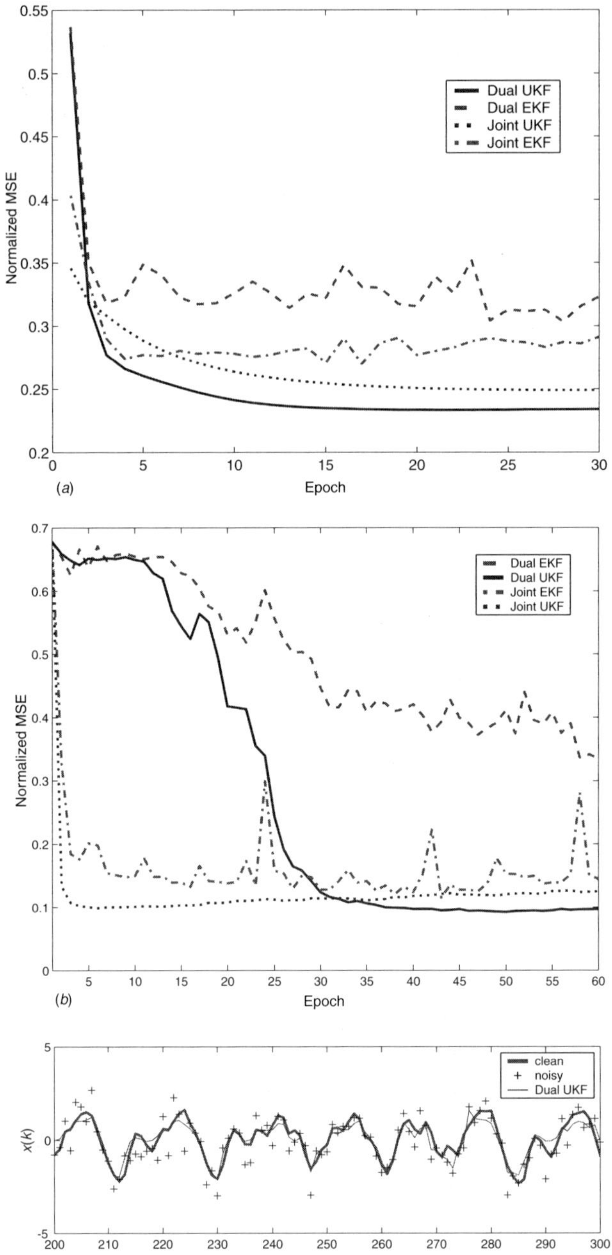

Figure 7.13 Comparative learning curves and results for the dual estimation experiments. Curves are averaged over 10 and 3 runs, respectively, using different initial weights. "Fixed" innovation covariances are used in the joint algorithms. "Annealed" covariances are used for the weight filter in the dual algorithms. (*a*) Chaotic AR neural network. (*b*) Mackey–Glass chaotic time series. (*c*) Estimation of Mackey–Glass time series: dual UKF.

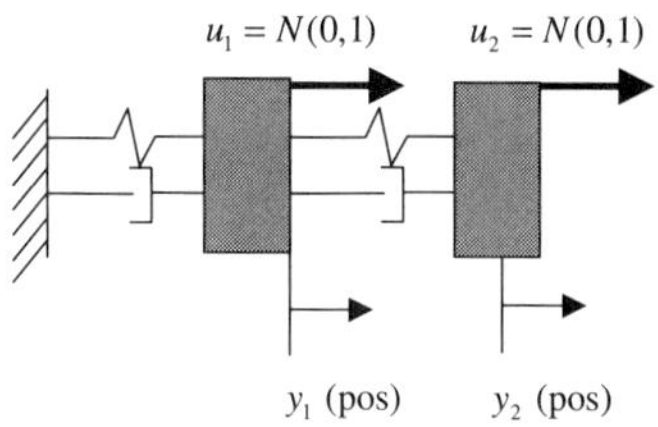

$$\begin{bmatrix} x_1 \\ x_1 \\ x_2 \\ x_2 \end{bmatrix} = \begin{bmatrix} 0 & 1 & 0 & 0 \\ -\omega_1^2 & -2\zeta_1\omega_1 & 0 & 0 \\ 0 & 0 & 0 & 1 \\ 0 & 0 & -\omega_2^2 & -2\zeta_2\omega_2 \end{bmatrix} \begin{bmatrix} x_1 \\ x_1 \\ x_2 \\ x_2 \end{bmatrix}$$

Figure 7.14 Mass-and-spring system.

Mode Estimation This example illustrates the use of the joint UKF for estimating the modes of a mass-and-spring system (see Fig. 7.14). This work was performed at the University of Washington by Mark Campbell and Shelby Brunke. While the system is linear, direct estimation of the natural frequencies ω_1 and ω_2 jointly with the states is a nonlinear estimation problem. Figure 7.15 compares the performance of the EKF and UKF. Note that the EKF does not converge to the true value for ω_2. For this experiment, the input process noise SNR is approximately 100 dB, and the measured positions y_1 and y_2 have additive noise at a 60 dB SNR (these settings effectively turn the task into a pure parameter-estimation

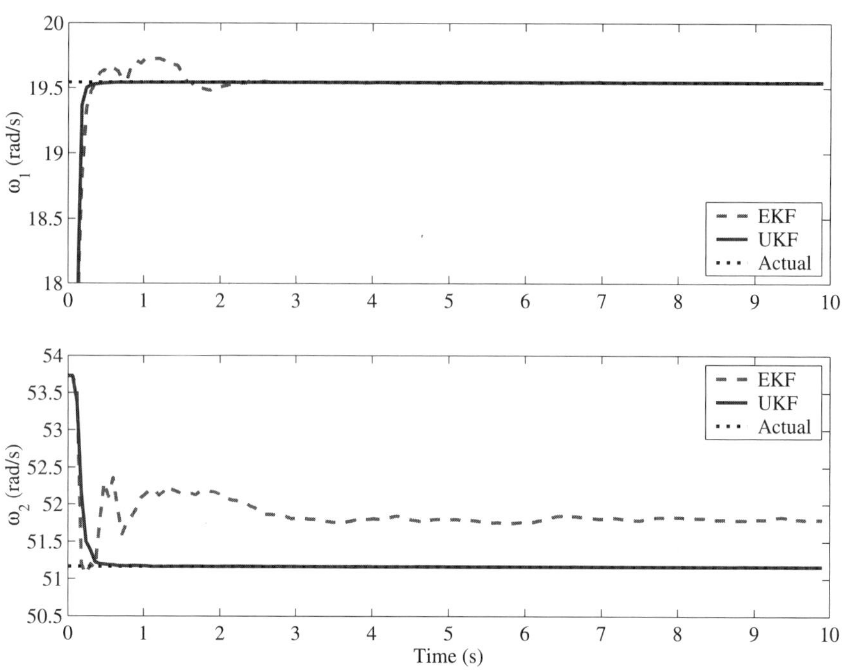

Figure 7.15 Linear mode prediction.

problem). A fixed innovations $\mathbf{R}^{\mathrm{r}}$ was used for the parameter estimation in the joint algorithms. Sampling was done at the Nyquist rate (based on ω_2), which emphasizes the effect of linearization in the EKF. For faster sampling rates, the performance of the EKF and UKF become more similar.

F15 Flight Simulation In this example (also performed at the University of Washington), joint estimation is done on an F15 aircraft model [31]. The simulation includes vehicle nonlinear dynamics, and engine and sensor noise modeling, as well as atmospheric modeling (densities, pressure, etc.) based on look-up tables. Also incorporated are aerodynamic forces based on data from Wright Patterson AFB. A closed-loop system using a gain-scheduled TECS controller is used to control the model [32]. A simulated mission was used to test the UKF estimator, and involved a quick descent, short tactical run, 180° turn, and ascent, with a possible failure in the stabilitator (horizontal control surface on the tail of the aircraft). Measurements consisted of the states with additive noise (20 dB SNR). Turbulence was approximately 1 m/s RMS. During the mission, the joint UKF estimated the 12 states (positions, orientations, and their derivatives) as well as parameters corresponding to aerodynamic forces and moments. This was done "off-line"; that is, the estimated states were not used within the control loop. Illustrative results are shown in Figure 7.16 for estimation of the altitude, velocity, and lift parameter (overall lift force on the aircraft). The left column shows the mission without a failure. The right column includes a 50% stabilitator failure at 65 seconds. Note that even with this failure, the UKF is still capable of tracking the state and parameters. It should be pointed out that the "black-box" nature of the simulator was not conducive to taking Jacobians necessary for running the EKF. Hence, implementation of the EKF for comparison was not performed.

Double Inverted Pendulum For the final dual estimation example, we again consider the double inverted pendulum, but this time we estimate both the states and system parameters using the joint UKF. Observations correspond to noisy measurements of the six states. Estimated states are then fed back for closed-loop control. In addition, parameter estimates are used at every time step to design the controller using the SDRE approach. Figure 7.17 illustrates performance of this adaptive control system by showing the evolution of the estimated and actual states. At the start of the simulation, both the states and parameters are unknown (the control system is unstable at this point). However, within one trial, the UKF

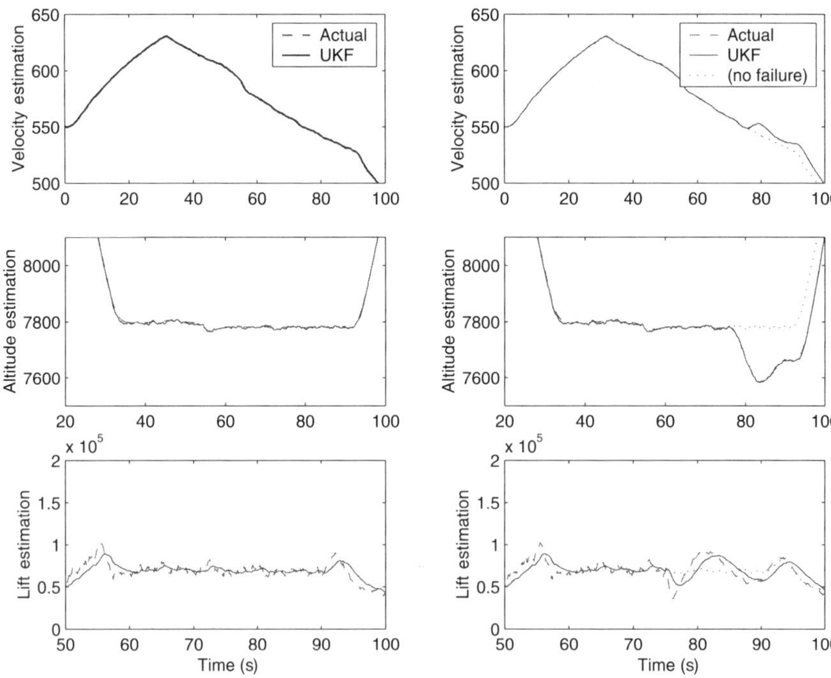

Figure 7.16 F15 model joint estimation (note that the estimated and true values of the state are indistinguishable at this resolution).

enables convergence and stabilization of the pendulum without a single crash.

7.6 THE UNSCENTED PARTICLE FILTER

The *particle filter* is a sequential Monte Carlo method that allows for a complete representation of the state distribution using sequential importance sampling and resampling [33–35]. Whereas the standard EKF and UKF make a Gaussian assumption to simplify the optimal recursive Bayesian estimation (see Section 7.2), particle filters make no assumptions on the form of the probability densities in question; that is, they employ full nonlinear, non-Gaussian estimation. In this section, we present a method that utilizes the UKF to augment and improve the standard particle filter, specifically through generation of the *importance proposal distribution*. This chapter will review the background fundamentals necessary to introduce particle filtering, and the extension based on the UKF. The

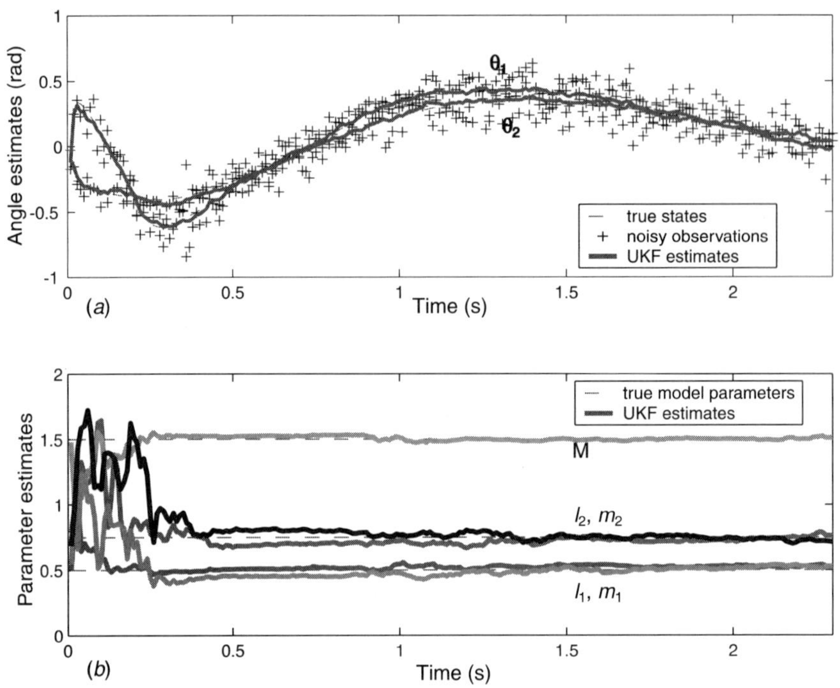

Figure 7.17 Double Inverted Pendulum joint estimation. Estimated states (*a*) and parameters (*b*). Only θ_1 and θ_2 are plotted (in radians).

material is based on work done by van der Merwe, de Freitas, Doucet, and Wan in [6], which also provides a more thorough review and treatment of particle filters in general.

Monte Carlo Simulation and Sequential Importance Sampling

Particle filtering is based on Monte Carlo simulation with sequential importance sampling (SIS). The overall goal is to directly implement optimal Bayesian estimation (see Eqs. (7.9)–(7.11)) by recursively approximating the complete posterior state density. In Monte Carlo simulation, a set of weighted particles (samples), drawn from the posterior distribution, is used to map integrals to discrete sums. More precisely, the posterior filtering density can be approximated by the following empirical estimate:

$$\hat{p}(\mathbf{x}_k|\mathbf{Y}_0^k) = \frac{1}{N}\sum_{i=1}^{N}\delta(\mathbf{x}_k - \boldsymbol{x}_k^{(i)}),$$

where the random samples $\{\boldsymbol{x}_k^{(i)};\ i = 1, \ldots, N\}$, are drawn from $p(\mathbf{x}_k|\mathbf{Y}_0^k)$ and $\delta(\cdot)$ denotes the Dirac delta function. The posterior *filtering* density $p(\mathbf{x}_k|\mathbf{Y}_0^k)$ is a marginal of the *full* posterior density given by $p(\mathbf{X}_0^k|\mathbf{Y}_0^k)$. Consequently, any expectations of the form

$$\mathbb{E}(\mathbf{g}(\mathbf{x}_k)) = \int \mathbf{g}(\mathbf{x}_k)p(\mathbf{x}_k|\mathbf{Y}_0^k)\, d\mathbf{x}_k \tag{7.96}$$

may be approximated by the following estimate:

$$\mathbb{E}(\mathbf{g}(\mathbf{x}_k)) \approx \frac{1}{N}\sum_{i=1}^{N} \mathbf{g}(\boldsymbol{x}_k^{(i)}). \tag{7.97}$$

For example, letting $g(\mathbf{x}) = \mathbf{x}$ yields the optimal MMSE estimate $\hat{\mathbf{x}}_k = \mathbb{E}[\mathbf{x}_k|\mathbf{Y}_0^k]$. The particles $\boldsymbol{x}_k^{(i)}$ are assumed to be independent and identically distributed (i.i.d) for the approximation to hold. As N goes to infinity, the estimate converges to the true expectation almost surely. Sampling from the filtering posterior is only a special case of Monte Carlo simulation, which in general deals with the complete posterior density $p(\mathbf{X}_0^k|\mathbf{Y}_0^k)$. We shall use this more general form to derive the particle filter algorithm.

It is often impossible to sample directly from the posterior density function. However, we can circumvent this difficulty by making use of importance sampling and alternatively sampling from a *known* proposal distribution $q(\mathbf{X}_0^k|\mathbf{Y}_0^k)$. The exact form of this distribution is a critical design issue, and is usually chosen in order to facilitate easy sampling. The details of this are discussed later. Given this proposal distribution, we can make use of the following substitution:

$$\begin{aligned}
\mathbb{E}(\mathbf{g}_k(\mathbf{X}_0^k)) &= \int \mathbf{g}_k(\mathbf{X}_0^k)\frac{p(\mathbf{X}_0^k|\mathbf{Y}_0^k)}{q(\mathbf{X}_0^k|\mathbf{Y}_0^k)}q(\mathbf{X}_0^k|\mathbf{Y}_0^k)\, d\mathbf{X}_0^k \\
&= \int \mathbf{g}_k(\mathbf{X}_0^k)\frac{p(\mathbf{Y}_0^k|\mathbf{X}_0^k)p(\mathbf{X}_0^k)}{p(\mathbf{Y}_0^k)q(\mathbf{X}_0^k|\mathbf{Y}_0^k)}q(\mathbf{X}_0^k|\mathbf{Y}_0^k)\, d\mathbf{X}_0^k \\
&= \int \mathbf{g}_k(\mathbf{X}_0^k)\frac{w_k(\mathbf{X}_0^k)}{p(\mathbf{Y}_0^k)}q(\mathbf{X}_0^k|\mathbf{Y}_0^k)\, d\mathbf{X}_0^k,
\end{aligned}$$

where the variables $w_k(\mathbf{X}_0^k)$ are known as the unnormalized importance weights,

$$w_k = \frac{p(\mathbf{Y}_0^k|\mathbf{X}_0^k)p(\mathbf{X}_0^k)}{q(\mathbf{X}_0^k|\mathbf{Y}_0^k)}. \tag{7.98}$$

We can get rid of the unknown normalizing density $p(\mathbf{Y}_0^k)$ as follows:

$$\begin{aligned}
\mathbb{E}(\mathbf{g}_k(\mathbf{X}_0^k)) &= \frac{1}{p(\mathbf{Y}_0^k)} \int \mathbf{g}_k(\mathbf{X}_0^k) w_k(\mathbf{X}_0^k) q(\mathbf{X}_0^k|\mathbf{Y}_0^k)\, d\mathbf{X}_0^k \\
&= \frac{\int \mathbf{g}_k(\mathbf{X}_0^k) w_k(\mathbf{X}_0^k) q(\mathbf{X}_0^k|\mathbf{Y}_0^k)\, d\mathbf{X}_0^k}{\int p(\mathbf{Y}_0^k|\mathbf{X}_0^k) p(\mathbf{X}_0^k) \dfrac{q(\mathbf{X}_0^k|\mathbf{Y}_0^k)}{q(\mathbf{X}_0^k|\mathbf{Y}_0^k)}\, d\mathbf{X}_0^k} \\
&= \frac{\int \mathbf{g}_k(\mathbf{X}_0^k) w_k(\mathbf{X}_0^k) q(\mathbf{X}_0^k|\mathbf{Y}_0^k)\, d\mathbf{X}_0^k}{\int w_k(\mathbf{X}_0^k) q(\mathbf{X}_0^k|\mathbf{Y}_0^k)\, d\mathbf{X}_0^k} \\
&= \frac{\mathbb{E}_{q(\cdot|\mathbf{Y}_0^k)}(w_k(\mathbf{X}_0^k)\mathbf{g}_k(\mathbf{X}_0^k))}{\mathbb{E}_{q(\cdot|\mathbf{Y}_0^k)}(w_k(\mathbf{X}_0^k))},
\end{aligned}$$

where the notation $\mathbb{E}_{q(\cdot|\mathbf{Y}_0^k)}$ has been used to emphasize that the expectations are taken over the proposal distribution $q(\cdot|\mathbf{Y}_0^k)$.

A *sequential* update to the importance weights is achieved by expanding the proposal distribution as $q(\mathbf{X}_0^k|\mathbf{Y}_0^k) = q(\mathbf{X}_0^{k-1}|\mathbf{Y}_0^{k-1})q(\mathbf{x}_k|\mathbf{X}_0^{k-1}, \mathbf{Y}_0^k)$, where we are making the assumption that the current state is not dependent on future observations. Furthermore, under the assumption that the states correspond to a Markov process and that the observations are conditionally independent given the states, we can arrive at the recursive update:

$$w_k = w_{k-1} \frac{p(\mathbf{y}_k|\mathbf{x}_k) p(\mathbf{x}_k|\mathbf{x}_{k-1})}{q(\mathbf{x}_k|\mathbf{X}_0^{k-1}, \mathbf{Y}_0^k)}. \tag{7.99}$$

Equation (7.99) provides a mechanism to sequentially update the importance weights given an appropriate choice of proposal distribution, $q(\mathbf{x}_k|\mathbf{X}_0^{k-1}, \mathbf{Y}_0^k)$. Since we can sample from the proposal distribution and evalute the likelihood $p(\mathbf{y}_k|\mathbf{x}_k)$ and transition probabilities $p(\mathbf{x}_k|\mathbf{x}_{k-1})$, all we need to do is generate a prior set of samples and iteratively compute the importance weights. This procedure then allows us to evaluate the expectations of interest by the following estimate:

$$\mathbb{E}(\mathbf{g}(\mathbf{X}_0^k)) \approx \frac{N^{-1} \sum_{i=1}^{N} \mathbf{g}(\boldsymbol{x}_{0:k}^{(i)}) w_k(\boldsymbol{x}_{0:k}^{(i)})}{N^{-1} \sum_{i=1}^{N} w_k(\boldsymbol{x}_{0:k}^{(i)})} = \sum_{i=1}^{N} \mathbf{g}(\boldsymbol{x}_{0:k}^{(i)}) \tilde{w}_k(\boldsymbol{x}_{0:k}^{(i)}), \tag{7.100}$$

where the normalized importance weights $\tilde{w}_k^{(i)} = w_k^{(i)} / \sum_{j=1}^{N} w_k^{(j)}$ and $\boldsymbol{x}_{0:k}^{(i)}$ denotes the ith sample trajectory drawn from the proposal distribution $q(\mathbf{x}_k|\mathbf{X}_0^{k-1}, \mathbf{Y}_0^k)$. This estimate asymptotically converges if the expectation and variance of $\mathbf{g}(\mathbf{X}_0^k)$ and w_k exist and are bounded, and if the support of the proposal distribution includes the support of the posterior distribution. Thus, as N tends to infinity, the posterior density function can be approximated arbitrarily well by the point-mass estimate

$$\hat{p}(\mathbf{X}_0^k|\mathbf{Y}_0^k) = \sum_{i=1}^{N} \tilde{w}_k^{(i)} \delta(\mathbf{X}_0^k - \boldsymbol{x}_{0:k}^{(i)}) \tag{7.101}$$

and the posterior *filtering* density by

$$\hat{p}(\mathbf{x}_k|\mathbf{Y}_0^k) = \sum_{i=1}^{N} \tilde{w}_k^{(i)} \delta(\mathbf{x}_k - \boldsymbol{x}_k^{(i)}). \tag{7.102}$$

In the case of filtering, we do not need to keep the whole history of the sample trajectories, in that only the current set of samples at time k is needed to calculate expectations of the form given in Eq. (7.96) and (7.97). To do this, we simply set, $\mathbf{g}(\mathbf{X}_0^k) = \mathbf{g}(\mathbf{x}_k)$. These point-mass estimates can approximate any general distribution arbitrarily well, limited only by the number of particles used and how well the above-mentioned importance sampling conditions are met. In contrast, the posterior distribution calculated by the EKF is a minimum-variance Gaussian approximation to the true distribution, which inherently cannot capture complex structure such as multimodalities, skewness, or other higher-order moments.

Resampling and MCMC Step The sequential importance sampling (SIS) algorithm discussed so far has a serious limitation: the variance of the importance weights increases stochastically over time. Typically, after a few iterations, one of the normalized importance weights tends to unity, while the remaining weights tend to zero. A large number of samples are thus effectively removed from the sample set because their importance weights become numerically insignificant. To avoid this degeneracy, a resampling or selection stage may be used to eliminate samples with low importance weights and multiply samples with high importance weights. This is often followed by a Markov-chain Monte Carlo (MCMC) move step, which introduces sample variety without affecting the posterior distribution they represent.

A selection scheme associates to each particle $\boldsymbol{x}_k^{(i)}$ a number of "children," N_i, such that $\sum_{i=1}^{N} N_i = N$. Several selection schemes have been proposed in the literature, including *sampling-importance resampling* (*SIR*) [36–38], *residual resampling* [25, 39], and *minimum-variance sampling* [34].

Sampling-importance resampling (SIR) involves mapping the Dirac random measure $\{\boldsymbol{x}_k^{(i)}, \tilde{w}_k^{(i)}\}$ into an equally weighted random measure $\{\boldsymbol{x}_k^{(j)}, N^{-1}\}$. In other words, we produce N new samples all with weighting $1/N$. This can be accomplished by sampling uniformly from the discrete set $\{\boldsymbol{x}_k^{(i)}; i = 1, \ldots, N\}$ with probabilities $\{\tilde{w}_k^{(i)}; i = 1, \ldots, N\}$. Figure 7.18 gives a graphical representation of this process. This procedure effectively replicates the original $\boldsymbol{x}_k^{(i)}$ particle N_i times (N_i may be zero).

In *residual resampling* [25, 39] a two-step process is used, which makes use of SIR. In the first step, the number of children are deterministicly set using the floor function, $N_i^A = \lfloor N\tilde{w}_t^{(i)} \rfloor$. Each $\boldsymbol{x}_k^{(i)}$ particle is replicated N_i^A times. In the second step, SIR is used to select the remaining $\bar{N}_t = N - \sum_{i=1}^{N} N_i^A$ samples, with new weights $w_t^{'(i)} = \bar{N}_t^{-1}(\tilde{w}_t^{(i)} N - N_i^A)$. These samples form a second set N_i^B, such that $\bar{N}_t = \sum_{i=1}^{N} N_i^B$, and are drawn as described previously. The total number of children of each particle is then set to $N_i = N_i^A + N_i^B$. This procedure is computationally cheaper than pure SIR, and also has lower sample variance. Thus, residual resampling is used for all experiments in Section 7.6.2 (in general, we have found that the specific choice of resampling scheme does not significantly affect the performance of the particle filter).

After the selection/resampling step at time k, we obtain N particles distributed approximately according to the posterior distribution. Since the selection step favors the creation of multiple copies of the "fittest"

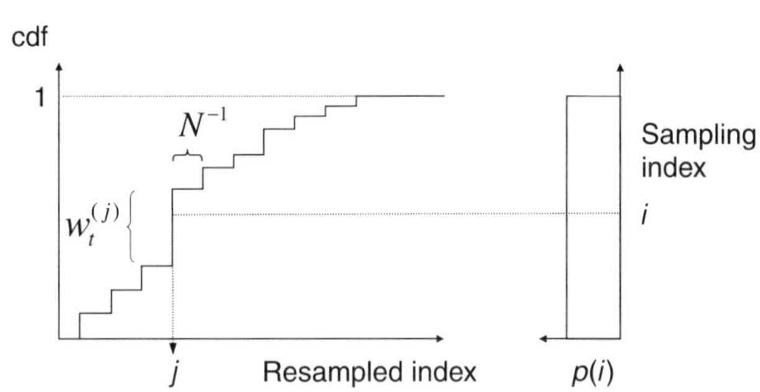

Figure 7.18 Resampling process, whereby a random measure $\{\mathbf{x}_k^{(i)}, \tilde{\mathbf{w}}_k^{(i)}\}$ is mapped into an equally weighted random measure $\{\mathbf{x}_k^{(j)}, N^{-1}\}$. The index i is drawn from a uniform distribution.

particle, many particles may end up having no children ($N_i = 0$), whereas others might end up having a large number of children, the extreme case being $N_i = N$ for a particular value i. In this case, there is a severe depletion of samples. Therefore, an additional procedure is often required to introduce sample variety after the selection step without affecting the validity of the approximation inferred. This is achieved by performing a single MCMC step on each particle. The basic idea is that if the particles are already distributed according to the posterior $p(\mathbf{x}_k|\mathbf{Y}_0^k)$ (which is the case), then applying a Markov-chain transition kernel with the same invariant distribution to each particle results in a set of new particles distributed according to the posterior of interest. However, the new particles may move to more interesting areas of the state space. Details of the MCMC step are given in [6]. For our experiments in Section 7.6.2, we found an MCMC step to be unnecessary. However, this cannot be assumed in general.

7.6.1 The Particle Filter Algorithm

The pseudo-code of a generic particle filter is presented in Table 7.6. In implementing this algorithm, the choice of the proposal distribution $q(\mathbf{x}_k|\mathbf{X}_0^{k-1}, \mathbf{Y}_0^k)$ is the most critical design issue. The optimal proposal distribution (which minimizes the variance on the importance weights) is given by [40–43]

$$q(\mathbf{x}_k|\mathbf{X}_0^{k-1}, \mathbf{Y}_0^k) = p(\mathbf{x}_k|\mathbf{X}_0^{k-1}, \mathbf{Y}_0^k), \tag{7.103}$$

that is, the true conditional state density given the previous state history and all observations. Sampling from this is, of course, impractical for arbitrary densities (recall the motivation for using importance sampling in the first place). Consequently, the transition prior is the most popular choice of proposal distribution [35, 44–47]:[10]

$$q(\mathbf{x}_k|\mathbf{X}_0^{k-1}, \mathbf{Y}_0^k) \stackrel{\circ}{=} p(\mathbf{x}_k|\mathbf{x}_{k-1}). \tag{7.104}$$

For example, if an additive Gaussian process noise model is used, the transition prior is simply

$$p(\mathbf{x}_k|\mathbf{x}_{k-1}) = \mathcal{N}(\mathbf{F}(\bar{\mathbf{x}}_{k-1}, 0), \mathbf{R}_{k-1}^{\mathbf{v}}). \tag{7.105}$$

[10]The notation $\stackrel{\circ}{=}$ denotes "chosen as," to indicate a subtle difference versus "approximation".

Table 7.6 Algorithm for the generic particle filter

1. *Initialization:* $k = 0$
   - For $i = 1, \ldots, N$, draw the states $\boldsymbol{x}_0^{(i)}$ from the prior $p(\mathbf{x}_0)$.
2. For $k = 1, 2, \ldots$
   (*a*) *Importance sampling step*
   - For $i = 1, \ldots, N$, sample $\boldsymbol{x}_k^{(i)} \sim q(\mathbf{x}_k | \mathbf{x}_{0:k-1}^{(i)}, \mathbf{Y}_0^k)$.
   - For $i = 1, \ldots, N$, evaluate the importance weights up to a normalizing constant:

$$w_k^{(i)} = w_{k-1}^{(i)} \frac{p(\mathbf{y}_k | \boldsymbol{x}_k^{(i)}) p(\boldsymbol{x}_k^{(i)} | \boldsymbol{x}_{k-1}^{(i)})}{q(\boldsymbol{x}_k^{(i)} | \boldsymbol{x}_{0:k-1}^{(i)}, \mathbf{Y}_0^k)}. \tag{7.106}$$

   - For $i = 1, \ldots, N$, normalize the importance weights:

$$\tilde{w}_k^{(i)} = w_k^{(i)} \left(\sum_{j=1}^{N} w_k^{(j)} \right)^{-1}.$$

   (*b*) *Selection step* (*resampling*)
   - Multiply/suppress samples $\boldsymbol{x}_k^{(i)}$ with high/low importance weights $\tilde{w}_k^{(i)}$, respectively, to obtain N random samples $\boldsymbol{x}_k^{(i)}$ approximately distributed according to $p(\mathbf{x}_k^{(i)} | \mathbf{Y}_0^k)$.
   - For $i = 1, \ldots, N$, set $w_k^{(i)} = \tilde{w}_k^{(i)} = N^{-1}$.

   (*c*) *MCMC move step* (*optional*)
   (*d*) *Output:* The output of the algorithm is a set of samples that can be used to approximate the posterior distribution as follows:

$$\hat{p}(\mathbf{x}_k | \mathbf{Y}_0^k) = \frac{1}{N} \sum_{i=1}^{N} \delta(\mathbf{x}_k - \boldsymbol{x}_k^{(i)}).$$

   The optimal MMSE estimator is given as

$$\hat{\mathbf{x}}_k = \mathbb{E}(\mathbf{x}_k | \mathbf{Y}_0^k) \approx \frac{1}{N} \sum_{i=1}^{N} \boldsymbol{x}_k^{(i)}.$$

   Similar expectations of the function $\mathbf{g}(\mathbf{x}_k)$ can also be calculated as a sample average.

The effectiveness of this approximation depends on how close the proposal distribution is to the true posterior distribution. If there is not sufficient overlap, only a few particles will have significant importance weights when their likelihood are evaluated.

The EKF and UKF Particle Filter An improvement in the choice of proposal distribution over the simple transition prior, which also address the problem of sample depletion, can be accomplished by moving the

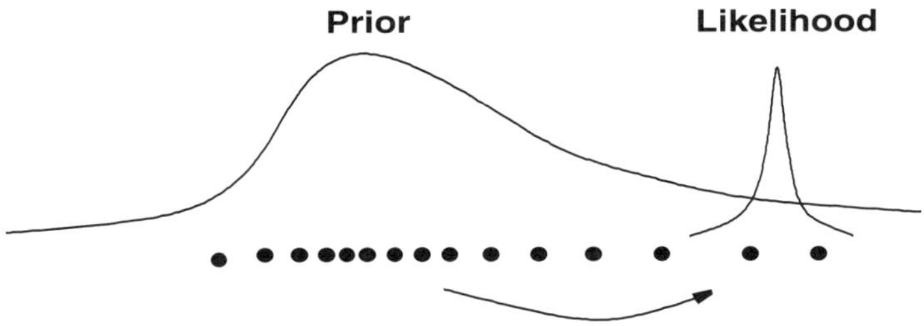

Figure 7.19 Including the most current observation into the proposal distribution, allows us to move the samples in the prior to regions of high likelihood. This is of paramount importance if the likelihood happens to lie in one of the tails of the prior distribution, or if it is too narrow (low measurement error).

particles towards the regions of high likelihood, based on the most recent observations $\mathbf{y}_k$ (see Fig. 7.19). An effective approach to accomplish this, is to use an EKF generated Gaussian approximation to the optimal proposal, that is,

$$q(\mathbf{x}_k|\mathbf{X}_0^{k-1}, \mathbf{Y}_0^k) \stackrel{\circ}{=} q_{\mathcal{N}}(\mathbf{x}_k|\mathbf{Y}_0^k), \tag{7.107}$$

which is accomplished by using a separate EKF to generate and propagate a Gaussian proposal distribution for each particle,

$$q_{\mathcal{N}}(\mathbf{x}_k^{(i)}|\mathbf{Y}_0^k) = \mathcal{N}(\bar{\mathbf{x}}_k^{(i)}, \mathbf{P}_k^{(i)}), \qquad i = 1, \ldots, N. \tag{7.108}$$

That is, at time k one uses the EKF equations, with the new data, to compute the mean and covariance of the importance distribution for each particle from the previous time step $k-1$. Next, we redraw the ith particle (at time k) from this new updated distribution. While still making a Gaussian assumption, the approach provides a better approximation to the optimal conditional proposal distribution and has been shown to improve performance on a number of applications [33, 48].

By replacing the EKF with the UKF, we can more accurately propagate the mean and covariance of the Gaussian approximation to the state distribution. Distributions generated by the UKF will have a greater support overlap with the true posterior distribution than the overlap achieved by the EKF estimates. In addition, scaling parameters used for sigma-point selection can be optimised to capture certain characteristic of the prior distribution if known; e.g. the algorithm can be modified to work with distributions that have heavier tails than Gaussian distributions such as Cauchy or Student-t distributions. The new filter that results from using a UKF for proposal distribution generation within a particle filter framework is called the *unscented particle filter* (UPF). Referring to the algorithm in Table 7.6 for the generic particle filter, the first item in the importance sampling step,

- For $i = 1, \ldots, N$, sample $\boldsymbol{x}_k^{(i)} \sim q(\mathbf{x}_k | \mathbf{x}_{0:k-1}^{(i)}, \mathbf{Y}_0^k)$,

is replaced with the following UKF update:

- For $i = 1, \ldots, N$:
  - Update the prior $(k-1)$ distribution for each particle with the UKF:
    * Calculate sigma points:

$$\boldsymbol{\mathcal{X}}_{k-1}^{(i)a} = [\bar{\boldsymbol{x}}_{k-1}^{(i)a} \quad \bar{\boldsymbol{x}}_{k-1}^{(i)a} + \gamma\sqrt{\mathbf{P}_{k-1}^{(i)a}} \quad \bar{\boldsymbol{x}}_{k-1}^{(i)a} - \gamma\sqrt{\mathbf{P}_{k-1}^{(i)a}}]. \tag{7.109}$$

    * Propagate particle into future (time update):

$$\boldsymbol{\mathcal{X}}_{k|k-1}^{(i)x} = \mathbf{F}(\boldsymbol{\mathcal{X}}_{k-1}^{(i)x}, \mathbf{u}_k, \boldsymbol{\mathcal{X}}_{k-1}^{(i)v}), \qquad \bar{\boldsymbol{x}}_{k|k-1}^{(i)} = \sum_{j=0}^{2L} W_j^{(m)} \boldsymbol{\mathcal{X}}_{j,k|k-1}^{(i)x}, \tag{7.110}$$

$$\mathbf{P}_{k|k-1}^{(i)} = \sum_{j=0}^{2L} W_j^{(c)} (\boldsymbol{\mathcal{X}}_{j,k|k-1}^{(i)x} - \bar{\boldsymbol{x}}_{k|k-1}^{(i)})(\boldsymbol{\mathcal{X}}_{j,k|k-1}^{(i)x} - \bar{\boldsymbol{x}}_{k|k-1}^{(i)})^T \tag{7.111}$$

$$\boldsymbol{\mathcal{Y}}_{k|k-1}^{(i)} = \mathbf{H}(\boldsymbol{\mathcal{X}}_{k|k-1}^{(i)x}, \boldsymbol{\mathcal{X}}_{k-1}^{(i)n}),$$

$$\bar{\mathbf{y}}_{k|k-1}^{(i)} = \sum_{j=0}^{2L} W_j^{(m)} \boldsymbol{\mathcal{Y}}_{j,k|k-1}^{(i)}. \tag{7.112}$$

    * Incorporate new observation (measurement update):

$$\mathbf{P}_{\tilde{\mathbf{y}}_k\tilde{\mathbf{y}}_k} = \sum_{j=0}^{2L} W_j^{(c)} (\boldsymbol{\mathcal{Y}}_{j,k|k-1}^{(i)} - \bar{\mathbf{y}}_{k|k-1}^{(i)})(\boldsymbol{\mathcal{Y}}_{j,k|k-1}^{(i)} - \bar{\mathbf{y}}_{k|k-1}^{(i)})^T, \tag{7.113}$$

$$\mathbf{P}_{\mathbf{x}_k\mathbf{y}_k} = \sum_{J=0}^{2L} W_j^{(c)} (\boldsymbol{\mathcal{X}}_{j,k|k-1}^{(i)} - \bar{\boldsymbol{x}}_{k|k-1}^{(i)})(\boldsymbol{\mathcal{Y}}_{j,k|k-1}^{(i)} - \bar{\mathbf{y}}_{k|k-1}^{(i)})^T, \tag{7.114}$$

$$\mathbf{K}_k = \mathbf{P}_{\mathbf{x}_k\mathbf{y}_k} \mathbf{P}_{\tilde{\mathbf{y}}_k\tilde{\mathbf{y}}_k}^{-1},$$

$$\bar{\boldsymbol{x}}_k^{(i)} = \bar{\boldsymbol{x}}_{k|k-1}^{(i)} + \mathbf{K}_k(\mathbf{y}_k - \bar{\mathbf{y}}_{k|k-1}^{(i)}), \tag{7.115}$$

$$\mathbf{P}_k^{(i)} = \mathbf{P}_{k|k-1}^{(i)} - \mathbf{K}_k \mathbf{P}_{\tilde{\mathbf{y}}_k\tilde{\mathbf{y}}_k} \mathbf{K}_k^T. \tag{7.116}$$

  - Sample $\boldsymbol{x}_k^{(i)} \sim q(\mathbf{x}_k^{(i)} | \mathbf{x}_{0:k-1}^{(i)}, \mathbf{Y}_0^k) \approx \mathcal{N}(\bar{\boldsymbol{x}}_k^{(i)}, \mathbf{P}_k^{(i)})$.

All other steps in the particle filter formulation remain unchanged.

7.6.2 UPF Experiments

The performance of the UPF is evaluated on two estimation problems. The first problem is a synthetic scalar estimation problem and the second is a real-world problem concerning the pricing of financial instruments.

Synthetic Experiment For this experiment, a time series was generated by the following process model:

$$x_{k+1} = 1 + \sin(\omega \pi t) + \phi_1 x_k + v_k, \tag{7.117}$$

where v_k is a Gamma $\mathcal{G}a(3, 2)$ random variable modeling the process noise, and $\omega = 0.04$ and $\phi_1 = 0.5$ are scalar parameters. A nonstationary observation model,

$$y_k = \begin{cases} \phi_2 x_k^2 + n_k, & t \leq 30, \\ \phi_3 x_k - 2 + n_k & t > 30, \end{cases} \tag{7.118}$$

is used, with $\phi_2 = 0.2$ and $\phi_3 = 0.5$. The observation noise, n_k, is drawn from a Gaussian distribution $\mathcal{N}(0, 0.00001)$. Given only the noisy observations y_k, the different filters were used to estimate the underlying clean state sequence x_k for $k = 1 \ldots 60$. The experiment was repeated 100 times with random re-initialization for each run. All of the particle filters used 200 particles and residual resampling. The UKF parameters were set to $\alpha = 1$, $\beta = 0$ and $\kappa = 2$. These parameters are optimal for the scalar case. Table 7.7 summarizes the performance of the different filters. The table shows the means and variances of the mean-square error (MSE) of the state estimates. Figure 7.20 compares the estimates generated from a

Table 7.7 State-estimation experiment results: the mean and variance of the MSE were calculated over 100 independent runs

	MSE	
Algorithm	Mean	Variance
Extended Kalman filter (EKF)	0.374	0.015
Unscented Kalman filter (UKF)	0.280	0.012
Particle filter: generic	0.424	0.053
Particle filter: MCMC move step	0.417	0.055
Particle filter: EKF proposal	0.310	0.016
Particle filter: EKF proposal and MCMC move step	0.307	0.015
Particle filter: UKF proposal (“*unscented particle filter*”)	0.070	0.006
Particle filter: UKF proposal and MCMC move step	0.074	0.008

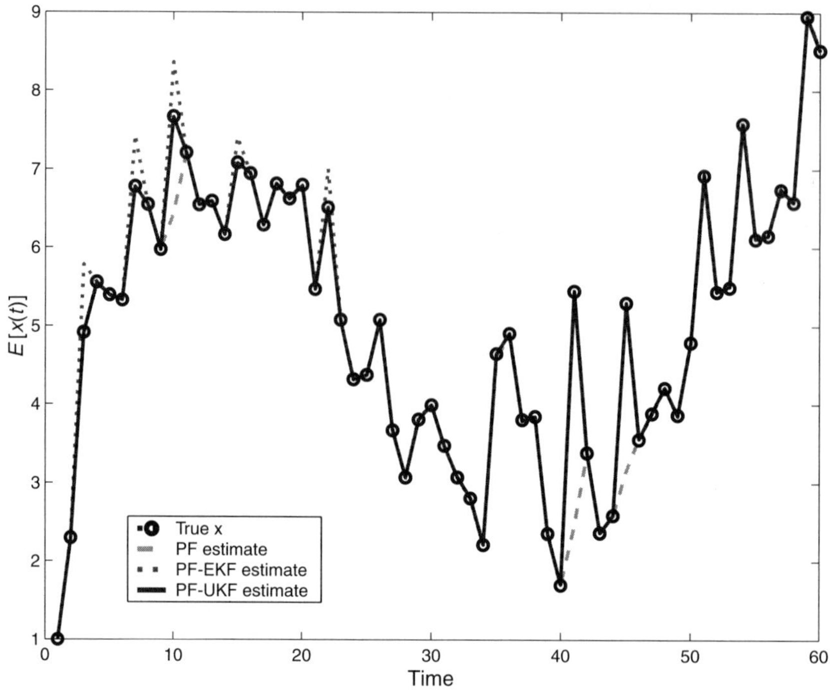

Figure 7.20 Plot of estimates generated by the different filters on the synthetic state-estimation experiment.

single run of the different particle filters. The superior performance of the *unscented particle filter* (UPF) is clear.

Pricing Financial Options Derivatives are financial instruments whose value depends on some basic underlying cash product, such as interest rates, equity indices, commodities, foreign exchange, or bonds [49]. A call option allows the holder to buy a cash product, at a specified date in the future, for a price determined in advance. The price at which the option is exercised is known as the strike price, while the date in which the option lapses is often referred to as the maturity time. Put options, on the other hand, allow the holder to sell the underlying cash product. In their seminal work [50], Black and Scholes derived the following industry standard equations for pricing European call and put options:

$$C = S\mathcal{N}_c(d_1) - Xe^{-rt_m}\mathcal{N}_c(d_2), \tag{7.119}$$

$$P = -S\mathcal{N}_c(-d_1) + Xe^{-rt_m}\mathcal{N}_c(-d_2), \tag{7.120}$$

where C denotes the price of a call option, P the price of a put option, S the current value of the underlying cash product, X the desired strike price, t_m the time to maturity, and $\mathcal{N}_c(.)$ the cumulative normal distribution, and d_1 and d_2 are given by

$$d_1 = \frac{\ln(S/X) + (r + \sigma^2/2)t_m}{\sigma\sqrt{t_m}},$$

$$d_2 = d_1 - \sigma\sqrt{t_m},$$

where σ is the (unknown) volatility of the cash product and r is the risk-free interest rate.

The volatility, σ, is usually estimated from a small moving window of data over the most recent 50–180 days [49]. The risk-free interest rate r is often estimated by monitoring interest rates in the bond markets. Our approach is to treat r and σ as the hidden states, and C and P as the output

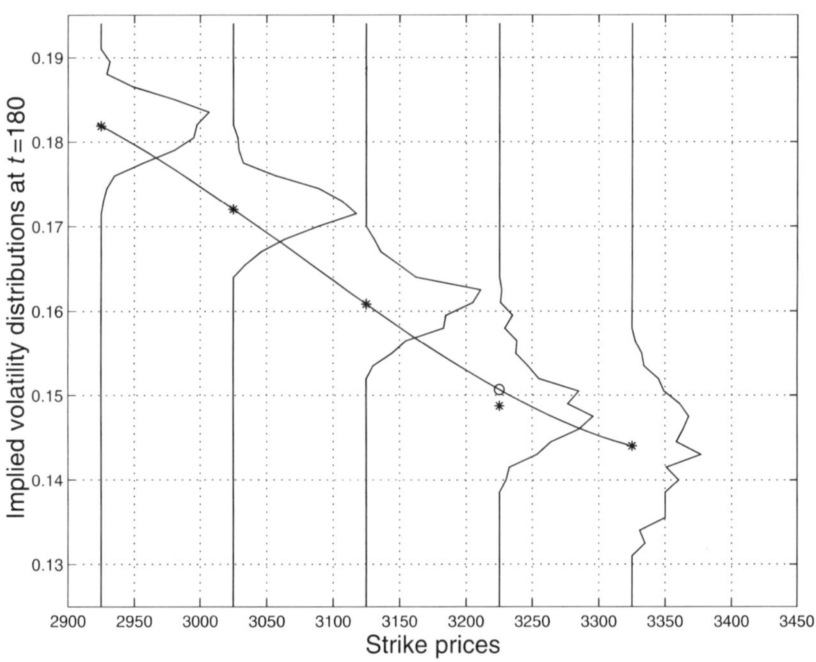

Figure 7.21 Probability smile for options on the FTSE-100 index (1994). Although the volatility smile indicates that the option with strike price equal to 3225 is underpriced, the shape of the probability gives us a warning against the hypothesis that the option is under-priced. Posterior mean estimates were obtained with the Black–Scholes model and particle filter ($*$), a fourth-order polynomial fit (–), and hypothesized volatility ($\circ$).

observations. S and t_m are treated as known control signals (input observations). This represents a parameter estimation problem, with the nonlinear observation given by Eqs. (7.119) or (7.120). This allows us to compute daily complete probability distributions for r and σ and to decide whether the current value of an option in the market is being either over-priced or under-priced. See [51] and [52] for details.

As an example, Figure 7.21 shows the implied probability density function of each volatility against several strike prices using five pairs of call and put option contracts on the British FTSE-100 index (from February 1994 to December 1994). Figure 7.22 shows the estimated volatility and interest rate for a contract with a strike price of 3225. In Table 7.8, we compare the one-step-ahead normalized square errors on a pair of options with strike price 2925. The square errors were only measured over the last 100 days of trading, so as to allow the algorithms to converge. The experiment was repeated 100 times with 100 particles in each particle filter (the mean value is reported; all variance were essentially zero). In this example, both the EKF and UKF approaches to improving the proposal distribution lead to a significant improvement over the standard particle filters. The main advantage of the UKF over the

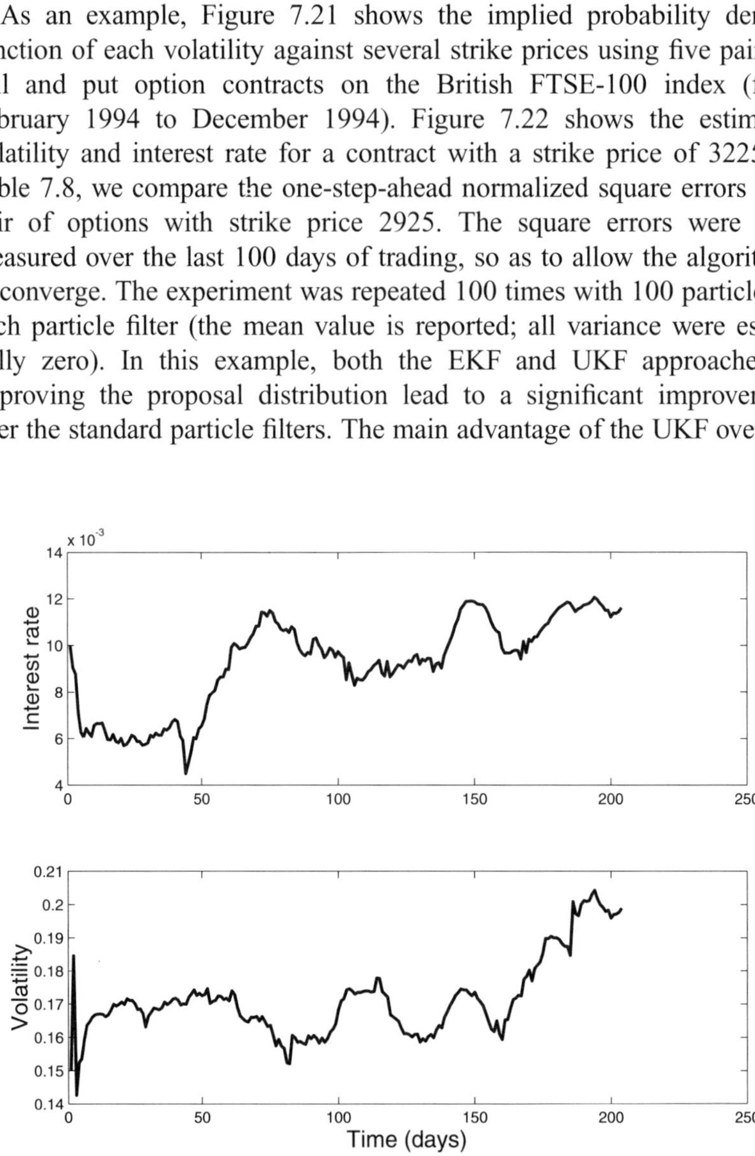

Figure 7.22 Estimated interest rate and volatility.

Table 7.8 One-step-ahead normalized square errors over 100 runs. The trivial prediction is obtained by assuming that the price on the following day corresponds to the current price

Option type	Algorithm	Mean NSE
Call	Trivial	0.078
	Extended Kalman filter (EKF)	0.037
	Unscented Kalman filter (UKF)	0.037
	Particle filter: generic	0.037
	Particle filter: EKF proposal	0.009
	Unscented particle filter	0.009
Put	Trivial	0.035
	Extended Kalman filter (EKF)	0.023
	Unscented Kalman filter (UKF)	0.023
	Particle filter: generic	0.023
	Particle filter: EKF proposal	0.007
	Unscented particle filter	0.008

EKF is the ease of implementation, which avoids the need to analytically differentiate the Black–Scholes equations.

7.7 CONCLUSIONS

The EKF has been widely accepted as a standard tool in the control and machine-learning communities. In this chapter, we have presented an alternative to the EKF using the unscented Kalman filter. The UKF addresses many of the approximation issues of the EKF, and consistently achieves an equal or better level of performance at a comparable level of complexity. The performance benefits of the UKF-based algorithms have been demonstrated in a number of application domains, including state estimation, dual estimation, and parameter estimation.

There are a number of clear advantages to the UKF. First, the mean and covariance of the state estimate is calculated to second order or better, as opposed to first order in the EKF. This provides for a more accurate implementation of the optimal recursive estimation equations, which is the basis for both the EKF and UKF. While equations specifying the UKF may appear more complicated than the EKF, the actual computational complexity is equivalent. For state estimation, both algorithms are in general of order L^3 (where L is the dimension of the state). For parameter estimation, both algorithms are of order L^2 (where L is the number of parameters). An efficient recursive square-root implementation (see Appendix B) was necessary to achieve the level of complexity in the parameter-estimation case. Furthermore, a distinct advantage of the UKF is its ease of implementation. In contrast to the EKF, no analytical derivatives (Jacobians or Hessians) need to be calculated. The utility of this is especially valuable in situations where the system is a "black box" model in which the internal dynamic equations are unavailable. In order to apply an EKF to such systems, derivatives must be found either from a principled analytical re-derivation of the system, or through costly and often inaccurate numerical methods (e.g., by perturbation). In contrast, the UKF relies on only functional evaluations (inputs and outputs) through the use of deterministically drawn samples from the prior distribution of the state random variable. From a coding perspective, this also allows for a much more general and modular implementation.

Even though the UKF has clear advantages over the EKF, there are still a number of limitations. As in the EKF, it makes a Gaussian assumption on the probability density of the state random variable. Often this assumption is valid, and numerous real-world applications have been successfully implemented based on this assumption. However, for certain problems (e.g., multimodal object tracking), a Gaussian assumption will not suffice, and the UKF (or EKF) cannot be applied with confidence. In such examples, one has to resort to more powerful, but also more computationally expensive, filtering paradigms such as particle filters (see Section 7.6). Finally, another implementation limitation leading to some uncertainty, is the necessity to choose the three unscented transformation parameters (i.e., α, β, and κ). While we have attempted to provide some guidelines on how to choose these parameters, the optimal selection clearly depends on the specifics of the problem at hand, and is not fully understood. In general, the choice of settings does not appear critical for state estimation, but has a greater affect on performance and convergence properties for parameter estimation. Our current work focuses on addressing this issue through developing a unified and adaptive way of calculating the optimal value of these parameters. Other areas of open research include utilizing the UKF for estimation of noise covariances, extension of the UKF to recurrent architectures that may require dynamic derivatives (see Chapter 2 and 5), and the use of the UKF and smoother in the expectation–maximization algorithm (see Chapter 6). Clearly, we have only begun to scratch the surface of the numerous applications that can benefit with use of the UKF.

APPENDIX A: ACCURACY OF THE UNSCENTED TRANSFORMATION

In this appendix, we show how the unscented transformation achieves second-order accuracy in the prediction of the posterior mean and covariance of a random variable that undergoes a nonlinear transformation. For the purpose of this analysis, we assume that all nonlinear transformations are analytic across the domain of all possible values of $\mathbf{x}$. This condition implies that the nonlinear function can be expressed as a multidimensional Taylor series consisting of an arbitrary number of terms. As the number of terms in the sum tend to infinity, the residual of the series tends to zero. This implies that the series always converges to the true value of the function.

If we consider the prior variable $\mathbf{x}$ as being perturbed about a mean $\bar{\mathbf{x}}$ by a zero-mean disturbance $\delta\mathbf{x}$ with covariance $\mathbf{P}_\mathbf{x}$, then the Taylor series expansion of the nonlinear transformation $f(\mathbf{x})$ about $\bar{\mathbf{x}}$ is

$$f(\mathbf{x}) = f(\bar{\mathbf{x}} + \delta\mathbf{x}) = \sum_{n=0}^{\infty}\left[\frac{(\delta\mathbf{x}\cdot\nabla_\mathbf{x})^n f(\mathbf{x})}{n!}\right]_{\mathbf{x}=\bar{\mathbf{x}}}. \tag{7.121}$$

If we define the operator $\mathbf{D}^n_{\delta\mathbf{x}} f$ as

$$\mathbf{D}^n_{\delta\mathbf{x}} f \triangleq [(\delta\mathbf{x}\cdot\nabla_\mathbf{x})^n f(\mathbf{x})]_{\mathbf{x}=\bar{\mathbf{x}}}, \tag{7.122}$$

then the Taylor series expansion of the nonlinear transformation $\mathbf{y} = f(\mathbf{x})$ can be written as

$$\mathbf{y} = f(\mathbf{x}) = f(\bar{\mathbf{x}}) + \mathbf{D}_{\delta\mathbf{x}} f + \frac{1}{2}\mathbf{D}^2_{\delta\mathbf{x}} f + \frac{1}{3!}\mathbf{D}^3_{\delta\mathbf{x}} f + \frac{1}{4!}\mathbf{D}^4_{\delta x} f + \cdots. \tag{7.123}$$

Accuracy of the Mean

The true mean of $\mathbf{y}$ is given by

$$\bar{\mathbf{y}} = \mathbb{E}[\mathbf{y}] = \mathbb{E}[f(\mathbf{x})] \tag{7.124}$$

$$= \mathbb{E}\left[f(\bar{\mathbf{x}}) + \mathbf{D}_{\delta\mathbf{x}} f + \frac{1}{2}\mathbf{D}^2_{\delta\mathbf{x}} f + \frac{1}{3!}\mathbf{D}^3_{\delta\mathbf{x}} f + \frac{1}{4!}\mathbf{D}^3_{\delta\mathbf{x}} f + \cdots\right]. \tag{7.125}$$

If we assume that $\mathbf{x}$ is a symmetrically distributed[11] random variable, then all odd moments will be zero. Also note that $\mathbb{E}[\delta\mathbf{x}\,\delta\mathbf{x}^T] = \mathbf{P}_\mathbf{x}$. Given this, the mean can be reduced further to

$$\bar{\mathbf{y}} = f(\bar{\mathbf{x}}) + \frac{1}{2}[(\nabla^T \mathbf{P}_\mathbf{x} \nabla) f(\mathbf{x})]_{\mathbf{x}=\bar{\mathbf{x}}} + \mathbb{E}\left[\frac{1}{4!}\mathbf{D}_{\delta\mathbf{x}}^4 f + \frac{1}{6!}\mathbf{D}_{\delta\mathbf{x}}^6 f + \cdots\right]. \quad (7.126)$$

The UT calculates the posterior mean from the propagated sigma points using Eq. (7.32). The sigma points are given by

$$\begin{aligned} \mathcal{X}_i &= \bar{\mathbf{x}} \pm (\sqrt{L+\lambda})\sigma_i, \\ &= \bar{\mathbf{x}} \pm \tilde{\sigma}_i \end{aligned}$$

where σ_i denotes the ith column[12] of the matrix square root of $\mathbf{P}_\mathbf{x}$. This implies that $\sum_{i=1}^{L}(\sigma_i \sigma_i^T) = \mathbf{P}_\mathbf{x}$. Given this formulation of the sigma points, we can again write the propagation of each point through the nonlinear function as a Taylor series expansion about $\bar{\mathbf{x}}$:

$$\mathcal{Y}_i = f(\mathcal{X}_i) = f(\bar{\mathbf{x}}) + \mathbf{D}_{\tilde{\sigma}_i} f + \frac{1}{2}\mathbf{D}_{\tilde{\sigma}_i}^2 f + \frac{1}{3!}\mathbf{D}_{\tilde{\sigma}_i}^3 f + \frac{1}{4!}\mathbf{D}_{\tilde{\sigma}_i}^4 f + \cdots.$$

Using Eq. (7.32), the UT predicted mean is

$$\begin{aligned} \bar{\mathbf{y}}_{UT} &= \frac{\lambda}{L+\lambda} f(\bar{\mathbf{x}}) + \frac{1}{2(L+\lambda)} \sum_{i=1}^{2L} \\ &\quad \times \left[f(\bar{\mathbf{x}}) + \mathbf{D}_{\tilde{\sigma}_i} f + \frac{1}{2}\mathbf{D}_{\tilde{\sigma}_i}^2 f + \frac{1}{3!}\mathbf{D}_{\tilde{\sigma}_i}^3 f + \frac{1}{4!}\mathbf{D}_{\tilde{\sigma}_i}^4 f + \cdots \right] \\ &= f(\bar{\mathbf{x}}) + \frac{1}{2(L+\lambda)} \sum_{i=1}^{2L} \left(\mathbf{D}_{\tilde{\sigma}_i} f + \frac{1}{2}\mathbf{D}_{\tilde{\sigma}_i}^2 f + \frac{1}{3!}\mathbf{D}_{\tilde{\sigma}_i}^3 f + \frac{1}{4!}\mathbf{D}_{\tilde{\sigma}_i}^4 f + \cdots \right). \end{aligned}$$

Since the sigma points are symmetrically distributed around $\bar{\mathbf{x}}$, all the odd moments are zero. This results in the simplification

$$\bar{\mathbf{y}}_{UT} = f(\bar{\mathbf{x}}) + \frac{1}{2(L+\lambda)} \sum_{i=1}^{2L} \left(\frac{1}{2}\mathbf{D}_{\tilde{\sigma}_i}^2 f + \frac{1}{4!}\mathbf{D}_{\tilde{\sigma}_i}^4 f + \frac{1}{6!}\mathbf{D}_{\tilde{\sigma}_i}^6 f + \cdots \right),$$

[11]This includes probability distributions such as Gaussian, Student-t, etc.
[12]See Section 7.3 for details of exactly how the sigma points are calculated.

and since

$$\begin{aligned}\frac{1}{2(L+\lambda)}\sum_{i=1}^{2L}\frac{1}{2}\mathbf{D}_{\tilde{\sigma}_i}^2 f &= \frac{1}{2(L+\lambda)}(\nabla f)^T\left[\sum_{i=1}^{2L}(\sqrt{L+\lambda}\sigma_i\sigma_i^T\sqrt{L+\lambda})\right](\nabla f)\\ &= \frac{L+\lambda}{2(L+\lambda)}(\nabla f)^T\frac{1}{2}\left[\sum_{i=1}^{2L}\sigma_i\sigma_i^T\right](\nabla f)\\ &= \frac{1}{2}[(\nabla^T\mathbf{P}_\mathbf{x}\nabla)f(\mathbf{x})]_{\mathbf{x}=\bar{\mathbf{x}}},\end{aligned}$$

the UT predicted mean can be further simplified to

$$\begin{aligned}\bar{\mathbf{y}}_{UT} &= f(\bar{\mathbf{x}}) + \frac{1}{2}[(\nabla^T\mathbf{P}_\mathbf{x}\nabla)f(\mathbf{x})]_{\mathbf{x}=\bar{\mathbf{x}}}\\ &\quad + \frac{1}{2(L+\lambda)}\sum_{i=1}^{2L}\left(\frac{1}{4!}\mathbf{D}_{\tilde{\boldsymbol{\sigma}}_i}^4 f + \frac{1}{6!}\mathbf{D}_{\tilde{\boldsymbol{\sigma}}_i}^6 f + \cdots\right).\end{aligned} \tag{7.127}$$

When we compare Eqs. (7.127) and (7.126), we can clearly see that the true posterior mean and the mean calculated by the UT agrees exactly to the third order and that errors are only introduced in the first and higher-order terms. The magnitudes of these errors depends on the choice of the composite scaling parameter λ as well as the higher-order derivatives of f. In contrast, a linearization approach calculates the posterior mean as

$$\bar{\mathbf{y}}_{\mathrm{LIN}} = f(\bar{\mathbf{x}}), \tag{7.128}$$

which only agrees with the true posterior mean up to the first order. Julier and Uhlman [2] show that, on a term-by-term basis, the errors in the higher-order terms of the UT are consistently smaller than those for linearization.

Accuracy of the Covariance

The true posterior covariance is given by

$$\mathbf{P}_\mathbf{y} = \mathbb{E}[(\mathbf{y}-\bar{\mathbf{y}}_T)(\mathbf{y}-\bar{\mathbf{y}}_T)^T] = \mathbb{E}[\mathbf{y}\mathbf{y}^T] - \bar{\mathbf{y}}\bar{\mathbf{y}}^T \tag{7.129}$$

where the expectation is taken over the distribution of $\mathbf{y}$. Substituting Eqs. (7.123) and (7.125) into (7.129), and recalling that all odd moments of $\delta\mathbf{x}$ are zero owing to symmetry, we can write the true posterior covariance as

$$\mathbf{P}_{\mathbf{y}} = \mathcal{A}_{\mathbf{x}}\mathbf{P}_{\mathbf{x}}\mathcal{A}_{\mathbf{x}}^T = \frac{1}{4}\{[(\nabla^T\mathbf{P}_{\mathbf{x}}\nabla)f(\mathbf{x})][(\nabla^T\mathbf{P}_{\mathbf{x}}\nabla)\mathbf{f}(\mathbf{x})]^T\}_{\mathbf{x}=\bar{\mathbf{x}}}$$
$$+ \underbrace{\mathbb{E}\left[\sum_{i=1}^{\infty}\sum_{j=1}^{\infty}\frac{1}{i!j!}\mathbf{D}_{\delta\mathbf{x}}^{i}f(\mathbf{D}_{\delta\mathbf{x}}^{j}f)^T\right]}_{ij>1}$$
$$- \underbrace{\left[\sum_{i=1}^{\infty}\sum_{j=1}^{\infty}\frac{1}{(2i)!(2j)!}\mathbb{E}[\mathbf{D}_{\mathbf{x}}^{2i}f]\mathbb{E}[\mathbf{D}_{\sigma\mathbf{x}}^{2j}f]^T],\right]}_{ij>1} \tag{7.130}$$

where $\mathcal{A}_{\mathbf{x}}$ is the Jacobian matrix of $f(\mathbf{x})$ evaluated at $\bar{\mathbf{x}}$. It can be shown (using a similar approach as for the posterior mean) that the posterior covariance calculated by the UT is given by

$$(\mathbf{P}_{\mathbf{y}})_{UT} = \mathcal{A}_{\mathbf{x}}\mathbf{P}_{\mathbf{x}}\mathcal{A}_{\mathbf{x}}^T - \frac{1}{4}\{[(\nabla^T\mathbf{P}_{\mathbf{x}}\nabla)f(\mathbf{x})][(\nabla^T\mathbf{P}_{\mathbf{x}}\nabla)f(\mathbf{x})]^T\}_{\mathbf{x}=\bar{\mathbf{x}}}$$
$$+ \frac{1}{2(L+\lambda)}\sum_{k=1}^{2L}\underbrace{\left[\sum_{i=1}^{\infty}\sum_{j=1}^{\infty}\frac{1}{i!j!}\mathbf{D}_{\tilde{\boldsymbol{\sigma}}_k}^{i}f(\mathbf{D}_{\tilde{\boldsymbol{\sigma}}_k}^{j}f)^T\right]}_{ij>1}$$
$$- \underbrace{\left[\sum_{i=1}^{\infty}\sum_{j=1}^{\infty}\frac{1}{(2i)!(2j)!4(L+\lambda)^2}\sum_{k=1}^{2L}\sum_{m=1}^{2L}\mathbf{D}_{\tilde{\boldsymbol{\sigma}}_k}^{2i}f(\mathbf{D}_{\tilde{\boldsymbol{\sigma}}_m}^{2j}f)^T\right]}_{ij>1}. \tag{7.131}$$

Comparing Eqs. (7.130) and (7.131), it is clear that the UT again calculates the posterior covariance accurately to the first two terms, with errors only introduced in the fourth- and higher-order moments. Julier and Uhlmann [2] show how the absolute term-by-term errors of these higher-order moments are again consistently smaller for the UT than for the linearized case that truncates the Taylor series after the first term, that is,

$$(\mathbf{P}_y)_{\text{LIN}} = \mathcal{A}_{\mathbf{x}}\mathbf{P}_{\mathbf{x}}\mathcal{A}_{\mathbf{x}}^T. \tag{7.132}$$

For this derivation, we have assumed the value of the β parameter in the UT to be zero. If prior knowledge about the shape of the prior distribution of $\mathbf{x}$ is known, β can be set to a non-zero value that minimizes the error in some of the higher (≥ 4) order moments. Julier [53] shows how the error in the kurtosis of the posterior distribution is minimized for a Gaussian $\mathbf{x}$ when $\beta = 2$.

APPENDIX B: EFFICIENT SQUARE-ROOT UKF IMPLEMENTATIONS

In the standard Kalman implementation, the state (or parameter) covariance $\mathbf{P}_k$ is recursively calculated. The UKF requires taking the matrix square-root $\mathbf{S}_k\mathbf{S}_k^T = \mathbf{P}_k$, at each time step, which is $\mathcal{O}(\frac{1}{6}L^3)$ using a Cholesky factorization. In the square-root UKF (SR-UKF), $\mathbf{S}_k$ will be propagated directly, avoiding the need to refactorize at each time step. The algorithm will in general still be $\mathcal{O}(L^3)$ for state estimation, but with improved numerical properties (e.g., guaranteed positive-semidefiniteness of the state covariances), similar to those of standard square-root Kalman filters [20]. However, for the special state-space formulation of parameter estimation, an $\mathcal{O}(L^2)$ implementation becomes possible (equivalent complexity to EKF parameter estimation).

The square-root form of the UKF makes use of three powerful linear-algebra techniques,[13] *QR decomposition*, *Cholesky factor updating*, and *efficient least squares*, which we briefly review below:

- *QR decomposition* The QR decomposition or factorization of a matrix $\mathbf{A} \in \mathbb{R}^{L\times N}$ is given by, $\mathbf{A}^T = \mathbf{QR}$, where $\mathbf{Q} \in \mathbb{R}^{N\times N}$ is orthogonal, $\mathbf{R} \in \mathbb{R}^{N\times L}$ is upper-triangular, and $N \geq L$. The upper-triangular part of $\mathbf{R}$, $\tilde{\mathbf{R}}$, is the transpose of the Cholesky factor of $\mathbf{P} = \mathbf{AA}^T$, that is, $\tilde{\mathbf{R}} = \mathbf{S}^T$, such that $\tilde{\mathbf{R}}^T\tilde{\mathbf{R}} = \mathbf{AA}^T$. We use the shorthand notation qr$\{\cdot\}$ to donate a QR decomposition of a matrix where only $\tilde{\mathbf{R}}$ is returned. The computational complexity of a QR decomposition is $\mathcal{O}(NL^2)$. Note that performing a Cholesky factorization directly on $\mathbf{P} = \mathbf{AA}^T$ is $\mathcal{O}(\frac{1}{6}L^3)$ plus $\mathcal{O}(NL^2)$ to form $\mathbf{AA}^T$.
- *Cholesky factor updating* If $\mathbf{S}$ is the original lower-triangular Cholesky factor of $\mathbf{P} = \mathbf{AA}^T$, then the Cholesky factor of the rank-1 update (or downdate) $\mathbf{P} \pm \sqrt{v}\mathbf{uu}^T$ is denoted by $\mathbf{S} =$ cholupdate$\{\mathbf{S}, \mathbf{u}, \pm v\}$. If $\mathbf{u}$ is a matrix and not a vector, then the result is M consecutive updates of the Cholesky factor using the M columns of $\mathbf{u}$. This algorithm (available in *Matlab* as `cholupdate`) is only $\mathcal{O}(L^2)$ per update.

[13] See [54] for theoretical and implementation details.

- *Efficient least squares* The solution to the equation $(\mathbf{AA}^T)\mathbf{x} = \mathbf{A}^T\mathbf{b}$ also corresponds to the solution of the overdetermined least-squares problem $\mathbf{Ax} = \mathbf{b}$. This can be solved efficiently using a QR decomposition with pivoting (implemented in *Matlab*'s "/" operator).

The complete specifications for the new square-root filters are given in Table 7.9 for state estimation and Table 7.10 for parameter estimation. Below we describe the key parts of the square-root algorithms, and how they contrast with the standard implementations. Experimental results and further discussion are presented in [7] and [55].

Square-Root State Estimation

As in the original UKF, the filter is initialized by calculating the matrix square root of the state covariance once via a Cholesky factorization, Eq. (7.133). However, the propagated and updated Cholesky factor is then used in subsequent iterations to directly form the sigma points. In Eq. (7.138) the *time update* of the Cholesky factor, $\mathbf{S}^-$, is calculated using a QR decomposition of the compound matrix containing the weighted propagated sigma points and the matrix square root of the additive process noise covariance. The subsequent Cholesky update (or downdate) in Eq. (7.137) is necessary since the zeroth weight, $W_0^{(c)}$, may be negative. These two steps replace the *time-update* of $\mathbf{P}^-$ in Eq. (7.55), and is also $\mathcal{O}(L^3)$.

The same two-step approach is applied to the calculation of the Cholesky factor, $\mathbf{S}_{\tilde{\mathbf{y}}}$, of the observation error covariance in Eqs. (7.142) and (7.143). This step is $\mathcal{O}(LM^2)$, where M is the observation dimension. In contrast to the way that Kalman gain is calculated in the standard UKF (see Eq. (7.61)), we now use two nested inverse (or *least-squares*) solutions to the following expansion of Eq. (7.60): $\mathcal{K}_k(\mathbf{S}_{\tilde{\mathbf{y}}_k}\mathbf{S}_{\tilde{\mathbf{y}}_k}^T) = \mathbf{P}_{\mathbf{x}_k\mathbf{y}_k}$. Since $\mathbf{S}_{\tilde{\mathbf{y}}}$ is square and triangular, efficient "back-substitutions" can be used to solve for $\mathcal{K}_k$ directly without the need for a matrix inversion.

Finally, the posterior measurement update of the Cholesky factor of the state covariance is calculated in Eq. (7.147) by applying M sequential Cholesky downdates to $\mathbf{S}_k^-$. The downdate vectors are the columns of $\mathbf{U} = \mathcal{K}_k\mathbf{S}_{\tilde{\mathbf{y}}_k}$. This replaces the posterior update of $\mathbf{P}_k$ in Eq. (7.63), and is also $\mathcal{O}(LM^2)$.

Square-Root Parameter Estimation

The parameter-estimation algorithm follows a similar framework to that of the state-estimation square-root UKF. However, an $\mathcal{O}(ML^2)$ algorithm, as

Table 7.9 Square-Root UKF for state estimation

Initialize with

$$\hat{\mathbf{x}}_0 = \mathbb{E}[\mathbf{x}_0], \qquad \mathbf{S}_0 = \text{chol}\left\{\mathbb{E}[(\mathbf{x}_0 - \hat{\mathbf{x}}_0)(\mathbf{x}_0 - \hat{\mathbf{x}}_0)^T]\right\}. \tag{7.133}$$

For $k \in \{1, \ldots, \infty\}$,

The sigma-point calculation and time update are given by

$$\mathcal{X}_{k-1} = [\hat{\mathbf{x}}_{k-1} \quad \hat{\mathbf{x}}_{k-1} + \gamma\mathbf{S}_k \quad \hat{\mathbf{x}}_{k-1} - \gamma\mathbf{S}_k], \tag{7.134}$$

$$\mathcal{X}^*_{k|k-1} = \mathbf{F}(\mathcal{X}_{k-1}, \mathbf{u}_{k-1}), \tag{7.135}$$

$$\hat{\mathbf{x}}_k^- = \sum_{i=0}^{2L} W_i^{(m)} \mathcal{X}^*_{i,k|k-1}, \tag{7.136}$$

$$\mathbf{S}_k^- = \text{qr}\left\{\left[\sqrt{W_1^{(c)}}(\mathcal{X}^*_{1:2L,k|k-1} - \hat{\mathbf{x}}_k^-) \quad \sqrt{\mathbf{R}^{\mathbf{v}}}\right]\right\} \tag{7.137}$$

$$\mathbf{S}_k^- = \text{cholupdate}\{\mathbf{S}_k^-, \mathcal{X}^*_{0,k} - \hat{\mathbf{x}}_k^-, W_0^{(c)}\}, \tag{7.138}$$

(augment sigma points)[14]

$$\mathcal{X}_{k|k-1} = [\mathcal{X}^*_{k|k-1} \quad \mathcal{X}^*_{0,k|k-1} + \gamma\sqrt{\mathbf{R}^{\mathbf{v}}} \quad \mathcal{X}^*_{0,k|k-1} - \gamma\sqrt{\mathbf{R}^{\mathbf{v}}}] \tag{7.139}$$

$$\mathcal{Y}_{k|k-1} = \mathbf{H}(\mathcal{X}_{k|k-1}) \tag{7.140}$$

$$\hat{\mathbf{y}}_k^- = \sum_{i=0}^{2L} W_i^{(m)} \mathcal{Y}_{i,k|k-1}, \tag{7.141}$$

and the measurement update equations are

$$\mathbf{S}_{\tilde{\mathbf{y}}_k} = \text{qr}\left\{\left[\sqrt{W_1^{(c)}}(\mathcal{Y}_{1:2L,k} - \hat{\mathbf{y}}_k) \quad \sqrt{\mathbf{R}_k^{\mathbf{n}}}\right]\right\}, \tag{7.142}$$

$$\mathbf{S}_{\tilde{\mathbf{y}}_k} = \text{cholupdate}\{\mathbf{S}_{\tilde{\mathbf{y}}_k}, \mathcal{Y}_{0,k} - \hat{\mathbf{y}}_k, W_0^{(c)}\} \tag{7.143}$$

$$\mathbf{P}_{\mathbf{x}_k\mathbf{y}_k} = \sum_{i=0}^{2L} W_i^{(c)}(\mathcal{X}_{i,k|k-1} - \hat{\mathbf{x}}_k^-)(\mathcal{Y}_{i,k|k-1} - \hat{\mathbf{y}}_k^-)^T, \tag{7.144}$$

$$\mathcal{K}_k = (\mathbf{P}_{\mathbf{x}_k\mathbf{y}_k}/\mathbf{S}_{\tilde{\mathbf{y}}_k}^T)/\mathbf{S}_{\tilde{\mathbf{y}}_k}, \tag{7.145}$$

$$\hat{\mathbf{x}}_k = \hat{\mathbf{x}}_k^- + \mathcal{K}_k(\mathbf{y}_k - \hat{\mathbf{y}}_k^-),$$

$$\mathbf{U} = \mathcal{K}_k\mathbf{S}_{\tilde{\mathbf{y}}_k}, \tag{7.146}$$

$$\mathbf{S}_k = \text{cholupdate}\{\mathbf{S}_k^-, \mathbf{U}, -1\}, \tag{7.147}$$

---

opposed to $\mathcal{O}(L^3)$, is possible by taking advantage of the *linear* state transition function. Specifically, the time update of the state covariance is given simply by $\mathbf{P}_{\mathbf{w}_k}^- = \mathbf{P}_{\mathbf{w}_{k-1}} + \mathbf{R}_{k-1}^{\mathbf{r}}$ (see Section 7.4 for a discussion on selecting $\mathbf{R}_{k-1}^{\mathbf{r}}$). In the square-root filters $\mathbf{S}_{\mathbf{w}_k}$ may thus be updated directly in Eq. (7.150) using one of two options: (1) $\mathbf{S}_{\mathbf{w}_k}^- = \lambda_{RLS}^{-1/2}\mathbf{S}_{\mathbf{w}_{k-1}}$, correspond-

[14]Alternatively, *redraw* a new set of sigma points that incorporate the additive process noise, i.e., $\mathcal{X}_{k|k-1} = [\hat{\mathbf{x}}_k^- \quad \hat{\mathbf{x}}_k^- + \gamma\mathbf{S}_k^- \quad \hat{\mathbf{x}}_k^- - \gamma\mathbf{S}_k^-]$.

Table 7.10 Square-root UKF for parameter estimation

Initialize with

$$\hat{\mathbf{w}}_0 = E[\mathbf{w}], \qquad \mathbf{S}_{\mathbf{w}_0} = \text{chol}\{E[(\mathbf{w} - \hat{\mathbf{w}}_0)(\mathbf{w} - \hat{\mathbf{w}}_0)^T]\}. \tag{7.148}$$

For $k \in \{1, \ldots, \infty\}$,
The time update and sigma point calculation are given by

$$\hat{\mathbf{w}}_k^- = \hat{\mathbf{w}}_{k-1}, \tag{7.149}$$

$$\mathbf{S}_{\mathbf{w}_k}^- = \lambda_{RLS}^{-1/2}\mathbf{S}_{\mathbf{w}_{k-1}} \quad \text{or} \quad \mathbf{S}_{\mathbf{w}_k}^- = \mathbf{S}_{\mathbf{w}_{k-1}} + \mathbf{D}_{\mathbf{r}_{k-1}}, \tag{7.150}$$

$$\mathcal{W}_{k|k-1} = [\hat{\mathbf{w}}_k^- \quad \hat{\mathbf{w}}_k^- + \gamma\mathbf{S}_{\mathbf{w}_k}^- \quad \hat{\mathbf{w}}_k^- - \gamma\mathbf{S}_{\mathbf{w}_k}^-], \tag{7.151}$$

$$\mathcal{D}_{k|k-1} = \mathbf{G}(\mathbf{x}_k, \mathcal{W}_{k|k-1}), \tag{7.152}$$

$$\hat{\mathbf{d}}_k = \sum_{i=0}^{2L} W_i^{(m)}\mathcal{D}_{i,k|k-1}, \tag{7.153}$$

and the measurement-update equations are

$$\mathbf{S}_{\mathbf{d}_k} = \text{qr}\left\{\left[\sqrt{W_1^{(c)}}(\mathcal{D}_{1:2L,k} - \hat{\mathbf{d}}_k) \quad \sqrt{\mathbf{R}^e}\right]\right\}, \tag{7.154}$$

$$\mathbf{S}_{\mathbf{d}_k} = \text{cholupdate}\{\mathbf{S}_{\mathbf{d}_k}, \mathcal{D}_{0,k} - \hat{\mathbf{d}}_k, W_0^{(c)}\}, \tag{7.155}$$

$$\mathbf{P}_{\mathbf{w}_k\mathbf{d}_k} = \sum_{i=0}^{2L} W_i^{(c)}(\mathcal{W}_{i,k|k-1} - \hat{\mathbf{w}}_k^-)(\mathcal{D}_{i,k|k-1} - \hat{\mathbf{d}}_k)^T, \tag{7.156}$$

$$\mathcal{K}_k = (\mathbf{P}_{\mathbf{w}_k\mathbf{d}_k}/\mathbf{S}_{\mathbf{d}_k}^T)/\mathbf{S}_{\mathbf{d}_k}, \tag{7.157}$$

$$\hat{\mathbf{w}}_k = \hat{\mathbf{w}}_k^- + \mathcal{K}_k(\mathbf{d}_k - \hat{\mathbf{d}}_k), \tag{7.156}$$

$$\mathbf{U} = \mathcal{K}_k\mathbf{S}_{\mathbf{d}_k}, \tag{7.158}$$

$$\mathbf{S}_{\mathbf{w}_k} = \text{cholupdate}\{\mathbf{S}_{\mathbf{w}_k}^-, \mathbf{U}, -1\}, \tag{7.159}$$

where

$$\mathbf{D}_{\mathbf{r}_{k-1}} = -\text{Diag}\{\mathbf{S}_{\mathbf{w}_{k-1}}\} + \sqrt{\text{Diag}\{\mathbf{S}_{\mathbf{w}_{k-1}}\}^2 + \text{Diag}\{\mathbf{R}_{k-1}^{\mathbf{r}}\}}.$$

ing to an exponential weighting on past data; (2) $\mathbf{S}_{\mathbf{w}_k}^- = \mathbf{S}_{\mathbf{w}_{k-1}} + \mathbf{D}_{\mathbf{r}_{k-1}}$, where the diagonal matrix $\mathbf{D}_{\mathbf{r}_{k-1}}$, is chosen to approximate the effects of annealing a diagonal process noise covariance $\mathbf{R}_k^{\mathbf{r}}$.[15] Both options avoid the costly $\mathcal{O}(L^3)$ QR and Cholesky-based updates necessary in the state-estimation filter.

[15]This update ensures that the main diagonal of $\mathbf{P}_{\mathbf{w}_k}^-$ is exact. However, additional off-diagonal cross-terms $\mathbf{S}_{\mathbf{w}_{k-1}}\mathbf{D}_{\mathbf{r}_{k-1}}^T + \mathbf{D}_{\mathbf{r}_{k-1}}\mathbf{S}_{\mathbf{w}_{k-1}}^T$ are also introduced (though the effect appears negligible).

REFERENCES

[1] S.J. Julier, J.K. Uhlmann, and H. Durrant-Whyte, "A new approach for filtering nonlinear systems," in *Proceedings of the American Control Conference*, 1995, pp. 1628–1632.

[2] S.J. Julier and J.K. Uhlmann, "A general method for approximating nonlinear transformations of probability distributions," Technical Report, RRG, Department of Engineering Science, University of Oxford, November 1996. http://www.robots.ox.ac.uk/siju/work/publications/letter_size/Unscented.zip.

[3] S.J. Julier and J.K. Uhlmann, "A new extension of the Kalman filter to nonlinear systems," in *Proceedings of AeroSense: The 11th International Symposium on Aerospace/Defence Sensing, Simulation and Controls, 1997.*

[4] E.A. Wan, R. van der Merwe, and A.T. Nelson, "Dual estimation and the unscented transformation," in S.A. Solla, T.K. Leen, and K.-R. Müller, Eds. *Advances in Neural Information Processing Systems 12*, Cambridge, MA: MIT Press, 2000, pp. 666–672.

[5] E.A. Wan and R. van der Merwe, "The unscented Kalman filter for nonlinear estimation, in *Proceedings of Symposium 2000 on Adaptive Systems for Signal Processing, Communication and Control (AS-SPCC), IEEE, Lake Louise, Alberta, Canada, October 2000.*

[6] R. van der Merwe, J.F.G. de Freitas, D. Doucet, and E.A. Wan, "The unscented particle filter," Technical Report CUED/F-INFENG/TR 380, Cambridge University Engineering Department, August 2000.

[7] R. van der Merwe and E.A. Wan, "Efficient derivative-free Kalman filters for online learning," in *Proceedings of European Symposium on Artificial Neural Networks (ESANN), Bruges, Belgium, April 2001.*

[8] S. Singhal and L. Wu, "Training multilayer perceptrons with the extended Kalman filter," in *Advances in Neural Information Processing Systems 1.* San Mateo, CA: Morgan Kauffman, 1989, pp. 133–140.

[9] G.V. Puskorius and L.A. Feldkamp, "Decoupled extended Kalman filter training of feedforward layered networks," in *Proceedings of IJCNN*, Vol. 1, International Joint Conference on Neural Networks, 1991, pp. 771–777.

[10] R.E. Kalman, "A new approach to linear filtering and prediction problems," *Transactions of the ASME, Ser. D, Journal of Basic Engineering*, **82**, 35–45 (1960).

[11] A. Jazwinsky, *Stochastic Processes and Filtering Theory.* New York: Academic Press, 1970.

[12] K. Ito and K. Xiong, "Gaussian filters for nonlinear filtering problems," *IEEE Transactions on Automatic Control*, **45**, 910–927 (2000).

[13] M. Nørgaard, N.K. Poulsen, and O. Ravn, "Advances in derivative-free state estimation for nonlinear systems," Technical Report IMM-REP-1998-15,

Department of Mathematical Modelling/Department of Automation, Technical University of Denmark, Lyngby, April 2000.

[14] J.R. Cloutier, C.N. D'Souza, and C.P. Mracek, "Nonlinear regulation and nonlinear H-infinity controls via the state-dependent Riccati equation technique: Part 1, Theory," in *Proceedings of the International Conference on Nonlinear Problems in Aviation and Aerospace, Daytona Beach, FL, May 1996.*

[15] M. Mackey and L. Glass, "Oscillation and chaos in a physiological control system," *Science*, **197**, 287–289 1977.

[16] A. Lapedes and R. Farber, "Nonlinear signal processing using neural networks: Prediction and system modelling," Technical Report LAUR 872662, Los Alamos National Laboratory, 1987.

[17] R.H. Shumway and D.S. Stoffer, "An approach to time series smoothing and forecasting using the EM algorithm," *Time Series Analysis*, **3**, 253–264 (1982).

[18] Z. Ghahramani and S.T. Roweis, "Learning nonlinear dynamical systems using an EM algorithm," in M.J. Kearns, S.A. Solla, and D.A. Cohn, Eds., *Advances in Neural Information Processing Systems 11: Proceedings of the 1998 Conference*. Cambridge, MA: MIT Press, 1999.

[19] F.L. Lewis, *Optimal Estimation*. New York: Wiley, 1986.

[20] A.H. Sayed and T. Kailath, "A state-space approach to adaptive RLS filtering," *IEEE Signal Processing Magazine*, pp. 18–60 (July 1994).

[21] S. Haykin. *Adaptive Filter Theory*, 3rd ed. Upper Saddle River, NJ: Prentice-Hall, 1996.

[22] A.T. Nelson, "Nonlinear estimation and modeling of noisy time-series by dual Kalman filtering methods," PhD Thesis, Oregon Graduate Institute, 2000.

[23] L. Ljung and T. Söderström, *Theory and Practice of Recursive Identification*. Cambridge, MA: MIT Press, 1983.

[24] D.J.C. MacKay, http://wol.ra.phy.cam.ac.uk/mackay/sourcedata. html.www.

[25] D.J.C. MacKay, "A practical Bayesian framework for backpropagation networks," *Neural Computation*, **4**, 448–472 (1992).

[26] K. Ikeda, "Multiple-valued stationary state and its instability of light by a ring cavity system," *Optics Communications* **30**, 257–261 (1979).

[27] H.H. Rosenbrock, "An automatic method for finding the greatest or least value of a function," *Computer Journal*, **3**, 175–184 (1960).

[28] G.V. Puskorius and L.A. Feldkamp, "Extensions and enhancements of decoupled extended Kalman filter training," in *Proceedings of ICNN*, Vol. 3, International Conference on Neural Networks 1997, pp. 1879–1883.

[29] E.A. Wan and A.T. Nelson, "Neural dual extended Kalman filtering: Applications in speech enhancement and monaural blind signal separation," in *Proceedings of Neural Networks for Signal Processing Workshop, IEEE*, 1997.

[30] M.B. Matthews, "A state-space approach to adaptive nonlinear filtering using recurrent neural networks," in *Proceedings of IASTED International Symposium on Artificial Intelligence Application and Neural Networks, 1990*, pp. 197–200.

[31] R.W. Brumbaugh, "An Aircraft Model for the AIAA Controls Design Challenge," PRC Inc., Edwards, CA.

[32] J.P. Dutton, "Development of a Nonlinear Simulation for the McDonnell Douglas F-15 Eagle with a Longitudinal TECS Control Law," Master Thesis, Department of Aeronautics and Astronautics, University of Washington, 1994.

[33] A Doucet, "On sequential simulation-based methods for Bayesian filtering," Technical Report CUED/F-INFENG/TR 310, Cambridge University Engineering Department, 1998.

[34] A. Doucet, J.F.G. de Freitas, and N.J. Gordon, "Introduction to sequential Monte Carlo methods," in A. Doucet, J.F.G. de Freitas, and N.J. Gordon, Eds. *Sequential Monte Carlo Methods in Practice*. Berlin: Springer-Verlag, 2000.

[35] N.J. Gordon, D.J. Salmond, and A.F.M. Smith, "Novel approach to nonlinear/non-Gaussian Bayesian state estimation," *IEE Proceedings, Part F*, **140**, 107–113 (1993).

[36] B. Efron, *The Bootstrap Jacknife and other Resampling Plans*. Philadelphia: SIAM, 1982.

[37] D.B. Rubin, "Using the SIR algorithm to simulate posterior distributions," in J.M. Bernardo, M.H. DeGroot, D.V. Lindley, and A.F.M. Smith, Eds., *Bayesian Statistics 3*. Oxford University Press, 1988, pp. 395–402.

[38] A.F.M. Smith and A.E. Gelfand, "Bayesian statistics without tears: a sampling–resampling perspective," *American Statistician*, **46**, 84–88, 1992.

[39] T. Higuchi, "Monte Carlo filter using the genetic algorithm operators," *Journal of Statistical Computation and Simulation*, **59**, 1–23 (1997).

[40] A. Kong, J.S. Liu, and W.H. Wong, "Sequential imputations and Bayesian missing data problems," *Journal of the American Statistical Association*, **89**, 278–288 (1994).

[41] J.S. Liu and R. Chen, "Blind deconvolution via sequential imputations," *Journal of the American Statistical Association*, **90**, 567–576 (1995).

[42] J.S. Liu and R. Chen, "Sequential Monte Carlo methods for dynamic systems," *Journal of the American Statistical Association*, **93**. 1032–1044 (1998).

[43] V.S. Zaritskii, V.V. Svetnik, and L.I. Shimelevich, "Monte-Carlo techniques in problems of optimal information processing," *Automation and Remote Control*, **36**, 2015–2022 (1975).

[44] D. Avitzour, "A stochastic simulation Bayesian approach to multitarget tracking," *IEE Proceedings on Radar, Sonar and Navigation*, **142**, 41–44 (1995).

[45] E.R. Beadle and P.M. Djurić, "A fast weighted Bayesian bootstrap filter for nonlinear model state estimation," *IEEE Transactions on Aerospace and Elecytronic Systems*, **33**, 338–343 (1997).

[46] M. Isard and A. Blake, "Contour tracking by stochastic propagation of conditional density," in *Proceedings of European Conference on Computer Vision, Cambridge, UK*, 1996, pp. 343–356.

[47] G. Kitagawa, "Monte Carlo filter and smoother for non-Gaussian nonlinear state space model," *Journal of Computational and Graphical Statistics*, **5**, 1–25 (1996).

[48] J.F.G. de Freitas, "Bayesian Methods for Neural Networks," PhD Thesis, Cambridge University Engineering Department, 1999.

[49] J.C. Hull, *Options, Futures, and Other Derivatives*, 3rd ed. Upper Saddle River, NJ: Prentice-Hall, 1997.

[50] F. Black and M. Scholes, "The pricing of options and corporate liabilities," *Journal of Political Economy*, **81**, 637–659 (1973).

[51] M. Niranjan, "Sequential tracking in pricing financial options using model based and neural network approaches," in M.C. Mozer, M.I. Jordan, and T. Petsche, Eds. *Advances in Neural Information Processing Systems* **8**, 1996, 960–966.

[52] J.F.G. de Freitas, M. Niranjan, A.H. Gee, and A. Doucet, "Sequential Monte Carlo methods to train neural network models," *Neural Computation*, **12**, 955–993 (2000).

[53] S.J. Julier, "The scaled unscented transformation." In preparation.

[54] W.H. Press, S.A. Teukolsky, W.T. Vetterling, and B.P. Flannery, *Numerical Recipes in C: The Art of Scientific Computing*, 2nd ed. Cambridge University Press, 1992.

[55] R. van der Merwe and E.A. Wan, "The square-root unscented Kalman filter for state and parameter-estimation," in *Proceedings of International Conference on Acoustics, Speech, and Signal Processing, Salt Lake City, UT, May 2001.*

INDEX